NUMERICAL SOLUTION OF PARTIAL DIFFERENTIAL EQUATIONS—III

SYNSPADE 1975

ACADEMIC PRESS RAPID MANUSCRIPT REPRODUCTION

Proceedings of the Third Symposium on the Numerical Solution of Partial Differential Equations, SYNSPADE 1975, Held at the University of Maryland, College Park, Maryland, May 19-24, 1975

Numerical Solution of Partial Differential Equations—III

SYNSPADE 1975

EDITED BY
BERT HUBBARD
Institute for Fluid Dynamics
and Applied Mathematics
University of Maryland
College Park, Maryland

ACADEMIC PRESS, INC.
New York • San Francisco • London • 1976
A Subsidiary of Harcourt Brace Jovanovich, Publishers

ACADEMIC PRESS, INC.
111 Fifth Avenue, New York, New York 10003

United Kingdom Edition published by
ACADEMIC PRESS, INC. (LONDON) LTD.
24/28 Oval Road, London NW1

LIBRARY OF CONGRESS CATALOG CARD NUMBER: 75-37403

AMS (MOS) 1970 Subject Classification 65-65
ISBN 0-12-358503-1

PRINTED IN THE UNITED STATES OF AMERICA

CONTENTS

CONTRIBUTORS

Numbers in parentheses indicate the pages on which authors' contributions begin.

J. D. Achenbach, (1) Northwestern University, Evanston, Illinois

Stuart S. Antman, (35) University of Maryland, College Park, Maryland

A. K. Aziz, (55) Naval Surface Weapons Center, Silver Spring, Maryland and University of Maryland, College Park, Maryland

Ivo Babuška, (89) University of Maryland, College Park, Maryland

Klaus-Jürgen Bathe, (117) Massachusetts Institute of Technology, Cambridge, Massachusetts

Louis Bauer, (443) Courant Institute of Mathematical Sciences, New York, New York

Alexandre Joel Chorin, (165) University of California, Berkeley, California

Ely M. Gelbard, (177) Argonne National Laboratory, Argonne, Illinois

Pierre Grisvard, (207) University of Nice, France

Antony Jameson, (275) Courant Institute of Mathematical Sciences, New York, New York

R. B. Kellogg, (321) University of Maryland, College Park, Maryland

Klaus Kirchgässner, (349) Universitat Stuttgärt, W. Germany

S. H. Leventhal, (55) Naval Surface Weapons Center, Silver Spring, Maryland

J. L. Lions, (373) College de France and Iria-Laboria, France

John E. Osborn, (393) University of Maryland, College Park, Maryland

Stanley Osher, (413) State University of New York at Stony Brook, Stony Brook, New York

Edward L. Reiss, (443) Courant Institute of Mathematical Sciences, New York, New York

Sergio Spagnolo, (469) University of Pisa, Italy

PREFACE

The Symposium on the Numerical Solution of Partial Differential Equations, SYNSPADE 1975, was the third in a series on this topic held on the campus of the University of Maryland, College Park, at 5-year intervals. During the week of 19-24 May 1975, researchers gathered at the Adult Education Center to listen to invited lectures and contributed papers and to discuss with each other the most recent developments in this field. This volume contains the invited addresses in full and a list of the contributed papers.

The emphasis of this symposium was on those difficult problems in partial differential equations exhibiting some type of singular behavior. This is a very broad topic that includes singular behavior of solutions induced by the geometry such as corners or the nature of the differential equation itself. Talks were given on the effects of nonlinearities, such as bifurcation, which occur in problems of nonlinear mechanics. Also discussed were equations of changing type and those with rapidly oscillating coefficients. The point of view of the symposium itself was to give equal weight to discussions of the mathematical models and their relation to experiment, behavior of solutions of the partial differential equations involved, and effective computational methods for their numerical solution. Thus this volume should have wide appeal to engineers and mathematicians alike.

An innovation in this symposium was a panel discussion on "Gaps between the Theory and Art of Scientific Computations" chaired by Dr. M.E. Rose (ERDA).

SYNSPADE 1975 was sponsored by the Institute for Fluid Dynamics and Applied Mathematics, University of Maryland, and was funded by a grant from the National Science Foundation. The generous financial assistance of the NSF and the hard work of many members of the University of Maryland faculty and staff made possible the success of this conference. Particular mention is due to the organizing committee: I. Babuška (Chairman), S. Antman, A.K. Aziz, B. Hubbard, and J. Osborn.

ELASTOSTATIC AND ELASTODYNAMIC STRESS SINGULARITIES*

J. D. Achenbach**

Summary

Elastostatic and elastodynamic stress singularities play a major part in considerations of fracture mechanics. In this paper some relatively well-known elastostatic results are briefly reviewed. The major part of the paper is, however, concerned with new elastodynamic results for stress intensity factors of a bifurcating crack.

1. Introduction

The singular behavior of solutions to partial differential equations at edges and corners has fascinated generations of analysts. A considerable body of literature dealing with the mathematical and physical aspects of singularities has accumulated over the years.

In several areas of the physical sciences and engineering the detailed study of singularities is motivated by specific physical considerations. From the physical point of view a flaw in a stressed body gives rise to a significant increase of the local stresses, particularly in the vicinity of a sharp edge of the flaw. If the local stresses reach a sufficiently high level the flaw becomes unstable and may grow into a sizeable fracture surface. Even if fracture does not occur immediately, the material is susceptible to stress

*This paper was prepared in the course of research sponsored by the National Science Foundation under Grant ENG70-01465 A02 to Northwestern University.

**Department of Civil Engineering, Northwestern University.

corrosion and fatigue damage in regions of stress concentration. The appearance of a singularity in analytical results, even though it is a consequence of mathematical idealizations, indicates a region in which certain field variables, such as stress components, are very large, and may give rise to fracture or other undesirable instabilities. Hence the preoccupation with stress singularities in the field of solid mechanics.

Fracture mechanics is primarily concerned with the analysis of the stability of flaws. Although fracture involves processes on the atomic level, it is a matter of wide experience that useful quantitative information can be obtained by basing computations on a continuum model for the medium. A continuum model, such as the homogeneous, isotropic linearly elastic continuum makes it feasible to carry out a mathematical analysis of the displacement and stress fields in the neighborhood of a flaw. A fracture criterion can subsequently be employed to determine the proclivity towards fracture.

Most of the analytical work in fracture mechanics is based on classical elasticity theory. For reference purposes we will summarize the equations governing the theory of linearized homogeneous elasticity. For a more detailed statement see e.g., Ref. [1]. Consider a rectangular cartesian coordinate system with coordinate axes x_i , $i = 1, 2, 3,$ and let $u_i(x_m,t)$ be the components of the displacement vector field. Within the restrictions of the linearized theory the deformation can then be described by the small-strain tensor, whose components are in indicial notation

(1.1) $$\varepsilon_{ij} = \frac{1}{2}(u_{i,j} + u_{j,i})$$

Here $u_{i,j}$ is $\partial u_i(x_1, x_2, x_3)/\partial x_j$. Denoting the components of the traction vector on a surface element by T_i , we have

$$T_i = \tau_{ij} n_j , \tag{1.2}$$

where n_j are the components of the outward normal and τ_{ij} are the components of the stress tensor. Balance of linear momentum subsequently leads to Cauchy's first law of motion,

$$\tau_{ij,j} = \rho \ddot{u}_i , \tag{1.3}$$

where ρ is the mass density, and the summation convention is invoked. Balance of angular momentum implies that the stress tensor is symmetric, i.e., $\tau_{ij} = \tau_{ji}$.

In the linearized theory the relation between the components of the stress tensor and the components of the strain tensor is

$$\tau_{ij} = C_{ijk\ell} \, \varepsilon_{k\ell} \tag{1.4}$$

where

$$C_{ijk\ell} = C_{jik\ell} = C_{k\ell ij} = C_{ij\ell k} \tag{1.5}$$

Thus, 21 of the 81 components are independent. The material is elastically isotropic when there are no preferred directions in the material, and the elastic constants are the same whatever the orientation of the cartesian coordinate system in which the components of τ_{ij} and ε_{ij} are evaluated. It can be shown that elastic isotropy implies that the constants $C_{ijk\ell}$ may be expressed as

$$C_{ijk\ell} = \lambda \delta_{ij} \delta_{k\ell} + \mu(\delta_{ik}\delta_{j\ell} + \delta_{i\ell}\delta_{jk}) \tag{1.6}$$

where δ_{ij} is the Kronecker delta. The stress-strain relation then assumes the well-known form

$$\tau_{ij} = \lambda\varepsilon_{kk}\delta_{ij} + 2\mu\varepsilon_{ij} \tag{1.7}$$

where λ and μ are Lamé's elastic constants.

Substitution of Eq. (1.1) into (1.7) and subsequent substitution of (1.7) in Eq. (1.3) produces the displacement equations of motion

$$\mu\, u_{i,jj} + (\lambda + \mu)u_{j,ji} = \rho\ddot{u}_i \tag{1.8}$$

Equations (1.8) form a somewhat awkward system of coupled partial differential equations.

The equations governing linearized elasticity were fully developed by the middle of the 19th century. It took another fifty years before the first mathematical analysis of elastostatic stress concentration based on elasticity theory was published. In 1898 G. Kirsch [2] investigated the two dimensional stress distribution around a small circular hole in a large body subjected to longitudinal tension. He showed that the peak circumferential stress at the hole is three times larger than the unperturbed stress. Fifteen years later stresses near an elliptic hole of semiaxes a,b $(a > b)$ in a field of longitudinal tension were analyzed by Inglis [3]. By taking the limit $b \to 0$ the work of Inglis provides the transition to a crack of width 2a. This limit which was apparently first computed by Griffith [4], produces a stress field with square-root singularities at the crack tips. By now the computation of stress singularities has filled numerous pages in technical journals.

Stress singularities are disturbing, not only because no real material can actually sustain singular stresses, but also because the basic premises of the linearized theory are violated by the appearance of singularities. Nevertheless the stress singularities, whose strengths are measured by stress intensity factors, have become accepted as useful mathematical idealizations in linear fracture mechanics. The usual rationalization is based on the concept of small scale yielding, whereby it is assumed that the large stresses are relieved in a small region of plastic flow. It is further assumed that the stresses outside of the small plastic zone are adequately represented by the dominant singular term of the elasticity solution. By this argument it is justifiable to base computations on elasticity solutions.

Square-root singularities play an important role in the energy criterion for fracture which was stated in 1921 by Griffith [4]. The use of energy considerations in a criterion for fracture is based on the observation that energy is "dissipated" during the fracture process, even if the material is perfectly elastic. The loss of mechanical energy as new fracture surface is formed becomes plausible if we consider the work of the internal (cohesive) forces that are removed as new fracture surface, which is free of tractions, is formed. Since the cohesive tractions that are removed are opposite in direction to the relative displacements of the newly formed fracture surfaces, their work is negative. Thus, in the course of crack propagation mechanical energy is extracted from the body. This energy is given out to whatever internal mechanism accounts for the cohesive forces. Generally the cohesive forces are implicitly identified with the atomic bonds which are broken when new free surface is formed. In the Griffith theory the work done in breaking these bonds is assigned to surface energy of the newly formed free surface.

The foregoing observations have led to the formulation of an energy criterion for fracture which states that crack growth can occur if the energy required to form fracture surface can just be delivered by the system. The energy delivered by the system is the negative of the work done as cohesive tractions are released. Within the idealized framework of linearized continuum mechanics the region over which cohesive tractions are released as the crack propagates is infinitesimal, namely, the tip of the propagating crack. Nevertheless the energy release rate has a finite value, because of the presence of the square-root terms in the field variables which enter in the energy release rate. A limit process which involves the shrinking of a region of release of cohesive tractions on the crack tip is shown in Ref. [5].

Now that the role of stress singularities in fracture mechanics has been discussed, the remainder of this article could conceivably be a survey of recent numerical and analytical techniques for analyzing the elastostatic and elastodynamic stress fields around cracks. There are, however, already several volumes of survey papers on elastostatic work. Among these we mention Volume II (Mathematical Fundamentals), of Fracture, [6], and a recent volume on methods of analysis [7]. The fundamentals of fracture mechanics have been discussed in two recently published texts, [8] and [9]. Rather than discuss elastostatic analysis in great detail, we will therefore highlight only a few key contributions, in the next section. In section 3 we proceed to discuss some recent work on elastodynamic analysis of stress fields around cracks.

2. Elastostatic Stress Singularities

Figure 1 shows the two-dimensional geometry of a slit in a body which is unbounded in the x_3-direction. For the two-dimensional case it is quite simple to analyze the asymptotic behavior of the fields near the crack tips by a technique which was proposed by Knein [10], Williams [11] and Karal and Karp [12]. One considers solutions, say for the displacements, of the general forms $r^p f_j(\theta)$. Substitution in the elastostatic displacement equations of equilibrium, i.e., Eq. (1.8) with zero right-hand side, yields in the limit $r \to 0$ a coupled system of ordinary differential equations for $f_j(\theta)$, which contains p as a parameter. Solution of these equations with the appropriate boundary conditions yields $p = 0.5$, as well as explicit expressions for $f_j(\theta)$. Separate angular distributions are obtained for in-plane displacements that are symmetric and antisymmetric with respect to the plane of the crack, respectively, and for anti-plane displacements. In the parlance of fracture mechanics the in-plane symmetric displacements correspond to Mode I fracture, in-plane antisymmetric displacements correspond to Mode II fracture, and anti-plane displacements correspond to Mode III fracture. These fracture modes are illustrated in Fig. 2. The corresponding variations with θ of the stress components are

In-plane symmetric: Mode I

$$\begin{Bmatrix} \tau_{11} \\ \tau_{12} \\ \tau_{22} \end{Bmatrix} = \frac{K_I}{(2\pi r)^{\frac{1}{2}}} \cos(\theta/2) \begin{Bmatrix} 1 - \sin(\theta/2)\sin(3\theta/2) \\ \sin(\theta/2)\cos(3\theta/2) \\ 1 + \sin(\theta/2)\sin(3\theta/2) \end{Bmatrix} \tag{2.1}$$

In-plane antisymmetric: Mode II

$$\begin{Bmatrix} \tau_{11} \\ \tau_{12} \\ \tau_{22} \end{Bmatrix} = \frac{K_{II}}{(2\pi r)^{\frac{1}{2}}} \begin{Bmatrix} -\sin(\theta/2)[2+\cos(\theta/2)\cos(3\theta/2)] \\ \cos(\theta/2)[1-\sin(\theta/2)\sin(3\theta/2)] \\ \sin(\theta/2)\cos(\theta/2)\cos(3\theta/2) \end{Bmatrix} \tag{2.2}$$

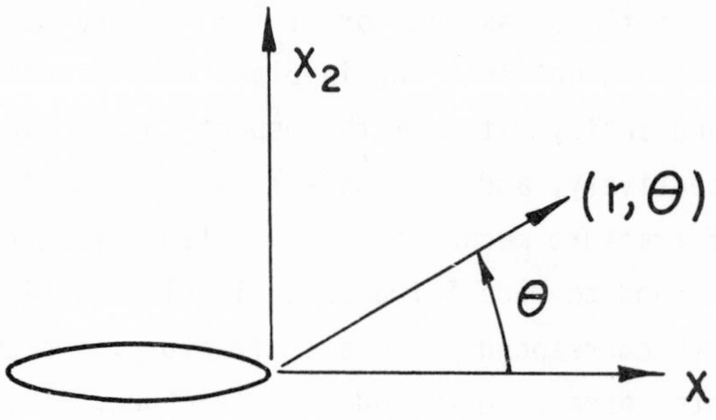

Fig. 1 Crack in a two-dimensional geometry.

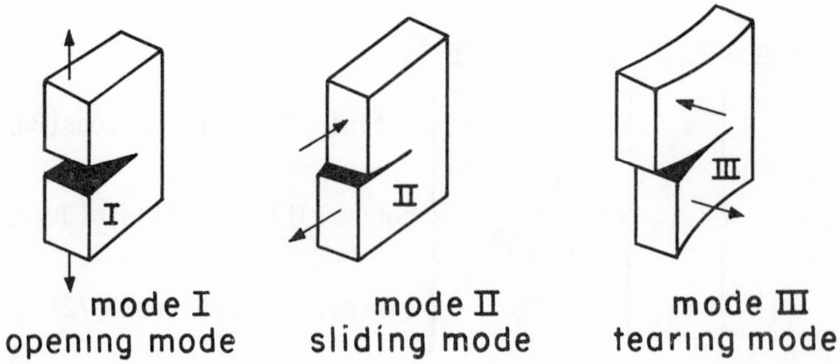

Fig. 2 Fracture modes.

Anti-plane: Mode III

$$(2.3) \qquad \begin{Bmatrix} \tau_{13} \\ \tau_{23} \end{Bmatrix} = \frac{K_{III}}{(2\pi r)^{1/2}} \begin{Bmatrix} -\sin(\theta/2) \\ \cos(\theta/2) \end{Bmatrix}$$

Here K_I , K_{II} and K_{III} are the stress-intensity factors. The results (2.1) - (2.3) also have applicability to three dimensional problems for cracks with a smoothly curved edge, as shown in Fig. 3. If we take a local cartesian coordinate system, the angular distributions in the x_1 x_2 - plane of the stress-intensity factors are as given by Eqs. (2.1) - (2.3). The geometrical details of the crack edge enter of course in the magnitudes of the stress-intensity factors.

Analytical and numerical methods to compute the stress intensity factors for two-dimensional cases have been explored in great detail, see e.g. Refs. [6] and [7]. Numerical analysis of stress fields around cracks, primarily by the finite element method, is taking an increasingly dominant role. In such work the singularities displayed in Eqs. (2.1) - (2.3) are usually accommodated by a special "singular element" which is embedded in conventional elements. The finite element computer program then calculates the strength of the singularity, i.e., the stress intensity factor. Recent work in this area is discussed in Ref. [13].

Some interesting computations of stress-intensity factors have been carried out in recent years by employing path-independent integrals. In a recent paper Eshelby [14] traces path-independent integrals back to a paper by Noether [15]. If we have a set of quantities $u_i(x_m)$ which depend on the variable x_m , and which satisfy the equations

(2.4) $$\frac{\partial W}{\partial u_i} - \left(\frac{\partial W}{\partial u_{i,j}}\right)_{,j} = 0 ,$$

where $W = W(u_i, u_{i,j})$ does not depend explicitly on x_m, then

(2.5) $$P_{\ell j,j} = 0 .$$

In Eq. (2.5)

(2.6) $$P_{\ell j} = W\delta_{\ell j} - \left(\frac{\partial W}{\partial u_{i,j}}\right) u_{i,\ell} .$$

It then follows from Gauss' theorem that

(2.7) $$F_\ell = \int_S P_{\ell j} n_j dS = 0 ,$$

provided that there are no singularities of the integrand within the surface S. If there are such singularities F_ℓ has the same value for all surfaces S which can be deformed into one another without passing any of the singularities of the integrand. In elastostatics, where $W = W(\varepsilon_{ij})$, and ε_{ij} is defined by Eq. (1.1), we have

(2.8) $$\tau_{ij} = \frac{1}{2}\left(\frac{\partial W}{\partial \varepsilon_{ij}} + \frac{\partial W}{\partial \varepsilon_{ji}}\right) ,$$

we find

(2.9) $$F_\ell = \int_S (W n_\ell - T_i u_{i,\ell}) dS .$$

In Eq. (2.9) T_i are the components of the stress vector, which are related to the components of the stress tensor by

(2.10) $$T_i = \tau_{ij} n_j \ .$$

Equation (2.9) is usually referred to as the J integral. It was discovered by Rice [16] that the J integral can be employed in a simple manner to compute stress-intensity factors. Examples are given in Refs. [16] and [17].

In conjunction with the simple deformation theory of plasticity, the J-integral has been used to calculate asymptotic results for so-called "small-scale" yielding near the tips of cracks loaded in Mode I, II, and III, see Refs. [17] - [20]. In these computations, the dominant part of the singular solution near the tip in the plastic range is determined to within a scalar factor by means of the method discussed previously in this section, in referring to Refs. [10] - [12]. The unknown factor is then found from the invariance of the J integral evaluated for paths around the crack tip at small and large radii, where conditions are respectively purely plastic and purely elastic.

In recent years path-independent integrals have been employed to compute stress-intensity factors for a nonlinear geometry, and for nonlinear material behavior by Knowles and Sternberg [21]. Path-independent integrals for elastodynamic problems have been discussed by Freund [22].

It is of interest to cite a few references for stress fields around a crack tip in an interface. For *elastostatic* problems the nature of the singularity near the tip of a crack along the interface of two bonded dissimilar half planes was analyzed by Williams [23]. Williams was apparently the first to discover that the singular parts of the stresses show intense oscillations. More complete elastostatic solutions were subsequently presented in Refs. [24] - [26]. Mathematically the oscillating singularities appear, because the factor

p , appearing in the asymptotic displacement expressions $r^p\ f_i(\theta)$, turns out to be a complex number for an interface flaw. For joined wedges of dissimilar materials the Mellin transform technique has been employed with advantage to analyze the fields near the vertices, see Refs. [27] and [28] for representative work.

Within the context of classical, linearized, isotropic, homogeneous elasticity, two-dimensional cracks have been considered in great detail. One class of problems that requires further attention concerns branching cracks. For a recent contribution we refer to Ref. [29]. Much work remains to be done for three-dimensional geometries, especially for surface flaws and through cracks in layers. The latter problem was recently attacked by Folias [30], with some surprising results, namely displacement distributions that are singular as a tip of the crack at a free face of the layer is approached.

3. Elastodynamic Stress Singularities

Elastodynamic effects become of importance if the propagation of a crack is fast, as for example in essentially brittle fracture, so that rapid motions are generated in the medium. The term "dynamic fracture" will be used to denote the effects of inertia resulting from the rapid propagation of a crack. Inertia effects in conjunction with fracture phenomena are also important if the external loads give rise to propagating mechanical disturbances (as for impact loads and explosive charges) which strike the crack and cause fracture. Spalling is an example of a fracture phenomenon caused by the rapid application of loads. The label "dynamic loading" is attached to the effects of inertia on fracture due to rapidly applied loads.

In the analysis of elastodynamic problems it is often found that at certain specific locations in a body the dynamic stresses are higher than the stresses computed from the corresponding problem of static equilibrium. The reflection of elastic waves provides an example of this effect. Reflection may give rise to an increase of the stress level, as is well known for reflection of a plane wave from a rigidly clamped boundary, when the stresses actually may be doubled in the vicinity of the boundary. A comparable effect occurs when a wave is diffracted by a crack. The dynamic stress "overshoot" in the vicinity of a crack tip may be as high as 30 percent. In view of the dynamic amplification of the stress level it is entirely conceivable that there are cases for which fracture does not occur under a gradually applied system of loads, but where the material does indeed fracture when the same system of loads is rapidly applied and gives rise to waves. In Reference [5] a number of observations on the influence of elastodynamic effects are presented on the basis of solutions of example problems. An extensive list of references is also presented in Ref. [5]. The proceedings of a recent meeting on dynamic fracture contain many contributions of interest, see Ref. [31].

A noteworthy feature of elastodynamic stress fields near a crack tip propagating in a homogeneous linearly elastic solid, is that maximums of stress intensity factors, which are in the plane of crack propagation at low speeds of the crack tip, move out of this plane when the speed of crack propagation exceeds a critical value. For the transient elastodynamic problem of a crack tip moving with a time-varying velocity along a rather arbitrary but smooth two-dimensional trajectory in an isotropic material, and along a principal plane in an orthotropic material, this effect was recently analyzed in Ref. [32]. The shift of maximum values of near-tip stresses out of the plane of crack propagation as the speed of the crack tip increases, suggests an elastodynamic explanation of bifurcation of a crack, which has been observed experimentally, see Refs. [33] - [35].

In this section we will first examine the nature of elastodynamic singularities near a propagating crack tip, by using an extension to elastodynamic problems of the technique employed by Knein [10], Williams [11], and Karal and Karp [12]. Then we will proceed to analyze the stress singularities for the case that a crack actually bifurcates.

Rapid crack propagation in the plane of the crack

Figure 4 shows a two-dimensional geometry, with a crack propagating in its own plane. A crack tip propagating along a rather arbitrary but smooth trajectory is discussed in Ref. [32]. The speed of the crack tip is $c(t)$, where $c(t)$ is an arbitrary function of time, subject to the conditions that $c(t)$ and dc/dt are continuous. A system of moving Cartesian coordinates (x,y) is centered at the crack tip, such that the x-axis is in the plane of the crack. Moving polar coordinates (r,θ) are attached to the moving crack tip.

In a plane two-dimensional geometry, the system of displacement equations of motion governing linearized elasticity separates into two uncoupled systems, for in-plane and anti-plane displacements, respectively, as is easily checked from Eq. (1.8).

Let the anti-plane displacement be instantaneously defined in terms of the moving coordinate system (x,y); i.e., $w = w(x,y,t)$. The second material derivative with respect to time, which is indicated by a superscript double dot, is then of the form

$$\ddot{w} = \frac{\partial^2 w}{\partial t^2} - \dot{c}(t)\,\frac{\partial w}{\partial x} - 2\,c(t)\,\frac{\partial^2 w}{\partial t\partial x} + [c(t)]^2\,\frac{\partial^2 w}{\partial x^2}\,. \tag{3.1}$$

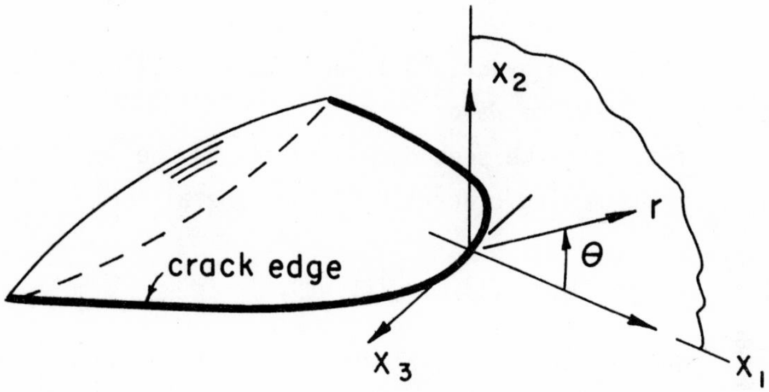

Fig. 3 Three-dimensional plane crack.

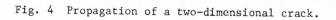

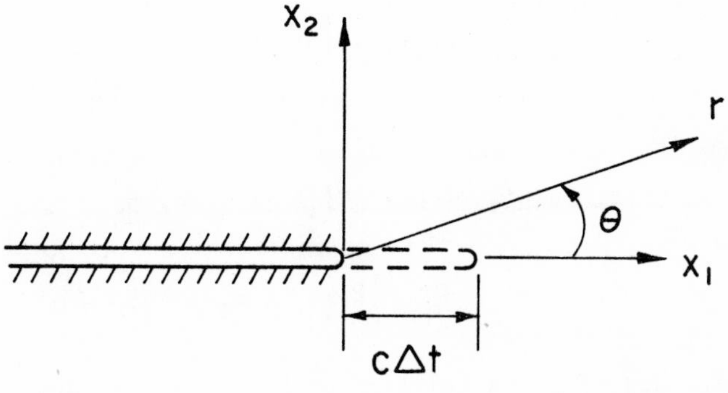

Fig. 4 Propagation of a two-dimensional crack.

Relative to the moving system of Cartesian coordinates the displacement equation of motion is

(3.2) $$\frac{\partial^2 w}{\partial x^2} + \frac{\partial^2 w}{\partial y^2} = \frac{1}{c_T^2}\ddot{w}$$

where $\ddot{w}$ is defined by Eq. (3.1) and $c_T = (\mu/\rho)^{\frac{1}{2}}$ is the velocity of transverse waves

In this paper we seek displacements in the immediate vicinity of the moving crack tip of the general form

(3.3) $$w(r,\theta,t) = r^q T(t) W(\beta,\theta)$$

where r and θ are the polar coordinates shown in Fig. 4, and β is defined as

(3.4) $$\beta = c(t)/c_T \quad .$$

The conditions of vanishing shear tractions on the crack surfaces and of antisymmetry of the displacement field require that

(3.5) $$\theta = \pm\pi: \quad \partial W/\partial\theta = 0 \ ,$$

(3.6) $$\theta = 0: \quad W = 0 \ .$$

Substituting Eq. (3.3) into Eq. (3.2), with $\ddot{w}$ according to Eq. (3.1), multiplying the result by r^{2-q}, and considering the limit $r \to 0$, the following equation for W is obtained

(3.7) $$(1-\beta^2\sin^2\theta)\frac{d^2W}{d\theta^2} - \beta^2(1-q)\sin(2\theta)\frac{dW}{d\theta} + q\left\{q + \beta^2[(2-q)\cos^2\theta-1]\right\}W = 0 \ .$$

Following Ref. [32], a solution of Eq. (3.7) is sought of the form

(3.8) $$W(\beta,\theta) = (1-\beta^2\sin^2\theta)^{q/2} W^*(\beta,\theta) \quad .$$

Substitution of Eq. (3.8) into Eq. (3.7) yields a much simpler equation for W^* which can, however, be further simplified by introducing the variable ω by

$$\tan \omega = (1-\beta^2)^{\frac{1}{2}} \tan \theta \; . \tag{3.9}$$

The resulting equation for W^* is

$$\frac{d^2W^*}{d\omega^2} + q^2W^* = 0 \; . \tag{3.10}$$

In view of the boundary condition at $\theta = 0$, the appropriate solution of Eq. (3.10) is $W^*(\beta,\omega) = A \sin q\omega$. The condition at $\theta = \pi$ gives $dW^*/d\omega = 0$, or $\cos q\pi = 0$, which yields as the lowest root $q = 0.5$. Thus the displacements are $O(r^{\frac{1}{2}})$ as $r \to 0$, and the shear stress $\tau_{\theta z}$ shows a square root singularity. As far as the dependence on θ is concerned it follows from Eq. (3.3) that $\tau_{\theta z} = (\mu/r)\partial w/\partial\theta$ is proportional to the following function, $T_{\theta z}$,

$$T_{\theta z} = (1-\beta^2)^{-\frac{1}{2}} \psi_1 \sin \theta + \psi_2 \cos \theta \tag{3.11}$$

Here ω has been eliminated by means of Eq. (3.9). The functions ψ_1 and ψ_2 are defined as

$$\psi_1 = \left[\frac{(1-\beta^2\sin^2\theta)^{\frac{1}{2}} - \cos\theta}{1 - \beta^2\sin^2\theta}\right]^{\frac{1}{2}} \tag{3.12}$$

$$\psi_2 = \left[\frac{(1-\beta^2\sin^2\theta)^{\frac{1}{2}} + \cos\theta}{1 - \beta^2\sin^2\theta}\right]^{\frac{1}{2}} . \tag{3.13}$$

For various values of c/c_T, the function $T_{\theta z}$ which governs the angular variation of the stress intensity factor is plotted in Fig. 5. It is noted that the maxima of $T_{\theta z}$ move out of the plane of the crack as the speed of the crack tip increases beyond $c/c_T \sim 0.5$.

For in-plane motions it is convenient to express the displacements $u_1(x_1,x_2,t)$ and $u_2(x_1,x_2,t)$ in terms of the displacement potentials φ and ψ by

$$(3.14a,b) \qquad u_1 = \frac{\partial\varphi}{\partial x_1} + \frac{\partial\psi}{\partial x_2} \quad ; \quad u_2 = \frac{\partial\varphi}{\partial x_2} - \frac{\partial\psi}{\partial x_1}$$

Equations (3.14a,b) satisfy the displacement equations of motion provided that φ and ψ are solutions of

$$(3.15a,b) \qquad \frac{\partial^2\varphi}{\partial x_1^2} + \frac{\partial^2\varphi}{\partial x_2^2} = \frac{1}{\varkappa^2 c_T^2}\ddot{\varphi} \quad ; \quad \frac{\partial^2\psi}{\partial x_1^2} + \frac{\partial^2\psi}{\partial x_2^2} = \frac{1}{c_T^2}\ddot{\psi} \ .$$

Here $\ddot{\varphi}$ and $\ddot{\psi}$ are defined analogously to Eq. (3.1), and

$$(3.16) \qquad \varkappa^2 = \frac{c_L^2}{c_T^2} = \frac{\lambda + 2\mu}{\mu} = \frac{2(1-\nu)}{1-2\nu} \ .$$

In Eq. (3.16) λ and μ are Lamé's elastic constants, and ν is Poisson's ratio.

In the vicinity of the crack tip, expressions for the potentials φ and ψ are sought in the forms

(3.17a,b)

$$\varphi(r,\theta,t) = r^p T(t)\Phi(\alpha,\theta) \ , \ \psi(r,\theta,t) = r^p T(t)\Psi(\beta,\theta)$$

respectively, where

$$(3.18) \qquad \alpha = \frac{c(t)}{c_L} = \frac{c(t)}{\varkappa c_T} = \frac{\beta}{\varkappa} \ .$$

In order that the strain energy density be integrable, we must have $\mathrm{Re}(p) > 1$. In the system of polar coordinates the boundary conditions on the surface of the propagating crack are $\tau_\theta = 0$ and $\tau_{r\theta} = 0$, for $\theta = \pm\pi$, $r > 0$, which yields at $\theta = \pm\pi$

$$(\beta^2-2)p\Phi - 2\frac{d\Psi}{d\theta} = 0 \quad , \quad 2\frac{d\Phi}{d\theta} + (\beta^2-2)p\Psi = 0 \quad . \tag{3.19a,b}$$

For Mode I fracture the displacements are symmetric with respect to $x_2 = 0$, and thus $\partial u_1/\partial\theta = 0$ and $u_2 = 0$, for $\theta = 0$, $r > 0$, while for Mode II fracture the displacement field is antisymmetric with respect to $x_2 = 0$, which implies $\partial u_2/\partial\theta = 0$ and $u_1 = 0$, for $\theta = 0$, $r > 0$. These conditions yield

For Mode I, $\theta = 0$: $d\Phi/d\theta = 0$, $\psi = 0$

For Mode II, $\theta = 0$: $\Phi = 0$, $d\Psi/d\theta = 0$.

The method of solution for $\Phi(\alpha,\theta)$ and $\Psi(\beta,\theta)$ is completely analogous to the method used for solving $W(\beta,\theta)$. Details can be found in Ref. [32]. For the symmetric fields the results are:

(3.20)

$$\Phi(\alpha,\theta) = \left\{(1-\alpha^2)^{\frac{1}{2}}\Phi_1 \sin 2\theta + \left[\left(1-\frac{\alpha^2}{2}\right)\cos 2\theta + \frac{\alpha^2}{2}\right]\Phi_2\right\}\frac{A}{\sqrt{2}}$$

where

$$\Phi_1 = \left[\frac{(1-\alpha^2\sin^2\theta)^{\frac{1}{2}} - \cos\theta}{1-\alpha^2\sin^2\theta}\right]^{\frac{1}{2}} \tag{3.21}$$

$$\Phi_2 = \left[\frac{(1-\alpha^2\sin^2\theta)^{\frac{1}{2}} + \cos\theta}{1-\alpha^2\sin^2\theta}\right]^{\frac{1}{2}} \tag{3.22}$$

and

(3.23)

$$\Psi(\beta,\theta) = \left\{-\left[\left(1 - \frac{\beta^2}{2}\right)\cos 2\theta + \frac{\beta^2}{2}\right]\Psi_1 + (1-\beta^2)^{\frac{1}{2}}\,\Psi_2\sin 2\theta\right\}\frac{B}{\sqrt{2}} .$$

Here Ψ_1 and Ψ_2 are defined by equations (3.12) and (3.13).

The functions Φ and Ψ governing the dependence on the angle θ of the potentials for the antisymmetric problem (Mode II fracture) can be obtained in the same manner. It is found that Φ is of the form given by equation (3.23) with β replaced by α , and Ψ is of the form (3.20) with α replaced by β .

The relevant stresses in the system of polar coordinates, r , θ , follow from Hooke's law. Results are presented in Ref. [32].

<u>Bifurcation of a crack</u>

The results obtained earlier in this section suggest an elastodynamic explanation of crack branching. We take the view that crack branching is a result of instability. A necessary condition for this instability should be determined on the basis of a comparison of states prior to branching and after branching has taken place. This comparison requires information on the elastodynamic near-tip fields for the case that a crack has actually bifurcated. In the remainder of this section the requisite mathematical methods are presented for analysis of the elastodynamic stress fields near a bifurcating crack. The analysis is carried out for the case of antiplane strain, and for instantaneous crack bifurcation. The in-plane case has not yet yielded to mathematical analysis.

In a stationary system of polar coordinates, r and θ , two-dimensional antiplane wave motions are governed by

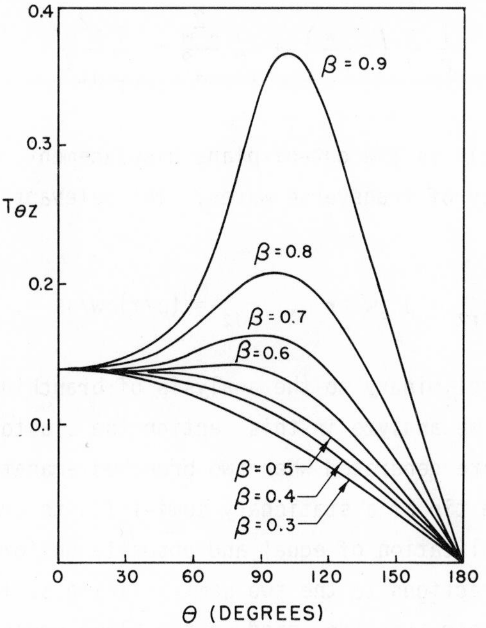

Fig. 5 $T_{\theta z}$ vs. θ for various values of $\beta = c/c_T$.

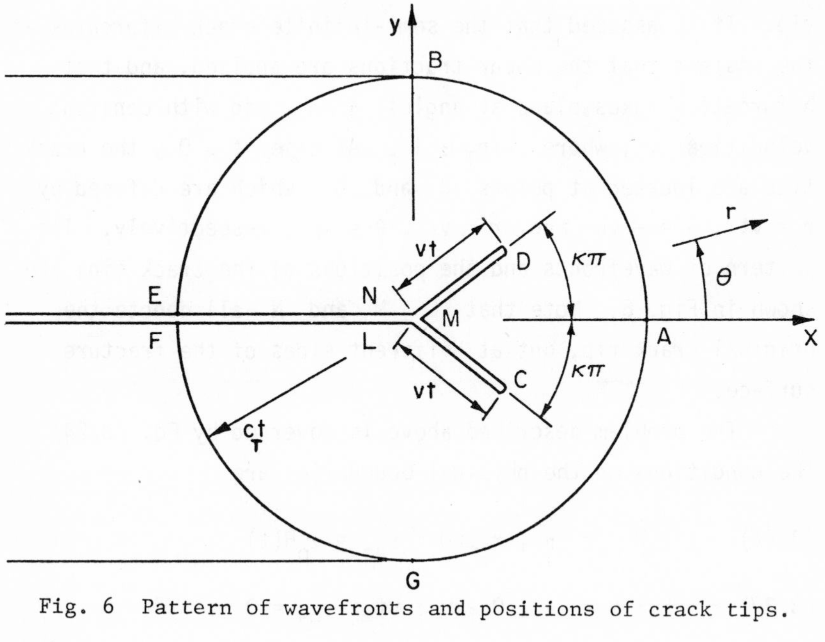

Fig. 6 Pattern of wavefronts and positions of crack tips.

(3.24) $$\frac{1}{r}\frac{\partial}{\partial r}\left(r\frac{\partial w}{\partial r}\right)+\frac{1}{r^2}\frac{\partial^2 w}{\partial \theta^2}=\frac{1}{c_T^2}\frac{\partial^2 w}{\partial t^2}$$

where $w(r,\theta,t)$ is the out-of-plane displacement, and c_T is the velocity of transverse waves. The relevant shear stresses are

(3.25) $$\tau_{rz}=\mu\,\partial w/\partial r \quad , \quad \tau_{\theta z}=(\mu/r)\partial w/\partial\theta \quad .$$

As a preliminary to the analysis of branching of a running crack we analyze in this section the elastodynamic fields which are generated when two branches emanate symmetrically from the tip of a stationary semi-infinite crack, upon the sudden application of equal and opposite uniform anti-plane shear tractions to the two semi-infinite surfaces of the crack. The shear tractions produce two plane waves, and a diffracted cylindrical wave with center at the original crack tip. It is assumed that the semi-infinite crack bifurcates at the instant that the shear tractions are applied, and that bifurcation takes place at angles $\pm\varkappa\pi$, and with constant velocities v, where $v/c_T < 1$. At time $t > 0$, the crack tips are located at points C and D, which are defined by $r = vt$, $\theta = -\varkappa\pi$ and $r = vt$, $\theta = \varkappa\pi$, respectively. The pattern of wavefronts and the positions of the crack tips are shown in Fig. 6. Note that L, M and N all denote the original crack tip, but at different sides of the fracture surface.

The problem described above is governed by Eq. (3.24). The conditions on the physical boundaries are

(3.26) $$\theta=\pm\pi \ , \ r>0 : \tau_{\theta z}=\tau_o H(t)$$

(3.27) $$\theta=\pm\pi \ , \ 0<r<vt : \tau_{\theta z}=0 \quad .$$

The uniform distributions of shear tractions at $\theta = \pm \pi$ generate plane waves, with constant particle velocities of magnitudes $\pm c_T \tau_0/\mu$. Along the segments BE and FG , see Fig. 6, the particle velocities then are

$$\frac{\pi}{2} < \theta \leq \pi \ , \ r = c_T t: \ \dot{w} = c_T \tau_0/\mu \tag{3.28}$$

$$-\pi \leq \theta < -\frac{\pi}{2} \ , \ r = c_T t: \ \dot{w} = -c_T \tau_0/\mu \ . \tag{3.29}$$

The material is undisturbed ahead of the segment BG , and thus

$$-\frac{\pi}{2} < \theta < \frac{\pi}{2} \ , \ r = c_T t: \ \dot{w} = 0 \ . \tag{3.30}$$

Equations (3.26) - (3.30) can be considered as boundary conditions on $\dot{w}(r,\theta,t)$ for the region of the cylindrical wave. It is evident from these conditions that the displacement field is antisymmetric with respect to $\theta = 0$, which implies that w , and thus $\dot{w}$ vanishes as $\theta = 0$.

$$\theta = 0 \ , \ 0 < r < c_T t: \ \dot{w} = 0 \tag{3.31}$$

It then suffices to consider only the region $0 \leq \theta \leq \pi$, $0 < r \leq c_T t$.

Let us consider the particle velocity as the dependent variable. For $t > 0$ it then follows from Eqs. (3.26) and (3.27) that

$$\theta = \pi \ , \ r > 0: \ \frac{\partial \dot{w}}{\partial \theta} = 0 \tag{3.32}$$

$$\theta = \varkappa\pi \ , \ 0 < r < vt: \ \frac{\partial \dot{w}}{\partial \theta} = 0 \ . \tag{3.33}$$

Equations (3.28) and (3.30) - (3.33) are the conditions on the particle velocity $\dot{w}$, on the boundaries of the region $0 < r < c_T t$, $0 \le \theta \le \pi$. These conditions suggest that the particle velocity $\dot{w}$ is self similar. The property of self-similarity implies that $\dot{w}$ depends on r/t and θ, rather than on θ and r and t separately. As discussed in Refs. [1, p. 154] and [36] it is then convenient to introduce the new variable $s = r/t$, whereupon the equation for $\dot{w}(s,\theta)$ easily follows. For $s < c_T$ Chaplygin's transformation $\beta = \cosh^{-1}(c_T/s)$ reduces the equation for $\dot{w}(s,\theta)$ to Laplace's equation

$$\text{(3.34a,b)} \qquad \frac{\partial^2 \dot{w}}{\partial \beta^2} + \frac{\partial^2 \dot{w}}{\partial \theta^2} = 0 \quad , \quad \beta = \cosh^{-1}(c_T/s) \ .$$

The real transformation given by Eq. (3.34b) maps the interior of the domain $0 \le \theta \le \pi$, $s < c_T$ into a semi-infinite strip in the θ-β plane containing a slit. The domain in the θ-β plane is shown in Fig. 7, where the boundary conditions corresponding to Eqs. (3.28) - (3.33) are also indicated.

In the θ-β plane the harmonic function $\dot{w}$ can be taken as the real part of an analytic function $G(\gamma)$

$$\text{(3.35)} \qquad \dot{w} = \operatorname{Re} G(\gamma) \quad , \quad \gamma = \beta + i\theta \quad ,$$

where $G(\gamma)$ can formally be obtained by conformal mapping techniques. The domain in the γ-plane can be related to the upper half of the ζ-plane by means of a Schwarz-Christoffel transformation $\gamma = \omega(\zeta)$, $\zeta = \xi + i\eta$. An appropriate transformation is

(3.36)

$$\omega(\zeta) = -\frac{\xi_M}{\xi_M+\xi_N}\frac{C_1}{(1-\xi_M^2)^{\frac{1}{2}}}\left\{\ell n\left[\left(1-\xi_M^2\right)^{\frac{1}{2}}\left(1-\zeta^2\right)^{\frac{1}{2}}+\zeta\xi_M+1\right]-\ell n\left(\zeta+\xi_M\right)\right\}$$

$$-\frac{\xi_N}{\xi_M+\xi_N}\frac{C_1}{(1-\xi_N^2)^{\frac{1}{2}}}\left\{\ell n\left[\left(1-\xi_N^2\right)^{\frac{1}{2}}\left(1-\zeta^2\right)^{\frac{1}{2}}-\zeta\xi_N+1\right]-\ell n\left(\zeta-\xi_N\right)\right]+C_2 .$$

Here C_1 and C_2 are complex constants. Equation (3.36) implies that the points A and E are mapped at $\zeta = -1$ and $\zeta = +1$, respectively, while the point D is mapped into the origin $\zeta = 0$. The mapping of the point E gives $C_2 = i\pi$. Considering the change in imaginary parts at M and N we obtain two relations. A comparison of the coordinates of point D in the γ- and ζ-planes, results in another relation, and still another relation is obtained from the mapping of the point B. For given values of $\varkappa$ and v/c_T, ξ_M, ξ_N and ξ_B can now be computed. These computations must be carried out on a digital computer. The values for limiting cases can, however, be obtained analytically, to yield

(3.37) $\qquad \varkappa = 0: \xi_M = 0 ; \xi_N = v/c_T ; \xi_B = c_T/v$

(3.38) $\qquad \varkappa = 0.5: \xi_M = v/c_T ; \xi_N = v/c_T ; \xi_B \to \infty$

(3.39)

$$\frac{v}{c_T} << 1: \xi_M = \left(\frac{\varkappa}{1-\varkappa}\right)^{1-\varkappa}\frac{v}{c_T}; \ \xi_N = \left(\frac{\varkappa}{1-\varkappa}\right)^{-\varkappa}\frac{v}{c_T}; \ \xi_B = \frac{\varkappa^{\varkappa}(1-\varkappa)^{1-\varkappa}}{1-2\varkappa}\frac{c_T}{v} .$$

Details of the computation of the pertinent points in the ζ-plane are given in Ref. [37].

The boundary conditions show in Fig. 7 transform into the following conditions on the real axis in the ζ-plane.

$$-\infty < \xi \leq -\xi_M: \quad \dot{w} = 0 \tag{3.40}$$

$$-\xi_M < \xi < 1: \quad \partial\dot{w}/\partial\eta = 0 \tag{3.41}$$

$$1 < \xi < \xi_B: \quad \dot{w} = c_T\tau_o/\mu \tag{3.42}$$

$$\xi_B \leq \xi < \infty: \quad \dot{w} = 0 \ . \tag{3.43}$$

A harmonic function which satisfies these conditions is

$$\dot{w} = A \ \mathrm{Im} \ \ell n \frac{F(\zeta)}{(\zeta-\xi_B)T} \tag{3.44}$$

where

$$A = c_T\tau_o/\mu\pi \ , \tag{3.45}$$

and

$$F(\zeta) = \left[\left(\zeta + \xi_M\right)^{\frac{1}{2}}\left(\xi_B - 1\right)^{\frac{1}{2}} + \left(\zeta - 1\right)^{\frac{1}{2}}\left(\xi_B + \xi_M\right)^{\frac{1}{2}}\right]^2 \tag{3.46}$$

$$T = \left[\left(\xi_B - 1\right)^{\frac{1}{2}} + \left(\xi_B + \xi_M\right)^{\frac{1}{2}}\right]^2 \ . \tag{3.47}$$

To obtain an explicit expression for $\dot{w}$ as a function of r/t and θ , ζ has to be solved in terms of γ from the mapping function $\gamma = \omega(\zeta)$ given by Eq. (3.36). This appears to be rather difficult. Without inverting the mapping function it is, however, possible to derive explicit expressions for the singular parts of the stresses and the particle velocities in the vicinity of the moving crack tip.

In the θ-β plane shown in Fig. 7, we consider a cylindrical coordinate system z, p, ψ centered at D. In the strip $\dot{w}$ is governed by Laplace's equation, and $\partial\dot{w}/\partial\theta$ vanishes on the surfaces of the slit. This implies that $\partial\dot{w}/\partial p$ and $(1/p)\partial\dot{w}/\partial\psi$, or $\dot{\tau}_{pz}$ and $\dot{\tau}_{\psi z}$, show square root singularities at the tip of the slit. The forms of $\dot{\tau}_{pz}$ and $\dot{\tau}_{\psi z}$ follow, from Eq. (2.3) as

$$\dot{\tau}_{pz} - i\dot{\tau}_{\psi z} \sim [K/(2p)^{\frac{1}{2}}]\exp(i\psi/2) . \tag{3.48}$$

It is not difficult to show that the intensity factor K can be expressed as

$$K = \mu\sqrt{2} \lim_{\zeta\to\xi_D} \frac{[\omega(\zeta)-\omega(\xi_D)]^{\frac{1}{2}}F'(\zeta)}{\omega'(\zeta)} . \tag{3.49}$$

For details we refer to Ref. [38]. In general the function $dF/d\zeta$ is continuous at $\zeta = \xi_D$, and $d\omega/d\zeta$ vanishes as ζ approaches ξ_D. In that case we use the expansions

$$\omega(\zeta) - \omega(\xi_D) = \frac{1}{2}\zeta^2 \omega''(\xi_D) + --- , \tag{3.50}$$

$$\omega'(\xi_D) = \zeta\,\omega''(\xi_D) + --- . \tag{3.51}$$

In the limit Eq. (3.49) then yields

$$K = \mu F'(\xi_D)/[\omega''(\xi_D)]^{\frac{1}{2}} . \tag{3.52}$$

For the problem at hand we find by using Eqs. (3.36) and (3.46)

$$K = -\frac{c_T\tau_o}{\pi}\,\frac{\xi_N(\xi_B-1)^{\frac{1}{2}}(\xi_B+\xi_M)^{\frac{1}{2}}}{(1-\varkappa)^{\frac{1}{2}}\xi_B(\xi_M+\xi_N)^{\frac{1}{2}}(1-\xi_N^2)^{\frac{1}{4}}} . \tag{3.53}$$

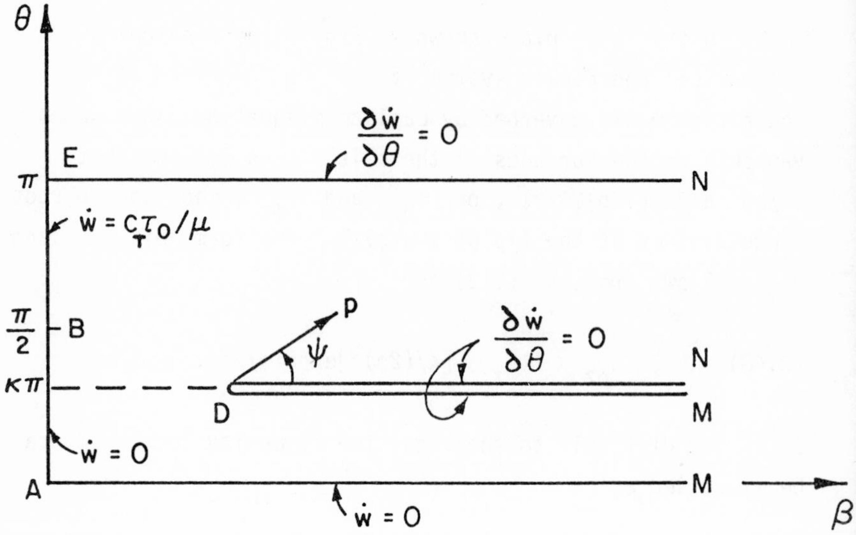

Fig. 7 Domain in the θ-β plane.

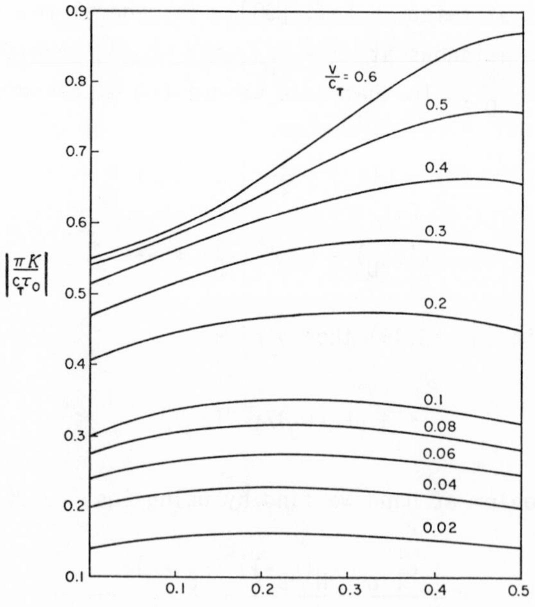

Fig. 8 Factor K versus κ for various values of v/c_T .

The factor K is plotted in Fig. 8.

The intensity factor K refers to fields in the β-θ plane. We still have to find the shear stresses in the vicinity of the crack tip in the physical plane, i.e., in terms of polar coordinates R and φ centered at point D. After a number of manipulations, which are described in some detail in Ref. [38] we obtain

$$\tau_{\varphi z} \sim k_\tau T_{\varphi z}\left(\frac{v}{c_T},\varphi\right)/R^{\frac{1}{2}} \tag{3.54}$$

where

$$k_\tau = \left(1 - \frac{v^2}{c_T^2}\right)^{\frac{1}{4}}\left(\frac{t}{v}\right)^{\frac{1}{2}} K \tag{3.55}$$

and $T_{\varphi z}$ follows from Eq. (3.11) by the substitution $\theta = \pi - \varphi$. It is, of course, not surprising that the function $T_{\varphi z}$ appears in this result, since it was shown earlier that the angular variation of the near tip stress field depends only on the instantaneous value of the speed of crack propagation.

The limitcase $\varkappa \to 0$ requires some special attention. By employing the results stated in Eq. (3.37) in the expression for K given by Eq. (3.53) we find in the limit $\varkappa \to 0$

$$K = -\frac{c_T\tau_0}{\pi}\,\frac{(1-v/c_T)^{\frac{1}{4}}}{(1+v/c_T)^{\frac{1}{4}}}\left(\frac{v}{c_T}\right)^{\frac{1}{2}} . \tag{3.56}$$

Equation (3.56) may be compared with the corresponding K for the case that the crack does not bifurcate, but propagates in its own plane. By examining the geometry of the mapping it is not difficult to see that the point M then coincides with the point D, while the positions of N and B are given by

Eq. (3.37). The corresponding expressions for $F'(\zeta)$ and $\omega'(\zeta)$ follow

$$(3.57) \qquad F'(\zeta) = \frac{c_T\tau_o}{\mu\pi}\,\frac{i}{\zeta - c_T/v}\,\frac{(c_T/v-1)^{\frac{1}{2}}(c_T/v)^{\frac{1}{2}}}{(\zeta-1)^{\frac{1}{2}}\zeta^{\frac{1}{2}}}$$

$$(3.58) \qquad \omega'(\zeta) = \frac{C_1}{(\zeta - v/c_T)(1-\zeta^2)^{\frac{1}{2}}}$$

where C_1 follows as

$$(3.59) \qquad C_1 = -\,(1-v^2/c_T^2)^{\frac{1}{2}} \quad .$$

The factor K now follows directly from Eq. (3.49) as

$$(3.60) \qquad K = -\,\frac{c_T\tau_o}{\pi}\,\frac{(1-v/c_T)^{\frac{1}{4}}}{(1+v/c_T)^{\frac{1}{4}}}\left(\frac{v}{c_T}\right)^{\frac{1}{2}}\sqrt{2} \quad .$$

The difference between Eqs. (3.56) and (3.60) is in the multiplying factor $\sqrt{2}$. Thus, when the crack bifurcates with infinitisimally small angle, the intensity factors drop considerably. For the quasi-static case it was found by Smith [Ref. 39, Eqs. (28) and (31)] that for small bifurcation angles the critical stress is $\sqrt{2}$ times as large as for propagation along a single segment in the plane of the crack, which is consistent with the results found here.

For various values of v/c_T the factor K given by Eq. (3.53) is plotted in Fig. 10. It is noted that the maximum of K shifts to a higher value of $\varkappa$ as v/c_T increases.

The results presented in this section can be used to examine the proclivity towards crack bifurcation by employing fracture mechanics considerations. For details we refer to

Ref. [37].

4. Concluding Remarks

The mathematical analysis of elastostatic and elastodynamic stress singularities is an essential part of the study of the stability of flaws with sharp edges in solid bodies. In this paper some analytical results for both elastostatic and elastodynamic problems were reviewed. Some new results for elastodynamic crack bifurcation were presented.

In the past the emphasis has been on purely analytical work. The limits of what can be done analytically may have been reached. There is an obvious trend towards semi-numerical techniques to deal with complicated geometries and more general constitutive behavior of materials.

REFERENCES

1. J. D. Achenbach, Wave Propagation in Elastic Solids, North-Holland Publ. Co./American Elsevier, 1973.
2. G. Kirsch, "Die Theorie der Elastizität und die Bedürfnisse der Festigkeitslehre," Zeitschrift des Vereins Deutscher Ingenieure 42, (1898), p. 797.
3. C. E. Inglis, "Stresses in a Plate due to the Presence of Cracks and Sharp Corners," Proc. Inst. Naval Architects 55, (1913), p. 219.
4. A. A. Griffith, "The Phenomena of Rupture and Flow in Solids," Phil. Trans. Roy. Soc. London A221, (1921), p. 163.
5. J. D. Achenbach, "Dynamic Effects in Brittle Fracture," Mechanics Today, Vol. I (S. Nemat-Nasser, editor), Pergamon, 1974, p. 1.
6. H. Liebowitz (editor), Fracture II, An Advanced Treatise, Academic Press, 1968.
7. G. C. Sih (editor), Mechanics of Fracture, Methods of Analysis and Solutions of Crack Problems, Noordhoff International Publishing, 1973.
8. J. F. Knott, Fundamentals of Fracture Mechanics, Halsted Press, 1973.
9. D. Broek, Elementary Engineering Fracture Mechanics, Noordhoff International Publishing, 1974.
10. M. Knein, "Zur Theorie des Druckversuchs," Abhandlungen aus dem aerodynamischen Institut an der T. H. Aachen 7, (1927), p. 62.

11. M. L. Williams, "Stress Singularities Resulting from Various Boundary Conditions in Angular Corners in Extension," J. of Appl. Mech. 19, (1952), p. 526.
12. S. N. Karp and F. C. Karel, Jr., "The Elastic-Field Behavior in the Neighborhood of a Crack of Arbitrary Angle," Com. P. and Appl. Math. XV, (1962), p. 413.
13. Pin Tong and T. H. H. Pian, "On the Convergence of the Finite Element Method for Problems with Singularity," Int. J. Solids & Structures 9, (1973), p. 313.
14. J. D. Eshelby, "The Calculation of Energy Release Rates," Prospects of Fracture Mechanics (G. C. Sih, H. C. van Elst and D. Broek, editors), Noordhoff Int. Publishing, 1975.
15. E. Noether, "Nachr. Ges. Wiss. Göttingen, math-phys. Klasse," (1918), p. 235, English translation by Tavel, M., Transport Theory and Statistical Physics, 1, (1971), p. 183.
16. J. R. Rice, "A Path Independent Integral and the Approximate Analysis of Strain Concentration by Notches and Cracks," J. Appl. Mech. 35, (1968), p. 379.
17. J. R. Rice, "Mathematical Analysis in the Mechanics of Fracture," in Fracture, An Advanced Treatise Vol. II, (H. Liebowitz, editor), Academic Press, 1968, p. 191.
18. J. W. Hutchinson, "Singular Behavior at the End of a Tensile Crack in a Hardening Material," J. Mech. Phys. Solids 16, (1968), p. 13.
19. J. W. Hutchinson, "Plastic Stress and Strain Fields at a Crack Tip," J. Mech. Phys. Solids 16, (1968), p. 337.
20. J. R. Rice and G. F. Rosengren, "Plane Strain Deformation near a Crack Tip in a Power-Law Hardening Material," J. Mech. Phys. Solids 16, (1968), p. 1.
21. J. K. Knowles and E. Sternberg, "Finite-Deformation Analysis of the Elastostatic Field near the Tip of a Crack: Reconsideration and Higher-Order Results," J. of Elasticity 4, (1974), p. 201.
22. L. B. Freund, "The Analysis of Elastodynamic Crack Tip Stress Fields," in Mechanics Today Vol. III (S. Nemat-Nasser, editor), in press.
23. M. L. Williams, "The Stresses around a Fault or Crack in Dissimilar Media," Bul. Seism. Soc. America 49, (1959), p. 199.
24. F. Erdogan, "Stress Distribution in a Nonhomogeneous Elastic Plane with Cracks," J. Appl. Mech. 30, (1963), p. 232.
25. A. H. England, "A Crack Between Dissimilar Media," J. Appl. Mech. 32, (1965), p. 400.
26. J. R. Willis, "Fracture Mechanics of Interfacial Cracks," J. Mech. Phys. Solids 19, (1971), p. 353.

27. D. B. Bogy, "Stress Singularities at Interface Corners in Bonded Dissimilar Isotropic Elastic Materials," Int. J. Solids Structures 7, (1971), p. 993.
28. D. B. Bogy, "Two Edge-Bonded Elastic Wedges of Different Materials and Wedge Angles under Surface Tractions," J. Appl. Mech 38, (1971), p. 377.
29. S. N. Chatterjee, "The Stress Field in the Neighborhood of a Branched Crack in an Infinite Elastic Sheet," Int. J. Solids Structures 11, (1975), p. 521.
30. E. S. Folias, "On the Three-Dimensional Theory of Cracked Plates," J. Appl. Mech, in press.
31. G. C. Sih (editor), Dynamic Crack Propagation, Noordhoff Int. Publishing, 1973.
32. J. D. Achenbach and Z. P. Bažant, "Elastodynamic Near-Tip Stress and Displacement Fields for Rapidly Propagating Cracks in Orthotropic Materials, J. Appl. Mech. 42, (1975), p. 183.
33. H. Schardin, "Velocity Effects in Fracture," in Fracture (B. L. Averbach, D. K. Felbeck, G. T. Hahn and D. A. Thomas, editors), John Wiley & Sons, 1959, p. 297.
34. F. Kerkhof, Bruchvorgänge in Gläsern, Verlag der Deutschen Glastechnischen Gesellschaft, Frankfurt (Main), 1970.
35. J. F. Kalthof, "On the Propagation Direction of Bifurcated Cracks," in Dynamic Crack Propagation (G. C. Sih, editor), Noordhoff International Publishing, 1974, p. 449.
36. J. D. Achenbach, "Shear Waves in an Elastic Wedge," Int. J. Solids & Structures 6, (1970), p. 379.
37. J. D. Achenbach, "Bifurcation of a Running Crack in Antiplane Strain," Int. J. Solids and Structures, in press.
38. J. D. Achenbach and V. K. Varatharajulu, "Skew Crack Propagation at the Diffraction of a Transient Stress Wave," Q. Appl. Math. 32, (1974), p. 123.
39. E. Smith, "Crack Bifurcation in Brittle Solids, J. Mech. Phys. Solids 16, (1968), p. 329.

FUNDAMENTAL MATHEMATICAL PROBLEMS IN THE THEORY OF NONLINEAR ELASTICITY*

Stuart S. Antman**

1. Introduction

During the last twenty-five years there has been a considerable clarification and development of the foundations of nonlinear continuum physics. Deformation, stress, and equations of motion can now be described simply and elegantly. There is a beautiful theory of constitutive relations. It has even been shown that there is more to classical thermodynamics than the chain rule of differentiation.

During this same time there has also been a remarkable development in nonlinear analysis and in numerical methods for studying nonlinear equations. It would seem high time to resolve the open problems of nonlinear continuum physics, either by subsuming them under the various nonlinear operator theories or by attacking them with effective numerical procedures. Unfortunately such a program does not yet seem viable because there are serious gaps in our knowledge of the physical problem and of mathematical methods.

In this paper we examine the delicate inter-relationship between mathematics and physics that is the source of these difficulties. We limit our attention to the field of nonlinear elasticity, which is the simplest exact theory of nonlinear solid mechanics. In Section 2, we describe the governing equations and examine their mathematical structure.

*This work was partially supported by N.S.F. Grant MPS 73-08587 A02.

**Department of Mathematics, University of Maryland.

In Section 3, we do the same for nonlinearly elastic rod and shell theories. In Section 4, we discuss a number of important specific problems. In Section 5, we conclude with a discussion of promising directions for research. We present a simplified version of the theory that purposely ignores certain subtle questions of analysis in the formulation of the problems. The interested reader can readily make explicit our tacit assumptions on the measurability of sets, the smoothness of functions, etc.

2. The Governing Equations

Let a body occupy a region B in a reference (or initial) configuration. We identify a particle of the body by its position $\underset{\sim}{z}$ in B. The position of particle $\underset{\sim}{z}$ at time t is denoted by $\underset{\sim}{r}(\underset{\sim}{z},t)$.

Having scarcely begun our description of the deformation, we confront our first and most serious obstacle. If $\underset{\sim}{r}$ is to represent the deformation of a body, then $\underset{\sim}{r}(\cdot,t)$ must be an invertible mapping so that two distinct particles of the body cannot simultaneously occupy the same point in space. This is a global restriction on the function $\underset{\sim}{r}$, which is to be characterized locally by partial differential equations. This global requirement does not seem readily handled by available methods of analysis. Moreover, this requirement demands that we prescribe a strategy for boundary conditions when two distinct parts of the body's boundary come into contact. These conditions would give rise to "variational inequalities" of a more vicious type than those presently under study. (Cf [15,16]).

We can retreat from this global condition by merely demanding that $\underset{\sim}{r}(\cdot,t)$ be locally invertible and orientation-preserving, at least almost everywhere. This is ensured by the requirement that the Jacobian be positive:

(1) $$\det \frac{\partial \underset{\sim}{r}}{\partial \underset{\sim}{z}} (\underset{\sim}{z},t) > 0 \quad \text{for (almost) all} \quad \underset{\sim}{z} \in B \; .$$

This local condition is to be imposed on candidates for the solution of the governing equations (which we have not yet exhibited). The collection of functions $\underset{\sim}{r}$ satisfying (1) is not convex since the average of the identity and a rotation of π about a fixed axis has a zero Jacobian. One can easily construct a pair of mappings with positive Jacobians whose average is negative. Moreover, by studying dislocations of a torus one can show that the class of functions satisfying (1) may have an infinite number of connected components, none of which is convex. (Cf. [7]). These geometrical properties of the admissible functions represent a serious obstacle to analysis.

If we wish to replace $\underset{\sim}{r}$ by a deformation variable that belongs to a convex set, then we can consider the Green deformation tensor $\underset{\sim}{C}$ defined by

(2) $$\underset{\sim}{C} = \left(\frac{\partial \underset{\sim}{r}}{\partial \underset{\sim}{z}}\right)^{*} \cdot \left(\frac{\partial \underset{\sim}{r}}{\partial \underset{\sim}{z}}\right) \quad .$$

Here the asterisk represents the transpose. When (1) holds, $\underset{\sim}{C}$ is positive definite and the collection of positive definite tensor functions is convex. Unfortunately, (2) represents an overdetermined system of six partial differential equations for the three components of $\underset{\sim}{r}$. (These equations must be solved in the course of the solution of most problems.) If $\underset{\sim}{C}$ is in $C^2(B)$ and if B is simply connected, then the necessary and sufficient condition for the existence of a solution $\underset{\sim}{r} \in C^3(B)$, unique to within a rigid body motion, is the vanishing of the Riemann-Christoffel curvature tensor based on $\underset{\sim}{C}$. Since this tensor involves second derivatives of $\underset{\sim}{C}$, this condition is extremely unpleasant for modern processes of analysis. I know of no suitable weakened versions of this condition. (See [29], however, for the linear version of (2).)

A guide to choosing between $\underset{\sim}{r}$ and $\underset{\sim}{C}$ as fundamental variables may be furnished by the following considerations. Solving the governing equations (in the static case) may be interpreted as finding the places where an infinite dimensional vector field vanishes on a class of admissible arguments. (The admissible class of $\underset{\sim}{r}$'s is defined by (1) and the admissible class of $\underset{\sim}{C}$'s consists of all positive definite tensor-valued functions.) The admissible class of $\underset{\sim}{r}$'s has unpleasant geometrical properties. Consider the paradigmatic problem of determining whether a continuous two-dimensional vector field that points outward on both boundaries of an annulus must vanish on the annulus. (The annulus is a set with unpleasant geometrical properties.) If the vector field is the gradient of a scalar, then elementary calculus shows that it must vanish on the interior of the annulus, whereas if it is not the gradient of a scalar the geometry of the annulus prevents us from making any conclusion. A sweeping generalization of this problem leads us to the conjecture that if a static problem is conservative so that it can be posed as a problem in the calculus of variations, then the detailed nature of the geometry of the admissible functions will not seriously hinder the analysis. Otherwise it might be necessary to tolerate the difficulties of compatibility conditions. This conjecture is supported by the analysis of nonlinearly elastic rod and shell problems [8].

The simplest way to present the equations of motion is in terms of the coordinate-free description that we have used so far. To lay the foundations of rod and shell theories it is however necessary to introduce curvilinear coordinates. We according let $\underset{\sim}{x} = (x^1, x^2, x^3)$ represent a triple of curvilinear coordinates for B . The collection of all such triples $\underset{\sim}{x}$ is assumed to be in one-to-one correspondence with the collection of particles $\underset{\sim}{z}$ in B . Let $\underset{\sim}{x}$ range over A as $\underset{\sim}{z}$ ranges over B . Let $\underset{\sim}{p}(\underset{\sim}{x},t) = \underset{\sim}{r}(\underset{\sim}{z},t)$. The volume element of B is $dv(\underset{\sim}{x}) = \sqrt{G(\underset{\sim}{x})}dx^1 dx^2 dx^3$, where G is the determinant of the metric tensor for $\underset{\sim}{x}$. Let E be a subset of A . If we make the usual assumption that the force acting on (the body corresponding to) E consists of a force distributed over the volume of E with intensity $\underset{\sim}{f}(\underset{\sim}{x},t)$ (force per unit volume at $\underset{\sim}{x}$ in the reference configuration) and a force distributed over the boundary ∂E with intensity $\underset{\sim}{\tau}(\underset{\sim}{x},t)$ (force per unit area at $\underset{\sim}{x}$ in the reference configuration), then the requirement that the resultant force on E equal the time rate of change of linear momentum of E can be expressed by

$$(3) \quad \int_E \underset{\sim}{f}(\underset{\sim}{x},t)dv(\underset{\sim}{x}) + \int_{\partial E} \underset{\sim}{\tau}(\underset{\sim}{x},t)da(\underset{\sim}{x}) = \int_E \rho(\underset{\sim}{x}) \frac{\partial^2}{\partial t^2} \underset{\sim}{p}(\underset{\sim}{x},t)dv(x) \quad .$$

Here $\rho(\underset{\sim}{x})$ represents the mass density at $\underset{\sim}{x}$ in the reference configuration and da represents the differential surface area on E . Relation (3) is to hold for all E in A . A similar relation describes the angular momentum principle, the consequences of which can be included in the constitutive relations described below. It may well happen that $\underset{\sim}{f}(\underset{\sim}{x},t)$ has the form $\hat{\underset{\sim}{f}}(\underset{\sim}{p}(\underset{\sim}{x},t),\underset{\sim}{x},\underset{\sim}{t})$ and that $\underset{\sim}{\tau}(\underset{\sim}{x},t)|_{x\in\partial A}$ has the form $\hat{\underset{\sim}{g}}(\underset{\sim}{p}(\underset{\sim}{x},t),\underset{\sim}{x},t)$. By a classical argument due to Cauchy or by a sophisticated generalization of it due to Noll (cf. [20]) it can be shown that

$$(4) \qquad \underset{\sim}{\tau}(\underset{\sim}{x},t) = \underset{\sim}{\tau}^k(\underset{\sim}{x},t)n_k(\underset{\sim}{x}) \ ,$$

where $\underset{\sim}{n} = (n_1,n_2,n_3)$ is the unit outer normal to ∂E . Here and below we sum repeated diagonal indices over the range 1,2, 3. (The vectors $\{\underset{\sim}{\tau}^k\}$ are called the Piola-Kirchhoff stress vectors.) The substitution of (4) into (3) allows us to apply Green's theorem to the resulting expression. If the integrands are continuous over A, then we immediately obtain the classical form of the equations of motion.

$$G^{-\frac{1}{2}}(G^{\frac{1}{2}}\underset{\sim}{\tau}^k)_{,k} + \underset{\sim}{f} = \rho \frac{\partial^2 \underset{\sim}{p}}{\partial t^2} . \tag{5}$$

Here $,k \equiv \dfrac{\partial}{\partial x^k}$.

We must also impose initial conditions and boundary conditions. The latter may be a complicated mixture of position and traction conditions. If we take the dot product of (5) with an arbitrary vector $\underset{\sim}{\eta}$ with components that vanish on ∂A at places determined by the position boundary conditions and if we then integrate by parts the resulting equation over A, we obtain the weak form of (5) (or the Principle of Virtual Work):

$$\int_A [-\underset{\sim}{\tau}^k \cdot \underset{\sim}{\eta}_{,k} + \underset{\sim}{f} \cdot \underset{\sim}{\eta}]dv + \int_{\partial A} \underset{\sim}{\sigma} \cdot \underset{\sim}{\eta}\, da = \int_A \rho \frac{\partial^2 \underset{\sim}{p}}{\partial t^2} \cdot \underset{\sim}{\eta}\, dv \quad \text{for all such } \underset{\sim}{\eta} . \tag{6}$$

Here $\underset{\sim}{\sigma}$ is the traction prescribed in a manner complementary to that of $\underset{\sim}{\eta}$. (We do not spell out the details of these boundary conditions.) Equations (3) and (6) make sense for functions $\underset{\sim}{\tau}^k, \underset{\sim}{f}, \underset{\sim}{\sigma}, \underset{\sim}{p}$ of quite general form, whereas the intermediate step (5) in this process leading from (3) to (6) uses a continuity hypothesis. This process is therefore philosophically unsound. It is also physically unsound because (5) could conceivably be transformed into an equivalent form that could have corresponding integral and weak forms that are not equivalent to (3) and (6). When solutions are not smooth as in the presence of shocks, these alternative versions may produce the wrong jump conditions. This difficulty cannot arise if we note that (3) and (6) are essentially equivalent. If we replace (3) by an integral over A of its integrand times the characteristic function of E, then we can obtain (6) by approximating $\underset{\sim}{\eta}$ by sums of characteristic functions. The derivation of (3) from (6) follows by letting $\underset{\sim}{\eta}$ approximate a characteristic function times a fixed vector.

A material is called <u>elastic</u> if there are functions $\hat{\underset{\sim}{\tau}}^k$ such that

$$(7) \qquad \underset{\sim}{\tau}^k(\underset{\sim}{x},t) = \hat{\underset{\sim}{\tau}}^k(\underset{\sim}{p},_\ell(\underset{\sim}{x},t),\underset{\sim}{x}) \quad .$$

The form of these functions is not arbitrary, but must be consistent with the angular momentum principle and with the requirement that the material properties of the body be unaffected by rigid motions. We do not indicate the explicit consequences of these requirements here. An elastic material is called <u>hyperelastic</u> if there is a scalar (<u>strain energy</u>) function ψ such that

$$(8) \qquad \hat{\underset{\sim}{\tau}}^k(\underset{\sim}{p},_\ell,x) = \frac{\partial\psi}{\partial\underset{\sim}{p},_k}(\underset{\sim}{p},_\ell,x) \quad .$$

If we substitute (7) into (5) (or(6)), we obtain a system of three second-order quasilinear partial differential equations in divergence form. Because the acceleration term occurs in such an innocuous way, the mathematical nature of the equation is essentially determined by the functions $\{\hat{\underset{\sim}{\tau}}^k\}$, which also describes the physical response of the material. A fundamental open question in the theory of nonlinear elasticity is to find a set of physically realistic restrictions on $\{\hat{\underset{\sim}{\tau}}^k\}$ that permit successful mathematical analysis of meaningful problems.

The simplest requirement on material response is roughly that an increase in strain should produce a concomitant increase in stress. This requirement of order-preservation is difficult to make precise because stress and strain are vectors (in our notation) and because there are many alternative ways of measuring these quantities. (Cf. [31].) Perhaps the simplest mathematical embodiment of this order-preservation is the monotonicity condition

$$(9) \qquad \underset{\sim}{\xi}_k \cdot \frac{\partial \hat{\underset{\sim}{\tau}}^k}{\partial \underset{\sim}{p}_{,\ell}} \cdot \underset{\sim}{\xi}_\ell > 0 \quad \forall (\underset{\sim}{\xi}_1, \underset{\sim}{\xi}_2, \underset{\sim}{\xi}_3) \neq (\underset{\sim}{0}, \underset{\sim}{0}, \underset{\sim}{0}) \quad .$$

For reasonable $\underset{\sim}{f}$, this condition leads to a complete existence and uniqueness theory for static boundary value problems. At first sight this seems to be a significant advantage of this condition, but this unqualified uniqueness actually implies that a rod subjected to compressive terminal thrusts would never buckle, however slender the rod and however large the load. This consequence makes (9) completely unacceptable because one of the few reasons for suffering the complexity of nonlinear elasticity is to describe buckling and related instabilities.

Another possible restriction on material behavior is the requirement that the equations be suitable for describing wave propagation. Precisely, the equations should be "totally hyperbolic" or have three real wave speeds. This leads to the strong ellipticity condition

$$(10) \qquad \mu_k \underset{\sim}{\lambda} \cdot \frac{\partial \hat{\underset{\sim}{\tau}}^k}{\partial \underset{\sim}{p}_{,\ell}} \cdot \underset{\sim}{\lambda} \mu_\ell > 0 \quad \forall (\underset{\sim}{\lambda}, \underset{\sim}{\mu}) \neq (\underset{\sim}{0}, \underset{\sim}{0}) \quad ,$$

i.e., (9) is to hold only for vectors $\underset{\sim}{\xi}_k$ of the restricted form $\mu_k \underset{\sim}{\lambda}$. When (8) holds, (10) is known as the strong Legendre-Hadamard condition. Recall that a solution of the Euler-Lagrange equations and natural boundary conditions for the potential energy function

$$\int_A [\psi(\underset{\sim}{p}_{,\ell},\underset{\sim}{x}) + f(\underset{\sim}{p}(\underset{\sim}{x}),\underset{\sim}{x})]dv \tag{11}$$

must satisfy the strong Legendre-Hadamard condition if it is to furnish (11) with a local minimum. Note that the Euler-Lagrange equations for (11) are just the equations of elastostatics. Unfortunately there is no existence theory for nonlinear static problems when (10) holds.

Students of nonlinear analysis may be aware of the <u>semi</u>- or <u>pseudo-monotoncity</u> condition, which is weaker than (9) but strong enough to ensure existence of static solutions without enforcing uniqueness. The effective way of generating this condition is to retain (9) but permit $\{\hat{\underset{\sim}{\tau}}^k\}$ to depend also on $\underset{\sim}{p}$ itself. But such a dependence would mean that the properties of the material would be altered by translating the body from one position to another. This absurd consequence renders this generalization useless.

There are another set of conditions, the <u>Coleman-Noll</u> conditions, whose physical consequences have been extensively studied (Cf. [30,33]) but whose mathematical consequences are unknown. Some of my work has indicated that these conditions are unduly restrictive. General discussions of the role of various classes of these material restrictions are given by [19, 21, 24, 31, 33]. (John Ball has shown me evidence that there might be some other kinds of mathematically effective assumptions that are physically reasonable.)

We account for the invertibility condition by demanding that some component of $\{\hat{\underset{\sim}{\tau}}^k\}$ become infinite as the Jacobian of (1) or as some principal subdeterminant of $\underset{\sim}{C}$ approaches zero. We schematically illustrate the corresponding behavior for the strain energy function in Fig. 1. This growth rate for small $\det(\partial\underset{\sim}{r}/\partial\underset{\sim}{z})$ is a source of analytic difficulty.

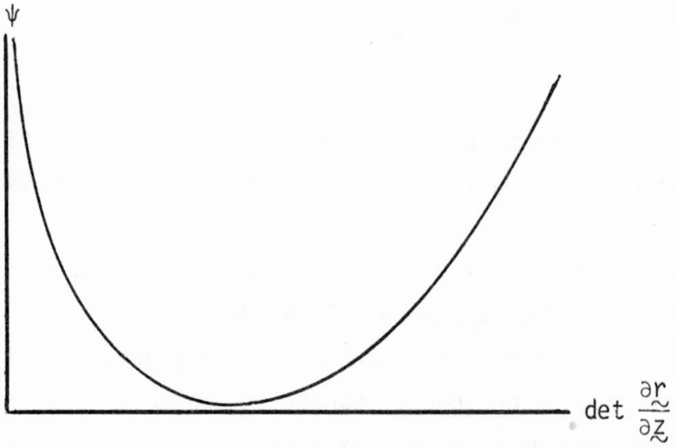

Fig. 1

But it successfully translates the invertibility condition into a constitutive restriction. It is clear that this behavior is inconsistent with commonly encountered stress-strain laws of the form

(12) some stress = linear function of some strain.

Nevertheless (12) is often justified by saying that under usual conditions the strains are never so critical as to cause the Jacobian to vanish. This fallacious argument puts the cart before the horse. The reason that strains are seldom critical is that growth conditions of the type illustrated in Fig. 1 make it exceedingly difficult to force the strains to their critical values.

If we let the Jacobian of (1) vanish, then (12) produces a corresponding <u>finite</u> stress (for usual measures of strain). In fact, if the Jacobian of (1) is negative, the corresponding stress would still be finite under (12). Thus by using (1) in a nonlinear problem, one could be in the ridiculous position of analyzing numerical data for a problem in which part of a body has changed it orientation.

When (12) is used, the nature and distribution of eigenvalues for static problems linearized about some base solution may differ markedly from their behavior when a stress-strain law consistent with Fig. 1 is used. This effect is most pronounced when parts of the body are in compression, which is the cause of buckling instabilitics.

Yet another difficulty attends the used of (12). There are many available measures of strain (cf. [31].) Two of the most useful for problems in solid mechanics are $\underset{\sim}{E} = \frac{1}{2}(\underset{\sim}{C}-\underset{\sim}{I})$ and $\underset{\sim}{H} = \underset{\sim}{C}^{\frac{1}{2}}-\underset{\sim}{I}$. These agree when $\underset{\sim}{C}$ is near the identity $\underset{\sim}{I}$ but have very different growth for large $\underset{\sim}{C}$. If (12) is used with the stress being the Piola-Kirchhoff stress, then the energy ψ is quadratic in $\underset{\sim}{H}$ when $\underset{\sim}{H}$ is the strain in (12) and cubic in $\underset{\sim}{H}$ when $\underset{\sim}{E}$ is the strain in (12). This distinction can be critical in problems for which there is a hydrostatic pressure because the potentials of such pressure loadings have comparable nonlinear rates of growth in $\underset{\sim}{r}$ and $\partial\underset{\sim}{r}/\partial\underset{\sim}{z}$. Consequently, the very existence of solutions under such loadings is affected by the choice of the strain appearing in (12). Cf. [2].

The final problem that would arise in the analysis of boundary value problems in nonlinear elastostatics in the choice of function space. Our first inclination would be to choose a Sobolev space W_p^1 of vector-valued functions with the index p characterizing the rate of growth of ψ for large values of $\partial\underset{\sim}{r}/\partial\underset{\sim}{z}$. Thus if $\underset{\sim}{w}$ were to represent something like $\partial\underset{\sim}{r}/\partial\underset{\sim}{z}$ or $\underset{\sim}{C}^{\frac{1}{2}}$, we would take $\underset{\sim}{w} \in L_p$:

$$(13) \qquad \int_B |\underset{\sim}{w}(\underset{\sim}{z})|^p d\underset{\sim}{z} < \infty .$$

This would suggest that the rate of growth of ψ is uniform in all directions $\underset{\sim}{w}$ for large $|\underset{\sim}{w}|$, i.e., the material is nearly homogeneous and isotropic. A possible generalization of this restrictive assumption is

$$(14) \qquad \sum_{k=1}^{3} \int_B |\underset{\sim}{w}(\underset{\sim}{z}) \cdot \underset{\sim}{e}_k(\underset{\sim}{z})|^{p_k(\underset{\sim}{z})} d\underset{\sim}{z} < \infty ,$$

where $\{\underset{\sim}{e}_k(\underset{\sim}{z})\}$ form a not necessarily orthormal basis and $\{p_k(\underset{\sim}{z})\}$ are ≥ 1 . It can be shown [7] that these $\underset{\sim}{w}$'s form a reflexive Banach space whose dual consists of functions $\underset{\sim}{w}^*$ satisfying

$$(15) \qquad \sum_{k=1}^{3} \int_B |\underset{\sim}{w}^*(\underset{\sim}{z}) \cdot \underset{\sim}{e}_k^*(\underset{\sim}{z})|^{p_k^*(\underset{\sim}{z})} d\underset{\sim}{z} < \infty ,$$

where $\{e_k^*(\underset{\sim}{z})\}$ is the basis dual to $\{\underset{\sim}{e}_k(\underset{\sim}{z})\}$ and $p_k^* = p_k/(p_k - 1)$. (Recently Kaniel [21] has shown how certain variational problems can be handled without the use of Banach spaces.)

3. Rod and Shell Theories

The difficulties presented by the three-dimensional theory of nonlinear elasticity have resulted in the development of innumerable, scarely distinguishable, theories of rods and shells, each endowed by its creator with special virtues that are not recognized by other such creators. In spite of this profusion of complex models, it is possible to present rather simply the essential mathematical structure of all geometrically exact theories of nonlinearly elastic rods and shells and then to use the resulting theories to study important, concrete problems.

Rod theories and shell theories characterize the behavior of bodies by a finite number of equations in just one and two independent spatial variables, respectively. For simplicity, we limit our attention to static problems in this section.

Let us denote the curvilinear coordinate x^3 by s and denote differentiation with respect to s by a prime. We suppose that $s \in [s_1,s_2]$ when $\underset{\sim}{x} \in A$. We set $A(s) = \{(x^1,x^2): (x^1,x^2,s) \in A\}$. Let $\underset{\sim}{b}$ be a function from $R^N \times A$ to R^3. Then we construct a rod theory by constraining the position field by

$$(16) \qquad \underset{\sim}{p}(\underset{\sim}{x}) = \underset{\sim}{b}(\underline{u}(s),\underset{\sim}{x}) .$$

We wish to obtain a suitable system of equations for the unknown $\underline{u}$. (Relation (16) may be interpreted as a holonomic constraint and $\underline{u}$ as a generalized coordinate vector.) A common form of (16) is

$$(17) \qquad \underset{\sim}{p}(\underset{\sim}{x}) = \sum_{k=1}^{N} \varphi_k(x^1,x^2)\underset{\sim}{u}_k(s) ,$$

where $\{\varphi_k\}_{k=1}^{\infty}$ are dense in $C(A(s))$. Corresponding to (16) we define test functions (or virtual displacements) by

$$(18) \qquad \underset{\sim}{\eta}(\underset{\sim}{x}) = \frac{\partial \underset{\sim}{b}}{\partial \underline{u}}(\underline{u}(s),\underset{\sim}{x})\cdot\underline{y}(s)$$

for arbitrary functions $\underline{y} : [s_1,s_2] \to R^N$. If we now substitute (16) and (18) into the equilibrium version of (6), we obtain the weak equilibrium equations (principle of virtual work) for rods:

$$(19) \qquad \int_{s_1}^{s_2} [\underline{m}\cdot\underline{y}' + \underline{n}\cdot\underline{y} - \underline{f}\cdot\underline{y}]ds + \underline{q}\cdot\underline{y}\,\Big|_{s_1}^{s_2} = 0 \quad \forall\underline{y} ,$$

where

$$(20) \qquad \underline{m}(s) = \int_{A(s)} \underset{\sim}{\tau}^3(\underset{\sim}{x})\cdot \frac{\partial \underset{\sim}{b}}{\partial \underline{u}} \sqrt{G}\, dx^1 dx^2 , \text{ etc.}$$

The corresponding constitutive relations are given by

$$(21)\quad \underline{m}(s) = \hat{\underline{m}}(\underline{u}(s),\underline{u}'(s),s) \equiv \int_{A(s)} \hat{\underset{\sim}{\tau}}^3(\underset{\sim}{b}_{,1}, \underset{\sim}{b}_{,2}, \underset{\sim}{b}_{,3}, \underset{\sim}{x}) \cdot \frac{\partial \underset{\sim}{b}}{\partial \underline{u}} \sqrt{G}\, dx^1 dx^2$$

and by a similar expression for $\underline{n}$. Thus the classical form of the equations for nonlinearly elastic rods is given by

$$(22)\quad \frac{d}{ds}\hat{\underline{m}}(\underline{u}(s),\underline{u}'(s),s) - \hat{\underline{n}}(u(s),u'(s),s) + \underline{f}(s) = \underline{0} \quad .$$

The essential mathematical structure of (22) is determined by the dependence of $\hat{\underline{m}}$ on $\underline{u}'$. An easy computation yields

$$(23)\quad \underline{a}\cdot \frac{\partial \hat{\underline{\underline{m}}}}{\partial \underline{u}'} \cdot \underline{a} = \int_{A(s)} \left(\frac{\partial \underset{\sim}{b}}{\partial \underline{u}} \cdot \underline{a}\right) \cdot \frac{\partial \underset{\sim}{\tau}^3}{\partial \underset{\sim}{p}_{,3}} \cdot \left(\frac{\partial \underset{\sim}{b}}{\partial \underline{u}} \cdot \underline{a}\right) \sqrt{G}\, dx^1 dx^2 \quad .$$

We observe that the integrand of (23) has the form (10) with $\underset{\sim}{\lambda} = \frac{\partial \underset{\sim}{b}}{\partial \underline{u}} \cdot \underline{a}$, $\mu_3 = 1$, $\mu_1 = \mu_2 = 0$, so that the strong ellipticity condition implies that (23) is positive definite, whence (22) is a semimonotone differential operator. This mathematical structure is the basis of a rather delicate existence and regularity theory for boundary value problems for (22), the difficulty largely caused by one-dimensional versions of the invertibility condition (1). Cf. [8].

The procedure leading to (19), (22), (23) is a generalization of the Galerkin-Kantorovich method. We used the general nonlinear representation (16) to show that all theories so obtained have the same mathematical structure. It is possible to obtain some preliminary results on the important question of the nature of the approximation, but definitive results must await further developments in the three-dimensional theory.

Shell theories are obtained in a completely analogous manner. Their development is somewhat easier because only one coordinate is to be eliminated. (Cf. [7]). In place of (23) we get a two-dimensional version of the strong ellipticity condition, however. But for axisymmetrical and cylindrical problems, these reduce to problems in just one space variable and the resulting equations have exactly the same form as (22), (23).

4. Special Problems

In spite of the generality of the theories described in Section 3, their common mathematical structure is powerful enough for them to provide detailed descriptions of the solutions to concrete, fundamental problems. Basically, all the classical problems of elastic stability of structures (Cf.[28]) can be readily treated in a fully nonlinear framework. In many cases, their solutions manifest a variety of new properties, which are observed in experiment but which cannot be described by the usual models of elastic stability theory. Here we briefly describe the nature of such results.

Consider an initially straight, doubly-symmetric rod of uniform cross-section that can suffer flexure, torsion, extension, and shear. The material of the rod is described by a set of six arbitrary constitutive relations consistent with the requirements of Section 3. The governing equations of equilibrium are a fifteenth-order system of quasilinear ordinary differential equations. Suppose that the rod is subjected only to terminal loads. Then it is possible to catalog precisely the qualitative behavior of every possible solution [10]. It is shown that the qualitative behavior for all such materials is like that of the solutions of the Kirchhoff kinetic analogy [24] except for the appearance of shear instabilities that may occur when parts of the rod are in tension. Moreover, one can describe the nature of all solutions with a helical axis (Kirchhoff's problem, cf. [5]). In this case, certain solutions exhibit effects akin to the Coulomb and Poynting effects of the three-dimensional theory.

Results used in the development of these papers can also be applied to treat the study of buckling of columns under compressive end thrusts. Even when the cross-section is not uniform, it is possible to get a nearly complete picture or the global behavior of nontrivial solution branches [11]. A complete local study of bifurcation can readily be made by following the method of Poincaré (cf. [23]) or the method of Lyapunov and Schmidt (cf. [32]).

Results of a related kind can be found for the deformation of an initially circular ring (or cylindrical tube) under the action of a hydrostatic pressure. If the ring is uniform and if it can suffer flexure, extension, and shear, then every solution for which the deformation is not "too large" has a configuration with n axes of symmetry with $n \geq 2$ [1,4,6]. The concept of "too large" depends upton the material; for certain materials it can be shown that these qualitative results hold as long as the deformed axis is a simple curve.

Work in progress indicates that comparable results should hold for buckling problems for circular plates and special caps.

Other kinds of instabilities, the tensile instabilities, are mathematically similar to, but mechanically distinct from the buckling instabilities. The simplest model exhibiting tensile instabilities is a rod endowed with enough kinematic structure to suffer longitudinal extension, traverse contraction, flexure, and shear. When such a rod is elongated by holding its ends at a fixed distance apart, then for appropriate boundary conditions, the equilibrium problem possesses a "trivial solution" representing a configuration with a straight axis, a uniform longitudinal extension, a uniform transverse contraction, and no shear. The problem may also posses other "trivial solutions" that are uniformly sheared. At certain critical values of the elongation, non-trivial solutions bifurcate from the trivial branches. These non-trivial solutions describe necking and shear (Luders band type) instabilities of the very sort that are observed in experimental results on the tension test [3,9]. The qualitative structure of the necking solutions can also be determined [4]. It is important to note that a nonlinear theory of elasticity predicts these effects, which are traditionally described by a theory of plastic yielding. Moreover, there is strong analytic evidence suggesting that these models of nonlinear elasticity can also exhibit hysteresis. These results indicate that there may be an intimate connection between nonlinear elasticity theory and plasticity theory.

In each of these static problems, the monotonicity and invertibility conditions play key roles; they are restrictive enough to permit critical steps in the analysis and yet not so restrictive as to prevent the appearance of an interesting variety of instabilities. This mathematical structure should also be helpful in the treatment of dynamical problems. The dynamical problems of nonlinear elasticity are governed by nonlinear hyperbolic equations, which can exhibit shocks. Equations manifesting the difficulties of our models have not been fully analyzed. For an overview of this theory consult [24]. Numerical studies of shocks are described by [27]. If a strong dissipative mechanism is introduced into the model of dynamical nonlinear elasticity, then the equations may become a nonlinear parabolic system describing nonlinear viscoelasticity. Here the mathematical structure should allow an effective analysis.

Such an analysis could lead to a study of problems of stability in the desirable context of stability of motion. Some typical work on simple models has been done by [13, 14, 18].

5. Conclusion

It is possible to formulate and treat a variety of important nonlinear problems in solid mechanics. Efforts to "simplify" such models by discarding certain terms, by ignoring the nature of large deformation, or by making mathematically convenient assumptions may deprive the model and its solutions of its physical significance and may frequently complicate the analysis. Well-formulated nonlinear theories of solid mechanics are analytically difficult, but they are designed to treat the physically difficult problems of buckling and related instabilities, formation of shocks, and fracture.

Although the results we have described in Section 4 are primarily qualitative, they do not answer all the physically important questions. Eg., knowledge of the global structure of multiple solutions for buckling problems and problems of tensile instabilities is prerequisite to an understanding of the nature of the instability. Available methods of global analysis are incapable of determining the detailed behavior of solutions. It is here that numerical methods can play a decisive role. An example of such a global numerical study of a buckling problem is given in [12]. The development of effective numerical procedures for such problems is a delicate art because these problems do not reduce to nonsingular matrix equations. A problem that must be faced in such analysis is the actual choice of constitutive relations that are physically reasonable and that elucidate qualitatively different sorts of behavior.

REFERENCES

1. S. S. Antman, "The Shape of Buckled Nonlinearly Elastic Rings," Z.A.M.P. 21 (1970), 422-438.
2. S. S. Antman, "The Theory of Rods," Handbuch der Physik Vol. VIa/2, Springer-Verlag, 1972, 641-703.
3. S. S. Antman, "Nonuniqueness of Equilibrium States for Bars in Tension," J. Math. Anal. Appl. 44 (1973), 333-349.
4. S. S. Antman, "Qualitative Theory of the Ordinary Differential Equations of Nonlinear Elasticity," in Mechanics Today 1972, edited by S. Nemat-Nasser, Pergamon Press, 1974, 58-101.
5. S. S. Antman, "Kirchhoff's Problem for Nonlinearly Elastic Rods," Quart. Appl. Math., 23 (1974), 221-240.
6. S. S. Antman, "Monotonicity and Invertibility Conditions in One-Dimensional Nonlinear Elasticity," Symposium on Nonlinear Elasticity, Mathematics Research Center, Univ. Wisconsin, edited by R. W. Dickey, Academic Press, 1973, 57-92.
7. S. S. Antman, "Boundary Value Problems of One-Dimensional Nonlinear Elasticity I : Foundations of the Theories of Nonlinearly Elastic Rods and Shells," to appear.
8. S. S. Antman, "Boundary Value Problems of One-Dimensional Elasticity II : Existence and Regularity Theory for Conservative Problems," in preparation.
9. S. S. Antman and E. Carbone, in preparation.
10. S. S. Antman and K. B. Jordan, "Qualitative Aspects of the Spatial Deformation of Nonlinearly Elastic Rods," Proc. Roy. Soc. Edinburgh, 73A (1975), 85-105.
11. S. S. Antman and G. Rosenfeld, in preparation.
12. L. Bauer, E. L. Reiss, and H. B. Keller, "Axisymmetric Buckling of Hollow Spheres and Hemispheres," Comm. Pure Applied Math. 23 (1970), 529-568.
13. J. Clements, "Existence Theorems for a Quasilinear Evolution," S.I.A.M. J. Appl. Math. 27 (1974), 745-752.
14. C. M. Dafermos, "The Mixed Initial-Boundary Value Problem for the Equations of Nonlinear One-Dimensional Viscoelasticity," J. Diff. Eqs. 6 (1969), 71-86.
15. G. Duvaut and J. L. Lions, "Les inéquations en mécanique et en physique," Dunod, 1972.
16. G. Fichera, "Boundary Value Problems of Elasticity with Unilateral Constraints," Handbuch der Physik Vol. VIa/2, Springer Verlag, Berlin, Göttingen, Heidelberg, 1972, 391-424.
17. J. M. Greenberg, "On the Equilibrium Configurations of Compressible Slender Bars," Arch. Ratl. Mech. Anal. 27 (1967), 181-194.

18. J. M. Greenberg, R. C. MacCamy, and V. J. Mizel, "On the Existence, Uniqueness and Stability of Solutions of the Equation $'(U_x)U_{xx} + U_{xtx} = U_{tt}$." J. Math. Mech. 17 (1968), 707-728.
19. M. Hayes, "Static Implication of the Strong Ellipticity Condition," Arch. Ratl. Mech. Anal. 33 (1969), 181-191.
20. M. E. Gurtin, "The Linear Theory of Elasticity," Handbuch der Physik VIa/2, Springer Verlag, Berlin 1972, 1-296.
21. R. Hill, "Constitutive Inequalities for Isotropic Elastic Solids under Finite Strain," Proc. Roy. Soc. (Sec. A) 314 (1970), 457-472.
22. S. Kaniel, "Convex Functionals," to appear.
23. J. B. Keller, "Bifurcation Theory for Ordinary Differential Equations," Bifurcation Theory and Nonlinear Eigenvalue Problems (J. B. Keller and S. S. Antman, editors), Benjamin, New York, 1969.
24. R. Knops and L. Payne, "Uniqueness Theorems in Linear Elasticity," Springer-Verlag, 1971.
25. P. D. Lax, "Hyperbolic Systems of Conservation Laws and the Mathematical Theory of Shock Waves," CBMS Regional Conf., S.I.A.M., 1973.
26. A. E. H. Love, A Treatise on the Mathematical Theory of Elasticity, 4th ed., Cambridge Univ. Press, 1927.
27. J. T. Oden, "Approximations and Numerical Analysis of Finite Deformation of Elastic Solids," Nonlinear Elasticity" (R. W. Dickey, editor), Academic Press, 1973, 175-228.
28. S. P. Timoshenko and J. M. Gere, Theory of Elastic Stability, McGraw-Hill, New York, 1961.
29. T. W. Ting, "St. Venant's Compatibility Conditions," Tensor N.S., 28 (1974), 5-12.
30. C. Truesdell and W. Noll, "The Nonlinear Field Theories of Mechanics," Handbuch der Physik III/3, Springer, Berlin, 1965.
31. C. Truesdell and R. Toupin, "The Classical Field Theories," Handbuch der Physik III/1, Springer, Berlin, 1960.
32. M. M. Vainberg and V. A. Trenogin, The Theory of Solution Branching of Nonlinear Equations (Russian), Nauka, Moscow, 1969.
33. C. C. Wang and C. Truesdell, Introduction to Rational Elasticity, Noordhoff, Leyden, 1973.

NUMERICAL SOLUTION OF LINEAR PARTIAL DIFFERENTIAL EQUATIONS OF ELLIPTIC-HYPERBOLIC TYPE*

A. K. Aziz**

and

S. H. Leventhal***

1. Introduction

The purpose of this paper is to survey the numerical methods for the solution of linear equations of mixed type. We consider a linear partial differential equation of the form

$$(1.1)\quad Lu = Au_{xx} + 2Bu_{xy} + Cu_{yy} + Du_x + Eu_y + Fu = G \quad ,$$

with real coefficients which are functions of the two independent variables x , y defined in a domain D of the x , y plane. If the discriminant $\Delta = B^2 - AC$ changes sign in the closure $\bar{D}$ of D (1.1) is called an equation of the mixed type.

The analytic study of equations of mixed type has been principally concerned with the following four kinds of problems.

(i) The Dirichlet problem in a domain D such that (1.1) is the elliptic in the interior of D and parabolic on part of the boundary ∂D of D .

(ii) The Cauchy problem where (1.1) is hyperbolic in

---

*Work partially supported by the Office of Naval Research under Contract NR044-453 and by the Naval Surface Weapons Center Independent Research Fund.

**Naval Surface Weapons Center, and University of Maryland.

***Naval Surface Weapons Center

D and parabolic on the curve (situated on ∂D) carrying the initial data.

(iii) Boundary value problems of Goursat type with (1.1) hyperbolic in D and parabolic on one of the curves carrying the boundary value.

(iv) Boundary value problems in D with (1.1) hyperbolic in one part of D, elliptic in another portion of D and parabolic on the curve or curves separating the hyperbolic and elliptic parts. We are concerned with the numerical solution of the fourth type of mixed problem.

According to the standard theory for (1.1) we may associate characteristic variables ξ and η such that $\xi(x,y)$ = constant and $\eta(x,y)$ = constant are the solutions of

$$\frac{dy}{dx} = \frac{-B \pm \sqrt{\Delta}}{A} \quad . \tag{1.2}$$

These variables are either real or complex conjugate. In terms of the new independent variables ξ and η equation (1.1) may be written as

$$u_{\xi\eta} + D_1 u_\xi + E_1 u_\eta + F_1 u = G_1 \quad . \tag{1.3}$$

A further change of variables ξ and η given by

$$\theta = \frac{1}{2}(\xi+\eta) \quad , \quad \sigma = \left(\frac{3}{4}\right)^{2/3} \{-(\xi-\eta)^2\}^{1/3} \tag{1.4}$$

yields the equation

$$\sigma u_{\theta\theta} + u_{\sigma\sigma} + a u_\theta + b u_\sigma + c u = g \quad . \tag{1.5}$$

We note that the variables θ and σ are always real. In analogy to normal forms for elliptic and hyperbolic equations, (1.5) may be considered as the normal form of the equations of mixed type. The equation

$$\sigma u_{\theta\theta} + u_{\sigma\sigma} = g \quad . \tag{1.6}$$

then corresponds to the Laplace and wave equations for $\sigma > 0$ and $\sigma < 0$ respectively. (1.6) is called the Tricomi equation and may be thought of as a natural completion of the usual classification of linear second order partial differential equations.

In the theory of partial differential equations there is a fundamental distinction between those of elliptic type, hyperbolic and parabolic type. Each type of equation has different requirements as to the boundary or initial data which must be specified to assure existence, uniqueness and continuous dependence on initial data, i.e., for the problem to be well posed. These requirements are well known for an equation of any particular type. Further many analytical and numerical techniques have been developed for solving the various types of partial differential equations, subject to suitable boundary conditions, including many nonlinear equations. However, for the equations of mixed type much less is known and it is usually difficult to know even what the proper boundary conditions are.

The question of existence and uniqueness of solutions for various categories of mixed problems has been dealt with in numerous papers; for a survey of the results and an extensive Bibliography the reader may consult the monograph of Bers [4], the book of Bitsadze [5] and the paper of Karatoprakliev [14]. The main body of results on the existence and uniqueness of solutions for the equations of mixed type have been primarily concerned with problems in two dimensions. There appear to be very few investigations of equations of mixed type in multi-dimensional regions. Recently G. D. Karatoprakliev [14] has formulated several boundary-value problems for equations of mixed type in bounded multidimensional regions. The theory of symmetric positive systems [8] is used in search for correctly posed boundary-value problems. [14] appears to be the first paper where the relation between multidimensional equations of mixed type and symmetric positive systems has been fully explored. As mentioned earlier the main objective of this paper is to review the existing numerical methods for the solution of linear equations of a elliptic-hyperbolic type. Although some of the considerations may be equally extended to other types of boundary-value problems for linear mixed equations, we shall not discuss these extensions in the present paper.

2. Numerical methods

The main numerical approaches for solving boundary-value problems of elliptic-hyperbolic type falls, in general, in one of the following categories: finite difference, finite element, or analytical approximative methods. In the following sections these approaches are reviewed and numerical results for some model problems are given.

2.1. Finite Difference Methods

The finite difference schemes used for the numerical solutions of (1.1) fall in two distinct categories. The first category comprises the finite difference methods which result from direct differencing of (1.1) in a suitable way, while the second category deals with difference analogues of first order systems to which (1.1) may be reduced.

2.1a. Direct Difference Methods

In applying finite difference methods directly to a second order linear mixed equation in general two approaches are used. One approach consists of first transforming by analytical methods the mixed problem to an equivalent purely elliptic boundary value problem and then solve the resulting problem by finite differences. In the second approach two different finite difference analogues of the given mixed equations are used, i.e., one difference scheme in the elliptic region and another difference scheme in the hyperbolic region.

(i) The earlier works on difference methods for linear equations of mixed type are all based on the reduction of the problem, by analytical methods, to a purely elliptic problem with complicated boundary conditions. In this connection theoretical results relating to the Lavrent'ev-Bitsadze equation are obtained in [11], [12], [15] and [18]. Along the same lines Vincenti and Wagoner give numerical results for the Tricomi equation, whose theoretical justification is given in [21].

(ii) Filippov [7] was the first to treat a special case of (1.1) by finite differences without first reducing the problem to an equivalent elliptic boundary value problem. He considered the Tricomi problem for the equation

$$(2.1.1)_{ii} \qquad Lu \equiv yu_{xx} + u_{yy} = f(x,y)$$

in a bounded domain D in the x, y plane. Ogawa [25] using techniques similar to those employed in [7] extended the results of [7] to the more general equation

$$(2.1.2)_{ii}$$

$$Lu \equiv K(y)u_{xx} + u_{yy} + a(x,y)u_x + b(x,y)u_y + c(x,y)u = f(x,y) \ ,$$

where $K(y)$ is a continuous function which is positive for y positive and negative for y negative. Thus the equation is elliptic, parabolic or hyperbolic according as $y > 0$, $y = 0$ or $y < 0$.

We briefly describe the finite difference scheme given in [25] which contains as a special case the finite difference procedure of Filippov [7].

For $y < 0$ the equation $(2.12)_{ii}$ has real characteristics given by the two families.

$$(2.1.3a)_{ii} \qquad \frac{dy}{dx} = (-K)^{-1/2}$$

$$(2.1.3b)_{ii} \qquad \frac{dy}{dx} = -(-K)^{-1/2}$$

Let A and B be two points on the x-axis with $x_A < x_B$. By D denote the open domain bounded by the characteristic Γ_1 of the family $(2.13b)_{ii}$ passing through A, the characteristic Γ_2 of the family $(2.13a)_{ii}$ passing through the point B, and by the simple arc $\Gamma^{\dagger}$ in the upper half-plane with end-points at A and B.

The Tricomi problem for the equation $(2.12)_{ii}$ on D consists of finding a function u, continuous in $\bar{D}$, which satisfies $(2.12)_{ii}$ in D, and which takes on the boundary values

$$(2.1.4)_{ii} \qquad u = \varphi_1 \text{ on } \Gamma^{\dagger}, \quad u = \varphi_2 \text{ on } \Gamma_2 ,$$

where φ_1 and φ_2 are prescribed functions.

The difference problem. Let K be of class $C^3(\bar{D})$ with $K(y) > 0$ for $y > 0$ and $K(y) < 0$ and $K'(y) > 0$ for $y < 0$. Furthermore assume that a and b are of class $C'(\bar{D})$ and c and f are of class $C(\bar{D})$.

Intergrating the equation $(2.13)_{ii}$ with respect to y, we have the characteristics of $(2.11)_{ii}$ in the form $x-x_0 = \pm G(y)$, where

$$(2.1.5)_{ii} \qquad G(y) = \int_y^0 [-K(\eta)]^{1/2} d\eta \quad , \quad y \leq 0 \quad .$$

Now divide the segment AB into N equal parts, each of length h , and through each of the points $x_K = x_A + Kh$ $(K = 1,2,\cdots,N-1)$ we draw the characteristics

$$x - x_K = \pm G(y) \quad .$$

These characteristics, together with the characteristics Γ_1 and Γ_2 , intersects at the points

$$(2.1.6)_{ii} \qquad (x_A + \frac{nh}{2} + Kh, -y_n) \; ;$$

$$K = 0,1,\cdots,N-n, n=1,2,\cdots,N \; ,$$

with the ordinate satisfying

$$(2.1.7)_{ii} \qquad G(-y_n) = \frac{nh}{2} \; , \; n=1,2,\cdots,N \; .$$

Taking for $y < 0$ the points given by $(2.1.6)_{ii}$, and for $y \geq 0$ the points of the form (x_A+Kh, mh) (K,m integers) which lie in $\bar{D}$, one obtains a mesh region $\bar{D}_h$. For $y > 0$ we take as the neighbors of the mesh point (x,y) the four (4) mesh points $(x+h,y), (x-h,y)$, $(x,y+h)$ and $(x,y-h)$. The boundary Γ_h of $\bar{D}_h$ consists of those points of D_h in the upper half plane for which not all four neighbors belong to $\bar{D}_h$, together with all points of $\bar{D}_h$ which lie on Γ_2 , and the point B . The totality of all points of $\bar{D}_h$ which are not boundary points is called the interior mesh region D_h . Let D_h^- and D_h^+ denote all the points of D_h for which $y < 0$ and $y > 0$ respectively, and let Γ_h^- and Γ_h^+ denote the points of Γ_h for $y < 0$ and $y \geq 0$ respectively, Finally let γ_h be the points in D_h for which $y = 0$. We now introduce a difference operator L_h which acts on any function u defined on $\bar{D}_h$. For any $(x,-y_n)$ of D_h^- let

$(2.1.8)_{ii}$

$$L_h u(x,-y_n) = \frac{1}{\lambda_n \lambda_{n+1}} \left\{ \frac{2\lambda_{n+1}}{\lambda_n + \lambda_{n+1}} u(x - \frac{h}{2}, -y_{n-1}) + \frac{2\lambda_n}{\lambda_n + \lambda_{n+1}} u(x - \frac{h}{2}, -y_{n+1}) - u(x-h,-y_n) - u(x,-y_n) \right\}$$

$$+ a(x,-y_n) \frac{1}{h} \left\{ u(x,-y_n) - u(x-h,-y_n) \right\}$$

$$+ b(x,-y_n) \frac{1}{\lambda_n + \lambda_{n+1}} \left\{ u(x - \frac{h}{2}, y_{n-1}) - u(x - \frac{h}{2}, -y_{n+1}) \right\}$$

$$+ c(x,-y_n) u(x,-y_n) \quad ,$$

where $\lambda_n = y_n - y_{n+1}$ and $y_0 = 0$. At a point $(x,y) \in D_h^+$ we let

$(2.1.9)_{ii}$

$$L_h u(x,y) = K(y) \frac{1}{h^2} \{u(x-h,y) - 2u(x,y) + u(x+h,y)\}$$

$$+ \frac{1}{h^2} \{u(x,y-h) - 2u(x,y) + u(x,y+h)\}$$

$$a(x,y) \frac{1}{h} \{u(x,y) - u(x-h,y)\} + b(x,y) \frac{1}{2h} \{u(x,y+h) - u(x,y-h)\} + c(x,y)u(x,y) \quad ,$$

at a point $(x,0)$ of γ_h we define

$(2.1.10)_{ii}$

$$L_h u(x,0) = \frac{2}{hy_2}\left\{\frac{y_2}{h+y_2}\, u(x,h) + \frac{h}{h+y_2}\, u(x,-y_2) - u(x,0)\right\}$$

$$+ a(x,0)\,\frac{1}{h}\,\{u(x,0) - u(x-h,0)\}$$

$$+ b(x,0)\,\frac{1}{h+y_2}\,\{u(x,h) - u(x,-y_2)\} + c(x,0)u(x,0)\ .$$

The problem of finding the solution of (2.1.1) in D subject to the boundary conditions $(2.1.4)_{ii}$ is replaced by the problem of finding the solution u of the difference equation

$(2.1.11)_{ii}$ $$L_h u = f$$

on the region D_h which satisfies the boundary conditions

$(2.1.12)_{ii}$ $$u = \varphi_1 \text{ on } \Gamma_h^+\ ,\ u = \varphi_2 \text{ on } \Gamma_h^-\ .$$

The equations $(2.1.11)_{ii}$ and $(2.1.12)_{ii}$ form a system of linear algebraic equations in the unknown values of u at the points of D_h, in which the number of equations is equal to the number of unknowns.

We note that when $K(y) = y$, $a = b = c = 0$ and $x_A = 0$ the operator L_h reduces to the difference operator given in [7] if in D_h^+ the mesh points (x,y) are taken as $(Kh,(\frac{3}{2}mh)^{2/3})$ (K and m integers). In this special case one has the error estimates $\|v - u\|_\infty \le ch^{2/3}$. It is known that if $v \in C^2(\bar{D})$. Then in D_h, $L_h v \to Lv$ uniformly as $h \to 0$ and if in addition the functions a, b, c and f satisfy certain suitable conditions (too lengthy to state here) the unique solution u of $(2.1.11)_{ii}$, $(2.1.12)_{ii}$ converges uniformly on D_h to the solution v of $(2.1.1)_{ii}$, $(2.1.4)_{ii}$ as $h \to 0$.

(iii) Another difference method which may be used to solve numerically linear elliptic-hyperbolic boundary value problems is the method of Murman and Cole [24] which was designed for numerical calculations concerning the steady transonic small disturbance flow where the governing equation is nonlinear. The basic idea in this method which consists of using different difference schemes in the different regions, is borrowed from Filippov [7]. In the elliptic region central differencing is used, while in the hyperbolic region one employs one sided differencing in the x-direction and central differencing in the y-direction. The overall order accuracy of the method is $O(h^K)$, $K \leq 1$. The derivatives are obtained by central differencing of the solution with an accuracy of less than $O(h)$. This method and variations of it are discussed in [13]. We shall compare this method applied to a linear problem with a finite element procedure which will be described in section 2.2 below.

2.1b. Difference Methods Based on Reduction to a First Order System

The second category of finite difference schemes used for the numerical solution of linear elliptic-hyperbolic boundary value problems is based on the theory of symmetric positive systems developed in [8].

We first summarize the fundamental results and identities for symmetric positive systems that are needed in the sequel. For the sake of simplicity of notation and for the later applications to second order elliptic-hyperbolic problems in the plane we confine our discussion to first order systems with two components.

Symmetric positive systems

Consider the first order system

(2.1.1b)

$$Lu = 2A_1(x,y)\frac{\partial u}{\partial x} + 2A_2(x,y)\frac{\partial u}{\partial y} + A_3 u = f \quad (x,y) \in \Omega$$

(2.1.2b) $$Mu = (\mu-\beta)u = 0 \quad (x,y) \in \partial\Omega ,$$

where Ω is a bounded open set in R^2, with a piecewise continously differentiable boundary $\partial\Omega$ and $\beta = n_i A_i$, with n_i's, $1 \le i \le 2$, being the components of the outer normal on $\partial\Omega$. The matrices $A_i \in L(R^2)$, $1 \le i \le 2$, are symmetric, Lipschitz continuous in $(x,y) \in \bar{\Omega}$. The entries of the matrix $A_3 \in L(R^2)$ are bounded in Ω. The matrix $\mu(x,y) \in L(R^2)$ is defined for $(x,y) \in \partial\Omega$. It is assumed that

(I) $\mu(x,y)$ is continuous for $(x,y) \in \partial\Omega$,

(II) $\mu+\mu' \ge 0$, $\operatorname{Ker}(\mu-\beta) + \operatorname{Ker}(\mu+\beta) = R^2$,

where μ' denote the adjoint of μ. We write the operator L in the form

$$Lu \equiv A_1\frac{\partial u}{\partial x} + A_2\frac{\partial u}{\partial y} + \frac{\partial}{\partial x}(A_1 u) + \frac{\partial}{\partial y}(A_2 u) + Ku$$

We shall say that L is positive if

(III) $C = K+K' > C_0 I$,

where

$$K = A_3 - \frac{\partial A_1}{\partial x} - \frac{\partial A_2}{\partial y} ,$$

C_0 is a positive constant, and I denotes the identity matrix. We define the formal adjoint L^* of L by

(2.1.3b) $$L^*u \equiv -\frac{\partial}{\partial x}(A_1 u) - \frac{\partial}{\partial y}(A_2 u) + A_3' u \quad .$$

We say that φ satisfies the adjoint boundary conditions if

(2.1.4b) $$(\mu'+\beta)\varphi = 0 \quad \text{for} \quad x,y \in \partial\Omega .$$

Let $E(\bar{\Omega})$ be the space of all (real) infinitely differentiable functions on Ω such that all the derivatives have continuous extensions to $\partial\Omega$. As usual let $L_2(\Omega)$ be the space of square integrable functions u on Ω with the norm

$$\|u\|^2_{L^2(\Omega)} = \int_\Omega u^2 \, dx \, dy .$$

The scalar product will be denoted by $(u,v)_{L^2(\Omega)}$. For $\ell \geq 1$ an integer the Sobolev space $H^\ell(\Omega)$ is defined as the closure of $E(\bar{\Omega})$ in the norm $\|\cdot\|_{H^\ell(\Omega)}$, where

$$\|u\|^2_{H^\ell(\Omega)} = \sum_{0 \leq |\alpha| \leq \ell} \|D^\alpha u\|^2_{L^2(\Omega)} ,$$

and

$$D^\alpha = \frac{\partial^{\alpha_1+\alpha_2}}{\partial x^{\alpha_1} \partial y^{\alpha_2}} , \quad \alpha = (\alpha_1,\alpha_2) , \quad |\alpha| = \sum_i \alpha_i .$$

(α_i are nonnegative integers.)

<u>Definition 2.1.1b</u> $u \in L^2(\Omega)$ is said to be a weak solution of (2.1.1b), (2.1.2b) if for all $\varphi \in H^1(\Omega)$ satisfying the boundary condition (2.1.4b) we have

$$(u, L^*\varphi)_{L^2(\Omega)} = (f,\varphi)_{L^2(\Omega)} .$$

<u>Definition 2.1.2b</u> We shall say that u is a strong solution of (2.1.1b), (2.1.2b) if there exists a sequence $\{u_j\} \in H^1(\Omega)$ satisfying the boundary condition (2.1.2b) and if

$$\lim_{j\to\infty} (\|u_j - u\|_{L^2(\Omega)} + \|f - Lu_j\|_{L^2(\Omega)}) = 0 .$$

There are several theorems for the existence of weak and strong solutions of symmetric positive systems, the interested reader may consult [8], [19] for the statements and proofs of these results.

Before describing the finite difference schemes for the symmetric positive systems we shall first illustrate the connection between the symmetric positive systems and the second order elliptic-hyperbolic boundary value problems. To this end we consider the model problem

$$yu_{xx} - u_{yy} = 0 \quad \text{in} \quad \Omega$$

$$u = \psi \text{ on } x_B , y_C , y_D$$

$$u \quad \text{unspecified on} \quad x_A ,$$

where the domain Ω is a parallelogram in the x-y plane, centered at the origin, the two sides $x = x_A$, $x = x_B$ are parallel to the y-axis, and the other two sides $y = y_C$ and $y_D(x)$ have slopes equal to 1 for simplicity. For brevity we denote the boundary $\partial\Omega$ by x_A , x_B , y_C , and y_D , remembering that y_C and y_D are functions of x and not constants. By changing variables in the usual manner (i.e. let $\varphi = u-v$, where v is a twice continuously differentiable function satisfying the boundary condition.) The above problem is transformed to

$$y\varphi_{xx} - \varphi_{yy} = f$$

$$\varphi = 0 \text{ on } x_B , y_C , y_D$$

$$\varphi \text{ unspecified on } \quad x_A .$$

Next we transform the equation into a first order system

$$\begin{pmatrix} y & 0 \\ 0 & 1 \end{pmatrix}\begin{pmatrix} \varphi_x \\ \varphi_y \end{pmatrix}_x + \begin{pmatrix} 0 & -1 \\ -1 & 0 \end{pmatrix}\begin{pmatrix} \varphi_x \\ \varphi_y \end{pmatrix}_y = \begin{pmatrix} f \\ 0 \end{pmatrix}$$

with boundary conditions

$$\varphi_y = 0 \text{ on } x_B \, , \; \varphi_x + \varphi_y = 0 \text{ on } y_C \text{ and } y_D$$

$$\varphi_x \, , \; \varphi_y \quad \text{unspecified on} \quad x_A$$

The coefficient matrices are symmetric but as yet do not satisfy condition III (positive definite K).

Following Freidrich [8], or Morawetz [22], [23] and Chu [6], we multiply the above system by a nonsingular matrix B of the form

$$B = \begin{pmatrix} b & cy \\ c & b \end{pmatrix} \quad , \quad (b \, , \, c \text{ functions of } x \text{ and } y) \, .$$

Letting $\psi = \{\varphi_x \, , \, \varphi_y\}$, $F = \{f \, , \, 0\}$, we obtain

$$2A_1\psi_x + 2A_2\psi_y = F \quad ,$$

where

$$2A_1 = \begin{pmatrix} by & cy \\ cy & b \end{pmatrix} \quad , \quad 2A_2 = \begin{pmatrix} -cy & -b \\ -b & -c \end{pmatrix}$$

and

$$K = \frac{-\partial A_1}{\partial x} - \frac{\partial A_2}{\partial y} = \frac{1}{2}\begin{pmatrix} (c_y - b_x)y + c & b_y - c_x y \\ b_y - c_x y & c_y - b_x \end{pmatrix} \quad ,$$

which can be made positive definite through the proper choice of b and c. This takes care of symmetry and positivity (condition (III)).

To show that the boundary condition satisfies hypotheses (II) we again follow the procedure of Freidrich's [8] or Chu [6] and Morawetz [22], [23]. In this procedure the fundamental observation is to note that $\psi\cdot\beta\psi$ can be written as the difference of two perfect square terms, i.e.,

$$\psi\cdot\beta\psi = \frac{1}{bn_x+cn_y}\,[(b^2-c^2y)(n_y\varphi_x-n_x\varphi_y)^2 - (n_y^2-yn_x^2)(b\varphi_x+c\varphi_y)^2]$$

It is known [6] that the domain shape, as well as the functions b and c can be chosen in such a way that

(a) $$n_y^2-n_x^2y < 0\ ,\quad bn_x+cn_y < 0\ ,\quad b^2-c^2y \geq 0$$

on the side x_A.

On the other three boundaries y_C, y_D and x_B:

(b) $$n_y^2-n_x^2 > 0\ ,\quad bn_x+cn_y < 0\ ,\quad b^2-c^2y \geq 0\ .$$

We define μ and M:

on x_A $\quad \mu = \beta$, $M = 0$.

Then $M\psi = 0$ implies that ψ is unspecified and $\psi\cdot\mu\psi = \psi\cdot\beta\psi > 0$. On y_C, y_D and x_B:

$$\psi\cdot\mu\psi = -\frac{1}{|bn_x+cn_y|}\,[(b^2-c^2y)(n_y\varphi_x-n_x\varphi_y)^2 + (n_y^2-yn_x^2)(b\varphi_x+c\varphi_y)^2]$$

$$\psi\cdot M\psi = -\frac{2}{bn_x+cn_y}\,[(b^2-c^2y)(n_y\varphi_x-n_x\varphi_y)^2]\ .$$

Then $M\psi = 0$ implies $n_y\varphi_x - n_x\varphi_y = \frac{d\varphi}{ds} = 0$ and $\psi \cdot \mu\psi > 0$. We observe that the operators $\mu - \beta$ and $\mu + \beta$ are, respectively 0 and 2μ. Their ranges are φ and the entire space respectively, and their null spaces are the entire space and φ respectively. Hence conditions II is satisfied. In section 2.4 below we give further examples of transformations and the corresponding boundary conditions for non-rectangular regions.

The Difference Problem

Since in the theory of symmetric positive systems the differential equations are characterized in a way independent of type, the difference equations derived from them will accordingly be type insensitive, i.e., the same scheme can be used in hyperbolic and elliptic regions. As other finite difference methods for such mixed equations use widely different techniques in each region, and necessitate rather delicate piecing together on the dividing line, a unified computational approach may offer significant simplifications, at least in principle.

K. O. Friedrichs [8] was the first to propose a finite difference procedure for the numerical solutions of symmetric positive systems in rectangular regions. Chu [6] further studied this method and extended it to curvilinear rectangular domains. Katsanis [16] gives a finite difference method for the solution of symmetric positive differential equations which is applicable to any region with piecewise smooth boundaries. We briefly describe the two difference schemes mentioned above.

(i) A finite difference scheme for curvilinear rectangular regions. It is shown in [6] that there exists a transformation of the dependent variables, coefficient matrices, the differential equations, and the boundary conditions such that when the domain is mapped from a curvilinear rectangle to a rectangle, the symmetric positive character of the equation is preserved. Thus without any loss of generality we may confine our considerations to rectangular domains.

Let R be the rectangle centered at the origin, with boundaries $x = x_A$, $x = x_B$, $y = y_C$ and $y = y_D$. Let R be partioned in a square grid of width h , a grid point is denoted by a pair of integers (σ,τ) with $(-s-2,-t-2) \leq (\sigma,\tau) \leq (s+2,t+2)$; the width h is selected so that s and t are even integers. The grid points $|\sigma| = s$, $|\tau| = t$ are called boundary points, those with $s < |\sigma| \leq s+2$, $t < |\tau| \leq t+2$ are extensions of the domain beyond the boundary and shall be called extension points.

We introduce the shift operators $S^x u(\sigma,\tau) = u(\sigma+1,\tau)$, $S^y u(\sigma,\tau) = u(\sigma,\tau+1)$. $S^{2x} \equiv S^x S^x$ and similarly for S^{2y} , S^{nx} and S^{ny} . $S^{-x} \equiv (S^x)^{-1}$, $S^{-y} \equiv (S^y)^{-1}$. All such operators commute.

The boundary operator B^0 and B^1 are defined such that $B^0 u$ is the value of u on the boundary and $B^1 u$ is the value of u one row beyond (into extension) respectively. B^{-1} is defined so that $B^{-1}u$ is the value one row within (into interior), and B^2, B^{-2} etc are similarly defined. (e.g. $B^1 = S^x B_0$ at $x = x_A$, $B^1 = S^{-x} B_0$ at $x = x_B$, etc.).

For each interior point we define the difference operator L_h :

$$L_h u \equiv \frac{(S^x A_1) S^{2x} u - (S^{-x} A_1) S^{-2x} u}{2h} + \frac{(S^y A_2) S^{2y} u - (S^{-y} A_2) S^{-2y} u}{2h} + Ku .$$

To each boundary point we assign the approximate boundary condition

$$M_h u = B^1 \mu \cdot B^0 u - B^1 \beta B^2 u = 0 \quad ,$$

where

$$B^1\beta = n_x B^1 A_1 + n_y B^1 A_2 \quad \text{(at each grid row } B^1\text{)}$$

$$B^1\mu = B^1\beta + M \quad .$$

The differential equation (2.1.1b) and the boundary conditions (2.1.2b) are replaced by the consistent finite difference operator L_h and the approximate boundary condition M_h for each even interior and boundary point, i.e.,

(2.1.5b)$_i$ $$L_h u = f \text{ in } R$$

(2.1.6b)$_i$ $$M_h u = 0 \text{ on } \partial R \quad .$$

In [8] and [6] various identities and convergence theorems are proved for the above finite difference approximation.

For the application and numerical implementation of this method to the model problem considered earlier we refer to [27].

(ii) A finite difference procedure for general domains.
In [16] a finite difference scheme for the symmetric positive system is given which is not restricted to curvilinear rectangular domains; in fact this procedure is applicable to any bounded domain with a piecewise smooth boundary.

The approach in this method follows the same line of argument as the method of Friedrich's [8] and Chu [6] described above. The main difference lies in the fact that in [16] first the differential equation is written in integral form, then the various integrals are approximated in a suitable manner to obtain a discrete analogue L_h of the differential operator L. The boundary condition is approximated in a similar way as in [6] with a slight modification. The order of accuracy of this scheme is $O(h^{\mu})$ with $\mu \le \frac{1}{2}$ in L^2 norm.

In [17] this method is applied to solve numerically the Tricomi problem for the model equation discussed earlier in this section where Ω is a bounded domain which consists of a rectangle for $y \leq 0$ and is bounded by two characteristics in the upper half plane $y > 0$. In section 2.4 these numerical results are compared with results obtained by a finite element procedure given in [1].

2.2. Finite Element Methods

As far as linear equations of mixed type are concerned, very little work on finite element methods has been done, except for [1], [28]. In [1] the results of the finite element method for the symmetric positive systems [20] are used to obtain the numerical solution of a boundary value problem for the Tricomi equation which has been dealt with in [17] by finite differences. In section 2.4 we shall show through examples that the technique used in [1] applies to more general equations, i.e., to equations of the form $(g(x,y)u_x)_x + u_{yy} = f$, where $g(x,y)$ changes sign in the domain under consideration. We observe that in this method the same discretization method is used for both the elliptic and hyperbolic regions.

In [28] a finite element method for the solution of Tricomi equation in the upper-half plane $y > 0$ is given. Error estimates are obtained for the elliptic region $y > 0$. It appears that the solution on the parabolic line $y = 0$ cannot be well approximated. Thus the solution in the hyperbolic region $y < 0$ may suffer by the error in the solution on the parabolic line, which must be used as part of the initial data.

The situation is very similar to what happens when two finite difference schemes of varying accuracy are used, one in the elliptic and another in the hyperbolic region.

We now briefly describe the finite element approach which shall be used for the numerical calculations in section

2.4 below.

Condider the first order system

$$\text{(2.2.1)} \qquad Lu \equiv A_1 \frac{\partial u}{\partial x} + A_2 \frac{\partial u}{\partial y} + A_3 u = f \quad \text{in} \quad \Omega$$

$$\text{(2.2.2)} \qquad Nu = 0 \quad \text{on} \quad \partial\Omega ,$$

where Ω is a bounded open set in the x,y plane with continuously differentiable boundary $\partial\Omega$, L is symmetric positive and N is admissible, i.e.

$$C = A_3 + A_3' - \frac{\partial A_1}{\partial x} - \frac{\partial A_2}{\partial y} \quad (A_3' = A_3 \text{ adjoint}) ,$$

is positive definite and

$$N+N^* > 0 , \ \text{Ker } N + \text{Ker}(N^*) = R^2 ,$$

$N^* = \beta+N'$, $\beta = n_x A_1 + n_y A_2$. ($n = (n_x, n_y)$ is the outer normal on $\partial\Omega$.) In the notation of section (2.1b) we have $N = \frac{\mu-\beta}{2}$. Define L^* as

$$L^*v = -\frac{\partial}{\partial x}(A_1 v) - \frac{\partial}{\partial y}(A_2 v) + A_3' v .$$

There are three possibilities for a variational formulation. The first is based on the weak formulation.

Find $\varphi \in L_2(\Omega)$ such that

$$(\varphi, L^*\psi)_{L^2(\Omega)} = (\psi, f)_{L^2(\Omega)} \quad \forall \ \psi \in H^1(\Omega) , \ N^*\psi = 0 \ \text{ on } \partial\Omega$$

This has the disadvantage that the trial functions must satisfy the adjoint boundary conditions. The second and third formulations are based on the basic identity

$$(L\varphi,\psi)_{L^2(\Omega)} + (N\varphi,\psi)_{L^2(\partial\Omega)} = (\varphi,L^*\psi)_{L^2(\Omega)} + (\varphi,N^*\psi)_{L^2(\partial\Omega)} \ , \ \forall\ \varphi,\psi \in H^1(\Omega) \ .$$

These formulations are:

if $\varphi \in H^1(\Omega)$ satisfies (2.2.1) , (2.2.2) then

$$(\varphi,L^*\psi)_{L^2(\Omega)} + (\varphi,N^*\psi)_{L^2(\partial\Omega)} = (f,\psi)\ \forall\ \psi \in H^1(\Omega)$$

and

$$(L\varphi,\psi)_{L^2(\Omega)} + (N\varphi,\psi)_{L^2(\partial\Omega)} = (f,\psi)\ \forall\ \psi \in H^1(\Omega) \ .$$

We observe that for $H^1(\Omega)$ solutions of (2.2.1),(2.2.2) the two formulations are equivalent.

Let V_h be a finite dimensional subspace of $H^1(\Omega)$, where h denotes a positive parameter that will tend to zero. Consider the discrete problem of finding $\varphi_h \in V_h$ such that

$$(2.2.3)\quad (L\varphi_h,\psi_h)_{L^2(\Omega)} + \int_{\partial\Omega} (N\varphi_h,\psi_h)ds = (f,\psi_h) \ , \ \forall\ \psi_h \in V_h \ .$$

Let $\{\varphi_i\}_{i=1}^n$ be a basis for V_h , i.e., $\varphi_h = \sum_{i=1}^n \alpha_i\varphi_i$, then (2.2.3) reduces to the linear system

where
$$K\alpha = f \quad ,$$

$$\alpha = \begin{pmatrix} \alpha_1 \\ \vdots \\ \alpha_n \end{pmatrix} \ , \quad F = \begin{pmatrix} (f,\varphi_1)_{L^2(\Omega)} \\ \vdots \\ (f,\varphi_n)_{L^2(\Omega)} \end{pmatrix} , \ K = \left[(L\varphi_i,\psi_j)_{L^2(\Omega)} + (N\varphi_i,\psi_j)_{L^2(\partial\Omega)} \right] .$$

From the fundamental identities and the admissibility of N it follows immediately that

$$\|\varphi_h\|_{L^2(\Omega)} \le C\|f\|_{L^2(\Omega)} \quad , \; C > 0 \quad ,$$

which implies the uniqueness of the solution of (2.2.3).

Convergence theorems and error estimates are given in [20]. In particular if the subspaces V_h are restricted to functions φ_h obtained by Lagrange interpolation polynomials of degree $\le K$ on triangular elements and if the solution u of (2.2.1), (2.2.2) are in $H^{K+1} \cap C^0(\bar{\Omega})$, then

(2.2.4) $$\|u-\varphi_h\|_{L^2(\Omega)} = O(h^K) \quad \text{as} \quad h \to 0 \quad .$$

The above estimate of Lesaint [20] does not appear to be optimal, in fact the numerical calculations indicate that

(2.2.5) $$\|u-\varphi_h\|_{L^2(\Omega)} = O(h^{K+1}) \quad \text{as} \quad h \to 0 \quad .$$

2.3 Analytical Approximative Methods

The finite difference and the finite element methods are only two of the possible approaches for numerical integration of equations of mixed type. Several authors, Bergman [2], [3], Guderley and Yoshihara [9], [10], use expansions in terms of particular solutions. Ovsiannikov [26] studied the representation of the solution of Tricomi's problem for the Tricomi equation by an expansion in terms of certain particular solutions. In some cases, this leads to an effective method for computing the solution.

2.4. Numerical Examples

I. We consider the Tricomi problem

(2.4.1) $$yu_{xx}-u_{yy} = f \quad \text{in} \quad \Omega$$

(2.4.2) $u = 0$ on Γ_2, Γ_3, Γ_4 ; u unspecified on Γ_1 and

$$2u_x + u_y = 0 \text{ on } \Gamma_5$$

where Ω is the domain in the x,y plane bounded by the characteristics Γ_1 and Γ_2 of (2.4.1) passing through the points (0,1) and (0,-1) respectively for $y \geq 0$ and bounded by the rectangle with sides Γ_3, Γ_4 and Γ_5 for $y < 0$ as shown in the figure I below.

By a change of independent variables:

$$\varphi = \begin{pmatrix} \varphi_1 \\ \varphi_2 \end{pmatrix}, \quad \varphi_1 = e^{-\lambda x} u_x, \quad \varphi_2 = e^{-\lambda x} u_y,$$

equation (2.4.1) may be written as the symmetric first order system

$$\begin{pmatrix} y & 0 \\ 0 & 1 \end{pmatrix} \varphi_x + \begin{pmatrix} 0 & -1 \\ -1 & 0 \end{pmatrix} \varphi_y + \begin{pmatrix} \lambda y & 0 \\ 0 & \lambda \end{pmatrix} \varphi = \begin{pmatrix} e^{-\lambda x} f \\ 0 \end{pmatrix} \tag{2.4.3}$$

However (2.4.3) is not positive. The multiplication of (2.4.3) by the matrix

$$T = \begin{pmatrix} a & by \\ b & a \end{pmatrix},$$

where a and b are functions of x,y yields the system

$$A_1\varphi_x + A_2\varphi_y + A_3\varphi = F, \tag{2.4.4}$$

where

$$A_1 = \begin{pmatrix} ay & by \\ by & a \end{pmatrix}, \quad A_2 = \begin{pmatrix} -by & -a \\ -a & -b \end{pmatrix}, \quad A_3 = \begin{pmatrix} \lambda ay & \lambda by \\ \lambda by & \lambda a \end{pmatrix},$$

$$F = \begin{pmatrix} a\, e^{-\lambda x} f \\ b\, e^{-\lambda x} f \end{pmatrix}, \quad C = A_3 + A_3' - \frac{\partial A_1}{\partial x} - \frac{\partial A_2}{\partial y} = \begin{pmatrix} 2\lambda ay+b & 2\lambda b \\ 2\lambda b & 2\lambda a \end{pmatrix}$$

It is easily seen that with the choice of $a = 2$, $b = 1$ and $\lambda = -\frac{1}{10}$ (2.4.3) becomes a symmetric positive system in Ω . Let $n = (n_x, n_y)$ be the outer normal to the boundary $\partial\Omega = \Gamma_1 \cup \Gamma_2 \cup \Gamma_3 \cup \Gamma_4 \cup \Gamma_5$. The boundary matrix β (see section 2.2) is defined by

$$\beta = n_x A_1 + n_y A_2$$

We recall that the boundary condition $N\varphi = 0$ is admissible if these exists a matrix μ such that: (i) $N = \frac{\mu-\beta}{2}$,
ii) $\mu(x) + \mu'(x) \geq 0$ (μ' - adjoint of μ) ,
iii) $\operatorname{Ker}(\beta-\mu) + \operatorname{Ker}(\beta+\mu) = R^2$. By writting the quadratic form $\varphi \cdot \beta\varphi$ as

$$\varphi \cdot \beta\varphi = \frac{(a^2-b^2y)(n_y\varphi_1 - n_x\varphi_2) - (n_y^2 - yn_x^2)(a\varphi_1 + b\varphi_2)^2}{an_x + bn_y}$$

μ may be defined so that N is admissible, namely,

$$\varphi \cdot \mu\varphi = \frac{|a^2-b^2y|(n_y\varphi_1 - n_x\varphi_2)^2 + |n_y^2 - yn_x^2|(a\varphi_1 + b\varphi_2)^2}{|an_x + bn_y|}$$

A simple calculation yields the following boundary matrices:

on $\Gamma_1 = \{(x,y) | y^3 = \frac{9}{4}(x-1)^2\}$ $N = \begin{pmatrix} 0 & 0 \\ 0 & 0 \end{pmatrix}$;

on $\Gamma_2 = \{(x,y) | y^3 = \frac{9}{4}(x+1)^2\}$ $N = \frac{2+\sqrt{y}}{\sqrt{1+y}} \begin{pmatrix} y & \sqrt{y} \\ \sqrt{y} & 1 \end{pmatrix}$;

on $\Gamma_3 = \{(x,y) | x = -1 \; , \; y \in [-1,0]\}$ $N = \begin{pmatrix} 0 & 0 \\ 0 & 2-\frac{y}{2} \end{pmatrix}$;

on $\Gamma_4 = \{(x,y) | y = -1 \quad x \in [-1,1]\}$ $N = \begin{pmatrix} 5 & 0 \\ 0 & 0 \end{pmatrix}$;

on $\Gamma_5 = \{(x,y) | x = 1 \quad y \in [-1,0]\}$ $N = \begin{pmatrix} -2y & -y \\ -y & -\frac{y}{2} \end{pmatrix}$.

Thus the boundary value problem (2.4.1), (2.4.2) is reduced to the symmetric positive system.

(2.4.4) $$A_1\varphi_x + A_2\varphi_y + A_3\varphi = F \text{ in } \Omega$$

(2.4.5) $$N\varphi = 0. \text{ on } \partial\Omega = \Gamma_1 \cup \Gamma_2 \cup \Gamma_3 \cup \Gamma_4 \cup \Gamma_5 \quad ,$$

where N is as described above.

Let f in (2.4.1) be given as

$$f(x,y) = y(y+1)[(4y^3 - 9(x+1)^2)(6x-2) - 18(x^2-1)(7x+1)]$$

$$-24(x+1)(x-1)^2 y(2y+1) \; ,$$

then

$$u(x,y) = (y+1)(x+1)(x-1)^2(4y^3 - 9(x+1)^2)$$

is the exact solution of (2.4.1), (2.4.2).

If we define

$$\varphi_1(x,y) = e^{-.1x} u_x \ , \ \varphi_2 = e^{-.1x} u_y \ ,$$

then

$$\varphi = (\varphi_1, \varphi_2)^t$$

is the exact solution of (2.4.4) and (2.4.5).

This is the numerical example that we shall consider. In order to apply the finite element procedure of section 2.2, we first triangulate the region Ω as in figure (I), where h refers to the length of the legs of the right triangles in the triangulation. For the space V_h we choose the set of piece-wise linear functions over these triangles. The convergence theorem in [20] guarantees $O(h)$ - rate of convergence in L_2. However numerical results indicate $O(h^2)$ - rate of convergence in L_2. The computation was carried out for $h = .2$ and $h = .1$. Table 1 gives the result of these computations.

ERRORS IN FINITE - ELEMENT COMPUTATION

MESH SIZE	H = .2	H = .1
MAX ERROR IN $\bar{\Omega}$	11.5666	1.44
MAX ERROR ON $\partial\Omega$	11.5666	1.44
MAX ERROR $y < 0$	1.01	.4248
MAX ERROR $y > 0$	3.545	.8734
L_2 - ERROR	1.88	.3506

Table I

A similar problem is solved in [17] by Katsanis using a finite difference method for positive symmetric systems with a rate of convergence of $O(h^{\mu})$, $\mu < \frac{1}{2}$ in L_2 . This difference scheme is pointwise divergent. Table II gives a comparison of the finite element and finite difference results.

FINITE DIFFERENCE - FINITE ELEMENT COMPARISON

METHOD	FINITE DIFFERENCE	FINITE ELEMENT
MAX ERROR H = .2	33.5	11.566
MAX ERROR H = .1	60.9	1.44
L_2 - ERROR H = .2	6.06	1.88
L_2 - ERROR H = .1	5.30	.3506
L_2 - RATE OF CONVERGENCE	.193	2.429

Table II

II. As a second example we consider a boundary value problem for the equation.

$$Lu \equiv (G(x,y)u_x)_x + u_{yy} = f \quad \text{in} \quad \Omega \quad , \tag{2.4.6}$$

where Ω is the unit square centered at the origin with the boundary $\partial\Omega = \Gamma_1 \cup \Gamma_2 \cup \Gamma_3 \cup \Gamma_4$ (see figure II). We prescribe the following boundary conditions:

$$\begin{cases} u = 0 \quad \text{on} \quad \Gamma_1 \quad \text{and} \quad \Gamma_4 \\ u \quad \text{unspecified on} \quad \Gamma_2 \quad \text{and} \;, \; au_x + bu_y = 0 \quad \text{on} \quad \Gamma_3 \;, \end{cases} \tag{2.4.7}$$

where a and b are given functions of x and y .

In order to apply the finite element procedure described in section 2.2 we first transform the above problem to a symmetric positive system with admissible boundary conditions. To this end let

$$\varphi_1 = au_x + bu_y \;,\; \varphi_2 = - bGu_x + au_y \quad ,$$

where a and b are functions of x and y .
Then

(2.4.8) $$L\varphi = A_1\varphi_x + A_2\varphi_y + A_3\varphi = f \quad ,$$

where

$$A_1 = \begin{pmatrix} \frac{aG}{\gamma} & -\frac{bG}{\gamma} \\ -\frac{bG}{\gamma} & -\frac{a}{\gamma} \end{pmatrix} , \quad A_2 = \begin{pmatrix} \frac{bG}{\gamma} & \frac{a}{\gamma} \\ \frac{a}{\gamma} & -\frac{b}{\gamma} \end{pmatrix}$$

$$A_3 = \begin{pmatrix} (\frac{aG}{\gamma})_x + (\frac{bG}{\gamma})_y & (\frac{a}{\gamma})_y - (\frac{bG}{\gamma})_x \\ (\frac{a}{\gamma})_y - (\frac{bG}{\gamma})_x & -[(\frac{a}{\gamma})_x + (\frac{b}{\gamma})_y] \end{pmatrix} , \quad \gamma = a^2+b^2G ,$$

$$C = A_3 + A_3' - \frac{\partial A_1}{\partial x} - \frac{\partial A_2}{\partial y} = A_3 \quad .$$

Let

$$r = (\frac{aG}{\gamma})_x + (\frac{bG}{\gamma})_y \quad , \quad s = (\frac{a}{\gamma})_y - (\frac{bG}{\gamma})_x \quad , \quad t = -[(\frac{a}{\gamma})_x + (\frac{b}{\gamma})_y] \ .$$

We see that C is positive definite if

(2.4.3) $$\Delta = rt-s^2 > 0 \quad \text{and} \quad T = r+t > 0 \quad .$$

If we suppose that $a = a(x)$ and $b = b(y)$, then (2.4.8) is equivalent to

(2.4.10) $$D = (a_x-b_y)(aG_x+bG_y-(a_x-b_y)G) > 0 \ .$$

The admissible boundary condition is determined, as in the previous example, by the quadratic forms

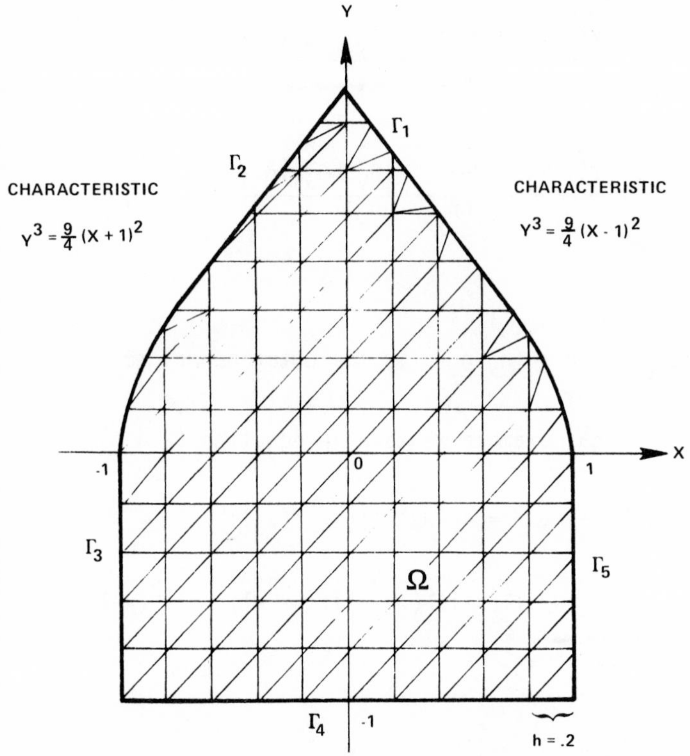
Y
CHARACTERISTIC
$Y^3 = \frac{9}{4}(X+1)^2$
CHARACTERISTIC
$Y^3 = \frac{9}{4}(X-1)^2$
Γ_1
Γ_2
Γ_3
Γ_4
Γ_5
Ω
X
-1
0
1
-1
h = .2

Fig. 1

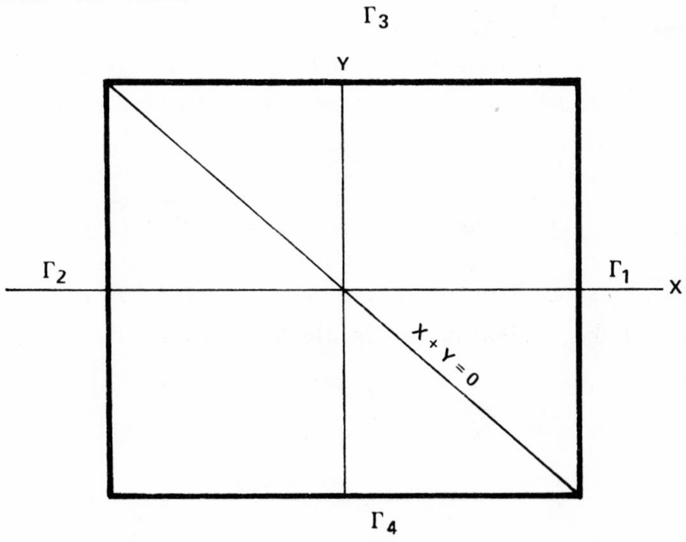
Γ_3
Y
Γ_2
Γ_1
X
X + Y = 0
Γ_4

Fig. 2

$$(2.4.10)\qquad \varphi\cdot\beta\varphi = \frac{\gamma(n_y\varphi_1 - n_x\varphi_2)^2 - (n_y^2+Gn_x^2)(a\varphi_1+b\varphi_2)^2}{\gamma(bn_y - an_x)}$$

$$(2.4.11)\qquad \varphi\cdot\mu\varphi = \frac{\gamma(n_y\varphi_1 - n_x\varphi_2)^2 + |n_y^2+Gn_x^2|(a\varphi_1+b\varphi_2)^2}{\gamma|bn_y - an_x|} ,$$

where

$\beta = n_x A_1 + n_y A_2$, $n = (n_x, n_y)$ is the outer normal to $\partial\Omega$.

Now for our numerical example we take in equation (2.4.6)

$$G(x,y) = x+y , \; f = 4[(3x^2-1)(x+y)(y+1)^2$$
$$+ (x-1)^2(x+1)^2(3y^2-1)+x(x^2-1)(y+1)^2(y-1)^2] .$$

With the above choices for G and f , the function

$$u(x,y) = (x+1)^2(x-1)^2(y+1)^2(y-1)^2$$

satisfies (2.4.6) and (2.4.7).

Now for $a = x+3$ and $b = 1$, the system (2.4.8) corresponding to $G(x,y) = x+y$ is symmetric positive. In fact since $|x| \le 1$, $|y| \le 1$

$$\gamma = (x+3)^2+(x+y) > 0 \quad \text{and} \quad D = (x+3)+1-(x+y) = 4-y > 0 .$$

For these choices of G , a and b (2.4.10) and (2.4.11) yield the following admissible boundary condition for (2.4.8).

$$\text{on } \Gamma_1 \; : \; N = \frac{\mu-\beta}{2} = \begin{pmatrix} 0 & 0 \\ 0 & 1/4 \end{pmatrix} ;$$

on Γ_2 : $N = \frac{\mu-\beta}{2} = \begin{pmatrix} 0 & 0 \\ 0 & 0 \end{pmatrix}$ (no boundary condition specified) ;

on Γ_3 : $N = \frac{\mu-\beta}{2} = \frac{1}{\gamma} \begin{pmatrix} (x+3)^2 & -(x+3) \\ -(x+3) & 1 \end{pmatrix}$;

on Γ_4 : $N = \frac{\mu-\beta}{2} = \begin{pmatrix} 1 & 0 \\ 0 & 0 \end{pmatrix}$.

As in example I, we apply the finite procedure of section 2.2, with the space V_h chosen as the set of piecewise linear functions over triangles. The numerical results indicate once again $O(h^2)$ rate of convergence in L_2 norm. The computation was carried out for $h = .2$ and $h = .1$. Table III gives the result of these computations. In table IV we compare the finite element results to the results obtained by a finite difference scheme proposed by Murman and Cole [24] (see section (2.1a)ii) for a nonlinear mixed problem in connection with small disturbance transonic flow.

ERROR IN FINITE - ELEMENT COMPUTATION

MESH SIZE	H = .2	H = .1
MAX ERROR IN $\bar{\Omega}$	.214915	.072785
MAX ERROR ON Γ	.183345	.056828
MAX ERROR - ELLIPTIC REGION	.214915	.072785
MAX ERROR - HYPERBOLIC REGION	.174956	.0687811
L_2 - ERROR	.1585	.0393554
L_2 - RATE OF CONVERGENCE	2.00984	

Table III

COMPARISON OF MURMAN'S FINITE DIFFERENCES METHOD
AND
FINITE ELEMENT SOLUTION

	L_2-ERROR SOLUTION	L_2-RATE OF SOLUTION	L_2-ERROR DERIVATIVE	L_2-RATE OF DERIVATIVE
FINITE DIFFERENCE MURMAN H = .2	.07562513	$H^{.57044}$	.3481913	$H^{.3843}$
FINITE DIFFERENCE MURMAN H = .1	.05095242		.2667630	
FINITE ELEMENT H = .2	-	-	.1585	$H^{2.0099}$
FINITE ELEMENT H = .1	-		.039355	

Table IV

REFERENCES

1. A. K. Aziz and S. Leventhal, "On Numerical Solutions of Equations of Hyperbolic-Elliptic Type." (To appear)
2. S. Bergman, "Methods or Determination and Computation of Flow Patterns of a Compressible Fluid," NACAT Technical Note No. 1018 (1946).
3. ______, "Operator Methods in the Theory of Compressible Fluids," Proceedings of Symposia in Appl. Math., 1 (1949) pp. 1-18.
4. L. Bers, "Mathematical Aspects of Subsonic and Transonic Gas Dynamics," Surveys In Appl. Math. III (1958).
5. A. V. Bitsadze, Equations of the Mixed Type, The MacMillan Company, New York (1964).
6. C. K. Chu, "Type-Insensitive Difference Schemes," Ph.D. Thesis, New York University (1959).
7. A. Filippov, "On Difference Methods for the Solution of the Tricomi Problem," Izv. Akad. Nauk SSR. Ser. Mat., 21 (1957), pp. 73-88.
8. K. O. Friedrichs, "Symmetric Positive Linear Differential Equations," Comm. Pure Appl. Math., 11 (1958), pp. 333-418.
9. G. Guderley and H. Yoshihara, "The Flow Over a Wedge Profile at Mach Number 1," Journal of the Aeronautical Sciences, 17 (1950), pp. 723-735.

10. ______________, "Two-dimensional Unsymmetric Patterns at Mach Number 1," Journal of the Aeronautical Sciences, 20 (1953), pp. 756-768.
11. Z. I. Halilov, "Solutions of a Problem for an Equation of Mixed Type by the Method of Grids," Akad. Nauk. Azerbaidzanskogo SSR, Trudy Inst. Fiz.-Mat., 6 (1953), pp. 5-13.
12. _________, "Solution of a Problem for an Equation of Mixed Type by the Method of Nets," Dokl. Akad. Nauk, Azerbaidzanskogo SSR, 9 (1953), pp. 189-194.
13. A. Jameson, "Numerical Solution of Nonlinear Partial Differential Equations of Mixed Type," These Proceedings.
14. G. D. Karatoprakliev, "Equation of Mixed Type and Degenerate Hyperbolic Equations in Multidimensional Regions," Differential'nye Uravnenija, No. 1. 8 (1973), pp. 55-67.
15. B. G. Karmanov, "On a Boundary Value Problem for an Equation of Mixed Types," Doklady, 95 (1954), pp. 439-442.
16. T. Katsanis, "Numerical Solution of Symmetric Positive Differential Equations," Math. Comp. 22 (1968), pp.763-783.
17. __________, "Numerical Solution of Tricomi Equation Using Theory of Symmetric Positive Differential Equations," SIAM J. Numer. Anal. No. 2 (1969), pp. 236-253.
18. O. A. Ladyzenskaya, "On one Method for Approximating the Solution to the Lavrent'ev-Bitsdze Problem," Uspehi Math. Nank, 9, 4 (1954), pp. 187-189.
19. P. D. Lax and R. S. Phillips, "Local Boundary Conditions for Dissipative Symmetric Linear Differential Operators," Comm. Pure Appl. Math. 13 (1960), pp. 427-455.
20. P. Lesaint, "Finite Element Methods for Symmetric Hyperbolic Equations," Numer. Math. 21 (1973), pp. 244-255.
21. D. Levey, "A Numerical Scheme for Solving a Boundary Value Problem for the Tricomi Equation," Ph.D. Thesis, New York University (1957).
22. C. Morawetz, "A Uniqueness Theorem for Frankl's Problem," Comm. Pure Appl. Math., 7 (1954), pp. 697-703.
23. ___________, "A Weak Solution for a System of Equations of Elliptic-Hyperbolic Type," Comm. Pure Appl. Math., 11 (1958), pp. 315-331.
24. E. M. Murman, and J. D. Cole, "Calculation of Plane Steady Transonic Flows," AIAA J. 9 (1971), pp. 114-121.
25. H. Ogawa, "On Difference Methods for the Solution of a Tricomi Problem," Tran. Amer. Math. Soc., 100 (1961), pp. 404-424.
26. L. V. Ovsiannikov, "Concerning the Tricomi Problem for One Class of Generalized Solutions of the Euler-Darboux Equation," Doklady, 91 (1953), pp. 457-460.
27. S. Schecter, "Quasi-Tridiagonal Matrices and Type-Insensitive Difference Equations," Quart. Appl. 18 (1960), pp. 285-295.

28. J. A. Trangenstein, "A Finite Element Method for the Tricomi Problem in the Elliptic Region," Ph.D. Thesis, Cornell University (1975).

HOMOGENIZATION AND ITS APPLICATION. MATHEMATICAL AND COMPUTATIONAL PROBLEMS*

Ivo Babuška**

1. The Homogenization, Its Content and Application

1.1. The content of the homogenization

Homogenization is an approach which studies the macro-behavior of a medium by its microproperties. The origin of this word is related to the question of a replacement of the heterogenous material by an "equivalent" homogenous one.

The content of the homogenization approach may be seen in the following three directions.

a) Applications in physics, mechanics, engineering, etc.

b) Mathematics

c) Numerical methods

There are many papers in direction a) but very few results are available in directions b) and c).

We will concentrate our attention mainly to the areas b) and c) with applications (and references) mostly in physics, mechanics and chemistry, although there are many more fields where homogenization is relevant. Instead of a general theory

---

* Research supported in part by the U.S. Energy Research and Development Administration under Contract #AEC AT (40-1)3443. Computer time for this project was supported in part through the facilities of the Computer Science Center of the University of Maryland.
** Institute for Fluid Dynamics and Applied Mathematics, University of Maryland.

we will discuss in this paper only a few characteristic examples. This will serve the expository character best.

1.2. The first example of a full dimensional homogenization and its history

Let us study the diffusion problem in a periodic heterogenous medium. Mathematically, we are interested in the analysis (in R_3) of the solution of the differential equation

$$(1)\qquad \sum_{i=1}^{3} \frac{\partial}{\partial x_i} a^H(x) \frac{du^H}{dx_i} = f(x)$$

where $x = (x_1, x_2, x_3)$ and

$$a^H(x) = a(\tfrac{x}{H})$$

with $0 < C_1 \le a(x) \le C_2 < \infty$ being a periodic function with the period 1 (generally discontinuous). When f has compact support and

$$\int_{R_3} f(x)dx = 0$$

then the solution u^H exists and is (up to a constant) uniquely determined.

Function $a^H(x)$ describes (micro) property of a periodic heterogenous medium. A question arises. Does the material behave "equivalently" as a homogenous material described by the differential equation

$$(2)\qquad \sum_{i,j=1}^{3} \frac{\partial}{\partial x_i} A_{i,j} \frac{\partial U}{\partial x_j} = f$$

with $A_{i,j}$ constants? These coefficients are often called bulk or effective diffusion coefficients. This and similar questions are very old. The first paper dealing with the problem of this type is [1]. See also [2] [3] [4] [5]. A very good survey of the results until 1925 is in [6].

The following table 1 (see [10]) shows the (constant) coefficients $A_{i,j}$,

$$A_{i,i} = K \ , \ i = 1,2,3 \ , \ A_{i,j} = 0 \quad \text{for} \quad i \neq j$$

for $k_A > 0$, $k_B > 0$, with

$$a(x) = k_A \quad \text{for} \quad x \in S = \{x | \sum_{i=1}^{3} x_i^2 \leq r^2 \ , \ r = .49237\}$$

and $a(x) = k_B$ for $x \in C - S$, with $C = [x| \ |x_i| < \frac{1}{2}]$ computed by different theories. In this case the sphere S takes 50% of the cell volume. Column a and f are lower and upper bounds for K derived in [7].

k_A / k_B	K/k_B					
	a	b	c	d	e	f
	[7]	[3]	[5]	[8]	[9]	[7]
∞	5.1	4.0	5.0	8.0	∞	66.7
10^2	4.5	3.8	4.7	6.7	27.1	19.2
10	2.72	2.8	3.06	3.3	4.0	4.1
2	1.40	1.43	1.43	1.44	1.44	1.46
.5	.71	.73	.72	.72	.73	.74
10^{-1}	.37	.47	.45	.44	.40	.48
0	.24	.40	.33	.35	.25	.40

Table 1

There are very many theories related to these problems obtained independently in Physics, Mechanics, Chemistry, etc. Table 1 shows a large and typical dispersion of the results stemming from different theories. The main reason for this dispersion is that the above equivalency is not precisely defined in a mathematical way.

Section 2 shows one possible definition of this equivalency and of the effective coefficients. Although the bulk coefficients are not well defined, there are lower and upper bounds for them, see e.g. [7]. In [11] the following bound is derived

$$\frac{1}{V_A/k_A+V_B/k_B} \le K \le V_A k_A + V_B k_B \tag{3}$$

where V_A resp V_B and k_A resp k_B are relative volumes $(V_A+V_B=1)$ resp diffusion coefficients for the two fractions of the material. In section 2 it will be shown that this formula has a precise sense.

So far we assumed that there is a regular array of the heterogeneities. A very impotant case is also when these heterogeneities have a random character. As a type of study devoted to this case we mention only [12], chapter 5. This lecture will not elaborate in this direction.

Eq. (1) depends on a parameter H. This parameter expresses the scale of the microbehavior (size of the cell) in comparison to the macroscale of f. Although the case $H \to 0$ has no real physical meaning, it is important to study this limiting process as a tool for the numerical solution. We will address this problem later in the paper.

Let us remark that there is not a unique way of introducing a "natural parameter". As an example, see Section 2.4. This of course contributes to the dispersion of the results.

1.3. The second example of a full dimensional homogenization

Another very important example of the homogenization approach may be found in the analysis of the composite materials. Linear theory of elasticity gives analogous equations to (1). Here, the Lamé equations for a vector function $u = (u_1, u_2, u_3)$ [1)] (the system is strongly elliptic) are the governing equations. Assuming a periodic heterogeneity (with period H) the same question as before arises. Is it possible to replace the heterogenous medium by an equivalent homogenous (in general anisotropic) one with some bulk (or effective) elastic properties? This and similar questions are discussed in many papers. For surveys, see e.g. [14], [15], [16].

As in previous examples of the diffusion equation, there is a large dispersion of results when different theories are used. [17] brings a comparison of results for a concrete case of composites (E-GLASS-EPOXY), with an extensive survey (109 references) of results until 1967. The large dispersion of results is caused by very different approaches to the problem with not precisely defining (mathematically) the principle notions as bulk modulae, etc. In comparison to the first example (diffusion), the case of composites brought a much larger variety of physical and engineering approaches into considerartion.

The oldest theory of composites is related to the so-called netting theory and the law of mixtures. One of these netting models was developed in [18]. [19] is an example of the application of the law of mixture, which is some simple averaging of the coefficients. A generalization using principles of the theory of strength of material was used in many papers. ([20] is a typical model of this category.)

As usual, the strength of material approach is more or less close to the use of the theory of elasticity. Many papers are in this direction. Some of the typical ones are e.g. [16]

1) See eq. [13], chapter 8.

and [21]. The variational approach and its refinement is the topic of many other papers, see e.g. [22]. [23] combines the variational and finite difference approach. There are many more different approaches and we refer the reader e.g. to <u>Journal of Composite Materials</u> where many additional references may be found. A large literature is devoted to the bounds of elastic constants (although these are not well defined), see e.g. [22], [24], [25], as typical references.

The derivation of the bulk elastic properties is not the only problem. E.g. there is the question about the relation of stresses in the original heterogenous media and in the homogeneous one. For a discussion and problems in this direction see e.g. [26]. This question will be addressed in section 2.2 from a mathematical point of view.

Let us remark that the example discussed in this section is a direct generalization of the problem of section 1.2. It is enough to realize the well known fact that the pure shear stress problem leads to equation (1).

We mentioned some problems and approaches for this solution. So far we did not quote any mathematical results. There are only a few which we will quote later together with mathematical analysis.

1.4. An example of a less dimensional homogenization

Let S be a two dimensional (for simplicity) smooth manifold in R_3. To any x in a neighborhood of S define $\gamma(x) \in S$ as the point which has shortest distance to x. [Because of smoothness of S, $\gamma(x)$ is well defined.] In addition let $\psi(x) > 0$ be defined on S (and smooth). Denote

$$S(H) = \{x \mid \|x-\gamma(x)\| \le \psi(\gamma(x))H\} .$$

For H sufficiently small, $S(H)$ is well defined. As an example of a less dimensional homogenization let us be interested in the solution u^H of (1) when

$$a^H(x) = a_o \ , \ x \notin S(H) \ ,$$

and

$$a^H(x) = \frac{b_o}{H} \ , \ x \in S(H) \ , \ 0 < b_o < \infty$$

with a_o , b_o constants.

A similar question arises as before. How does the solution u^H behave for small H ? This and similar problems have many applications. We mention as an example the problem of hydrolic fracturing in oil problems where S and ψ are respectively the geometry and thickness of an intrusion [see e.g. [27], [28]]. There are also problems when

$$a^H(x) = b_o H \ , \text{ etc.}$$

1.5. Nonstationary problems of homogenization

So far we introduced only stationary problems (time independent). Of course it is obvious that all examples we have shown may be naturally generalized for nonstationary cases. Especially important is e.g. the case of dynamics and viscoelasticity for composite media and problems related to these questions; (see e.g. [29]). In chemistry the transient problem for diffusion equation is a very important one, etc.

2. Mathematics of the Homogenization

2.1. Introduction

Let us study now more precisely the problems introduced in the previous section and give an answer to some problems hinted there. For simplicity, only two dimensional case will be treated. This restriction is not essential. The results will be formulated for diffusion equation. Analogous results are valid for the example of composite material explained in section 1.3. This example will be used in the illustration of the numerical approach in section 3.

Let us formulate our goals. For given H [H is given in the problem and cannot be changed] we have to find the solution u^H and the fluxes

$$t_i^H = a^H \frac{\partial u^H}{\partial x_i} , \quad i = 1,2 ,$$

of equation (1). Realizing that we have thousands of cells, in practice it is impossible to find u^H directly from the microstructure. The only promising approach is the homogenization. It is important to get some information about the error too, because as we said <u>H is</u> (physically) <u>fixed</u>. Let us underline once more that there is no unique way to imbed the given problem into a one parametric system.

2.2. Homogenization problem for infinite domain

We will assume that $\Omega = R_2$. Section 2.3 removes this restriction and shows additional problems. The first obvious question is to study the behavior of u^H and t_i^H for $H \to 0$. This problem is related to the problem of the weak convergence of the coefficients a^H . The convergence of u^H was studied in an abstract way in [30], [31], [32], and in more concrete form in [33]. For computational reasons and applications more detailed analysis is necessary. In [34], [35], [36], (see also [37], [38], [39]), the following theorem (in more general form) has been proven.

<u>THEOREM 1</u>. <u>Let</u> f <u>be smooth with compact support and in addition let</u>

$$\int_{R_2} f\,dx = 0 .$$

Assume that $a(x)$ in (1) is piecewise smooth and u^H is normalized so that $\int_K u^H dx = 0$ where $K = \{x \mid |x_i| < \frac{1}{2}, i=1,2\}$. Then

1) there exists $U \in L^{(1)}(R_2)$,[2] $\int_K U\,dx = 0$ such that

$$\|u^H - U\|_{L_2(\tilde{\Omega})} \leq CH \tag{4}$$

where $\tilde{\Omega}$ is any bounded domain [C depends on $\tilde{\Omega}$, f, a(x) , but is independent of H],

2) function U satisfies (elliptic) differential equation

$$\sum_{i,j=1}^{2} A_{i,j} \frac{\partial^2 U}{\partial x_i \partial x_j} = f \tag{5}$$

where

$$A_{m,n} = \int_K \left(\sum_{i=1}^{2} a(x) \frac{\partial w_m}{\partial x_i} \frac{\partial w_n}{\partial x_i} \right) dx, \quad m, n = 1, 2 \tag{6}$$

with

$$w_k = x_k - \eta_k(x), \ \eta_k(x) \in H^1_{per}(K),^{3)} \ \int_K \eta_k(x) dx = 0, \ k = 1, 2$$

and $\eta_k(x)$ is such that

---

2) $$\|u\|^2_{L^{(1)}(R_2)} = \int_{R_2} \left(\sum_{i=1}^{2} \left(\frac{\partial U}{\partial x_i} \right)^2 \right) dx$$

3) $H^1_{per}(K)$ is subspace of $H^1(K)$ of periodic functions.

(7) $$\int_K \left[\sum_{i=1}^{2} a(x) \frac{\partial w_k}{\partial w_i} \frac{\partial \xi}{\partial x_i} \right] dx = 0 \quad \underline{\text{for every}} \quad \xi \in H^1_{per}(K) ,$$

3) <u>denoting</u>

(8) $$T_i = \sum_{j=1}^{2} A_{i,j} \frac{\partial U}{\partial x_j} , \; i = 1, 2$$

<u>there exist periodic functions</u> $\chi_j^{[i]}(x)$ <u>such that</u>

(9) $$\| t_i^H - \sum_{j=1}^{2} T_j \chi_j^{[i]}(\tfrac{x}{H}) \|_{L_2(\tilde{\Omega})} \le CH , \; i = 1, 2 .$$

Functions $\chi_i^{[i]}$ are composed of partial derivatives of w_i and of the function $a(x)$,

4)

(10) $$\int_{R_2} \left[\sum_{i=1}^{2} a^H(x) \left(\frac{\partial u^H}{\partial x_i} \right)^2 - \sum_{i,j=1}^{2} A_{i,j} \frac{\partial U}{\partial x_i} \frac{\partial U}{\partial x_j} \right] dx \le CH$$

Theorem 1 gives an answer to some problems raised in section 1.2. (resp 1.3).

1) The bulk properties (coefficients $A_{i,j}$) may be uniquely determined by the condition $u^H \to U$ for $H \to 0$ and any f smooth.

2) The bulk properties may also be uniquely determined by the condition that accumulated energy for the original and homogenized medium will be equal when $H \to 0$ for any f . [Accumulated energy condition is used e.g. in [25].]

3) Because coefficients $A_{i,j}$ are scalar products (in energy), usual two sided estimates (using direct and complementary principles) may be used for diagonal terms $A_{j,j}$ (or

eigenvalues of the matrix $\{A_{i,j}\}$). Formula (3) may be easily derived in this way.

4) Although U is close to u^H fluxes T_i and t_i^H are far apart. Nevertheless functions $\chi_j^{[i]}$ serve as magnifying factors which brings the expression

$$\sum_{j=1}^{2} T_j \chi_j^{[i]}$$

close to t_i^H .

5) Theorem 1 also gives as a corollary a simple interpretation of the homogenized fluxes T_i . Define

$$Z_i^H(x) = \frac{1}{H^2} \int_{(\xi-x)/H \in K} a(\xi) \frac{\partial u^H}{\partial x_i}(\xi) d\xi .$$

$Z_i^H(x)$ is an average flux over an area of the cell and leads to an estimate

$$\|T_i - Z_i^H\|_{L_2(\tilde{\Omega})} \leq CH .$$

So T_i are the average fluxes (when $H \to 0$) .

6) The functions w_k and $\chi_j^{[i]}$ may be obtained by analysis of a single cell only (which is numerically quite feasible). We shall now give a simple example. Assume

$$a(x) = p_1 > 0 \quad \text{for} \quad -\tfrac{1}{2} < x_1 < -\tfrac{1}{4}$$

$$\tfrac{1}{4} < x_1 \leq \tfrac{1}{2}$$

$$a(x) = p_2 > 0 \quad \text{for} \quad |x_1| \leq \tfrac{1}{4}$$

(i.e., a(x) be independent of x_2). In this case it is easy to determine all functions needed in theorem 1. We obtain

$\eta_2 = 0$ and $\eta_1(x)$ is independent of x_2, with

$$\frac{\partial \eta_1}{\partial x_1} = 1 - \frac{C}{a}, \quad C^{-1} = \tfrac{1}{2}\left(\frac{1}{p_1} + \frac{1}{p_2}\right).$$

Hence $A_{1,2} = A_{2,1} = 0$ and

$$A_{11} = \int_{-\frac{1}{2}}^{+\frac{1}{2}} \left(\frac{C^2}{a}\right) dx = C = \frac{2p_1p_2}{p_1+p_2}$$

$$A_{22} = \tfrac{1}{2}(p_1+p_2).$$

In addition

$$\chi_1^{[1]} = 1, \quad \chi_2^{[1]} = 0$$

$$\chi_2^{[2]} = \frac{a(x)}{A_{22}}, \quad \chi_1^{[2]} = 0.$$

The case just introduced is the problem of a layered medium. [For some additional information about the problems of heterogeneity in diffusion, see e.g. [40].]

The bulk properties are relevant only for small H, which means small in comparison with scale of changes of f. It seems that the only way to check whether H is small enough is to get a solution with error of order H^2 and compare the results. To find such an approximation is more laborious (see [34], [35]). We have to solve additional periodic problems on unit cell and find in addition to U another function $\tilde{U}$ which satisfies eq. (11)

$$(11) \qquad \sum_{i,j=1}^{2} A_{i,j} \frac{\partial^2 \tilde{U}}{\partial x_1 \partial x_j} = HLU$$

where L is a differential operator with constant coefficients involving (only) third derivatives. The coefficient may be determined with the help of the periodic solution mentioned above. Such an analysis leads to the approximate solution for u^H and t_i^H with error of order H^2. For the precise formulation see [34], [35], [39]. Many questions still remain open here. E.g. eq. (11) may be interpreted as the equation for the term of order H when the equation

$$- HLV + \sum_{i,j=1}^{2} A_{i,j} \frac{\partial^2 V}{\partial x_1 \partial x_2} = f \tag{12}$$

is solved by asymptotic expansions and

$$V = U + \tilde{U} . \tag{13}$$

There is an open question whether it is not better to solve (12) instead of (11) and (5). Some preliminary simple numerical experiments suggest this possibility. Formal analysis of some expansions have been used by different authors. Let us mention here as typical only [41], [42].

Quite analogous results to those mentioned above are valid also for the example of composite materials discussed in section 1.3.

So far we were interested in eq. (1). Similar results hold also for eq.

$$\sum_{i,j=1}^{2} \frac{\partial}{\partial x_i} a^H(x) \frac{\partial u^H}{\partial x_j} - g^H(x)u^H = f \tag{14}$$

where $g^H(x) = g(\frac{x}{H})$. For more, see [33]. The homogenized equation (analogous to (5)) is

(15) $$\sum_{i,j=1}^{2} A_{i,j} \frac{\partial^2 U}{\partial x_i \partial x_j} - QU = f$$

where $A_{i,j}$ are the same coefficients as before and $Q = \int_K g(x)dx$.

2.3. The problem for bounded domain

Let us start with the case $\Omega = R^+ = \{x|x_2>0\}$ and assume the boundary condition $u^H(x_1,0) = 0$. Then the following theorem holds.

THEOREM 2

Let $A_{i,j}$ and $\chi_i^{(i)}$ be the same as in Theorem 1. In addition let U be the solution of (5) with $U(x_1,0) = 0$. Then

1)

(16) $$\|u^H - U\|_{L_2(\tilde{\Omega})} \leq CH$$

(17) $$\|t_i^H(x) - \sum_{j=1}^{2} T_j(x)\chi_j^{[i]}(\tfrac{x}{H})\|_{L_2(\Omega(Z))} \leq$$

$$\leq C\left[H^{\frac{1}{2}}e^{-\frac{Z\bar{C}}{H}} + H\right]$$

Here $\Omega(Z) \subset R^+$ is any bounded domain with distance Z to ∂R^+, $Z \geq 0$.

2) There exists function $\kappa_i^{[k]}(x)$ defined on $P = \{x|\ |x_1| < \frac{1}{2}\ ,\ x_2 \geq 0\}$ periodic in x_1 with period 1 and exponentially decreasing with x_2 so that

(18)
$$\|t_i^H(x) - \sum_{j=i}^{2} T_j(x)\chi_j^{[i]}(\tfrac{x}{H}) - \sum_{j=i}^{2} T_j(x_1,0)\kappa_j^{[i]}(\tfrac{x}{H})\|_{L_2(\Omega)} \le CH$$

For the proof see [34].

Expression (17) and (18) shows the existence of a boundary layer. Functions $\kappa_j^{[i]}$ are practically computable because of the exponential decay. Using this fact we may restrict the analysis to a string of a few cells only.

Similar boundary layer behavior holds for the approximation of order H^2 and other types of boundary conditions. For smooth domain (16) holds but instead of (17) only estimate with $CH^{\frac{1}{2}}$ has been proven. The boundary layer behavior is in general an open problem (e.g. also for half plane with irrational angle to the array of cells).[4)]

2.4. Homogenization of the problem with grained boundary

Let $\psi_r(x)$ be periodic function with period 1 such that

$$\psi_r(x) = 0 \quad \text{for} \quad x \in S_r = \{x|\ \|x\| \le r\ ,\ r < \tfrac{1}{2}\}$$

$$\psi_r(x) = 1 \quad \text{for} \quad x \in K - S_r\ .$$

Denote

$$\Omega_r^H \subset R_2\ ,\ \Omega_r^H = \{x|\ \psi_r(\tfrac{x}{H}) = 1\}$$

(a domain with holes). An example of the problem with grained boundary is the problem of finding u^H defined on Ω_r^H so that

$$\Delta u^H = f$$

4) Remark that the boundary layer behavior was experimentally observed and is the reason for a rather unusual failure of fibrous composite laminates. See e.g. [43].

and $u^H = 0$ on the boundary $\partial\Omega_r^H$ for a given H_o and r_o. In addition assume that f is smooth and has compact support. This example will show that there are many ways to homogenize.

Assume that r is a function of H such that $r(H_o) = r_o$ and choose two special forms.

Case I, $r_1(H) = r_o$

Case II, $r_2(H) = e^{-H_o^2 H^{-2} |\ell n\, r_o|}$

THEOREM 3

If $r(H) = r_1(H)$, *then*

(19) $$\|u^H\|_{L_2(\Omega)} \le CH^2 .$$

In addition there exists a periodic function (with period 1) $\chi(x)$ *so that*

(20) $$\|u^H - U_1\|_{L_2(\Omega)} \le CH^3$$

where

(21) $$U_1 = H^2 f(x)\chi(\tfrac{x}{H}) .$$

If $r(H) = r_2(H)$ *then*

(22) $$\|u^H - U_2\|_{L_2(\Omega)} \to 0$$

where

(23) $$\Delta U_2 - |\ell n\, r_o|^{-1} H_o^{-2} U_2 = f .$$

For more precise statement of (19), (20), see [39]. pressions (21) and (22) are following from [44].

When r_0 and H_0 are given, then an obvious question arises. Is it better to use $u^H \approx U_1$ or $u^H \approx U_2$? It is intuitively clear that if

$$|\ell n\, r_0|^{-1} H_0^{-2}$$

is a small number, then U_2 is better to use. If

$$|\ell n\, r_0|^{-1} H_0^{-2}$$

is large, then it is better to use function U_1 for practical reasons. Unfortunately a rigorous mathematical analysis of such problems is not available. We remark that similar problems are arising in the analysis of the cases introduced earlier.

2.5. The case of less dimensional homogenization

Let us study the case introduced in section 1.4. As before assume that f is smooth, has compact support, and $\int_{R_2} f\, dx = 0$ and solve the problem

$$\frac{\partial}{\partial x_1} a^H \frac{\partial u^H}{\partial x_1} + \frac{\partial}{\partial x_2} a^H \frac{\partial u^H}{\partial x_2} = f$$

when $a^H(x) = 1$ for $x \notin S(H)$ and $a^H = \frac{1}{H}$ for $x \in S(H)$ where $S(H)$ is a neighborhood of a smooth curve S as explained in section (1.4). Denote by U the solution of the following problem

$$\Delta U = f \quad \text{on} \quad R_2 - S$$

where U is continuous on R_2, and on S the jump $J(U)$ in the normal derivatives satisfies

$$J(U) = \frac{d}{ds} \psi(S) \frac{\partial U}{\partial s} .$$

It is possible to show that U minimizes the quardratic

functional

$$\int_{R_2} \left[\left(\frac{\partial U}{\partial x_1}\right)^2 + \left(\frac{\partial U}{\partial x_2}\right)^2\right] dx + \oint_S \psi(S)\left(\frac{\partial U}{\partial s}\right)^2 ds - 2\int_{R_2} fU\, dx$$

This problem belongs to a class of "not normal" interface conditions and is closely related to the case of "not normal" boundary conditions, which was studied in [45]. This paper proves the existence and regularity of the solution in the dependence on input data.

The following theorem can be proved.

THEOREM 4

$$\|u^H - U\|_{L^{(1)}(\Omega)} \leq CH^{\frac{1}{2}} \quad {}^{5)}$$

where u^H resp U are functions introduced above. For the proof of a sharper version of Theorem 4 with higher order convergence see [46]. The other problem mentioned in section 1.4 is also analyzed in [46].

There are problems analogous to this which we discussed in section 2.5, namely, the problem with grained boundary located on a line. For this, see e.g. [47], [48].

2.6. Additional results

Homogenization results can be generalized in the same directions.

1) Assume that $a^H(x) = a(\frac{x}{H}, x)$ with $a(\xi, x)$ be periodic in ξ with periodicity 1 and smooth. Then an analogous theorem to theorem 1 holds. Instead of (5) we have to solve differential equation

---

5) For $L^{(1)}(\Omega)$ see footnote.

$$\sum_{i,j=1}^{2} \frac{\partial}{\partial x_i} A_{i,j}(x) \frac{\partial U}{\partial x_j} = f \tag{24}$$

where coefficients $A_{i,j}$ are computed as in (6) where integration is over the domain $K(\xi) = \{\xi | \xi - x \in K\}$ and w_k satisfies (7) with second variable (in the function a) being fixed.

2) A similar theorem holds also when the problem is nonlinear. E.g. let

$$a^H\left(x, \frac{\partial u^H}{\partial x_i}\right) = a\left(\frac{\partial u^H}{\partial x_1}, \frac{\partial u^H}{\partial x_2}\right)$$

Then U satisfies (24) with $A_{i,j}$ being a function of $\frac{\partial U}{\partial x_i}$, $i = 1,2$. The function

$$A_{i,j}\left(\frac{\partial U}{\partial x_1}, \frac{\partial U}{\partial x_2}\right)$$

is computed out of similar eq. to (6). For more about it see [36].

In one dimension the analysis of the (full dimensional) homogenization may be easily made by using theorems for continuous dependence on a parameter. For linear problems a very general analysis may be obtained by applying results of [49]. In the one dimensional case there are no major problems with nonlinear problems either. E.g. the results of [50] may be used very effectively for such an analysis.

So far we have analyzed only stationary problems. Using ideas explained above many results may be obtained for nonstationary solutions too, see e.g. [51], but the majority of the most important problems still remain open.

3. Applications

3.1. Formulation of an example

We will be interested in the analysis of the composite material consisting of two isotropic phases characterized by Young's module of elasticity E_F resp E_M and Poisson ratio ν_F resp ν_M. The structure of the cell and the particular data are shown in Fig. 1. The plane stress state is assumed. We are interested in the analysis of the infinite region and behavior around the boundary of the halfplane with relative structure shown in Fig. 3. We assume that only volume forces are given and the boundary is free of stresses.

3.2. Computational procedure and results

In the vein as explained in section 2 the procedure is as follows.

1) Computation of the bulk properties.

The computation goes analogously as in Theorem 1. First we have to determine functions analogous to ω_k. Here we need to solve three periodic elasticity problems on the cell. [Because of symmetry our case consists only of two such solutions.] This analysis was performed by use of program NONSAP - for the description, see e.g. [52]. NONSAP is a finite element program and (square) bilinear elements have been used. For accuracy purposes three computations have been made with mesh sizes $h = \frac{1}{4}, \frac{1}{8}, \frac{1}{16}$. Bulk modulae have been computed using these functions. The homogenized material is an orthotropic one. The stress $(\sigma_x, \sigma_y, \nu_{xy})$ strain $(\ell_x, \ell_y, \ell_{xy})$ relation is as follows.

a) Original materials.

$$\sigma_x = \frac{E}{1-\nu^2}(\ell_x+\nu\ell_y) = C_{11}\ell_x + C_{12}\ell_y$$

$$\sigma_y = \frac{E}{1-\nu^2}(\ell_y+\nu\ell_x) = C_{21}\ell_x + C_{22}\ell_y$$

$$\tau_{xy} = 2G\ \ell_{xy}$$

with

(25) $$G = \frac{E}{2(1+\nu)}$$

(26) $$C_{11} = C_{22}\ , \quad C_{12} = C_{21}$$

b) Homogenized material.

$$C\sigma_x = C_{11}\ell_x + C_{12}\ell_y$$

$$\sigma_y = C_{12}\ell_x + C_{22}\ell_y$$

$$\tau_{xy} = 2G\ \ell_{xy}\ ,$$

and

(27) $$\nu = \frac{C_{12}}{C_{11}}\ .$$

Because the homogenized material is an orthotropic one, relation (25) does not hold. In Table 2 the elasticity constant for the matrix, fibers, and homogenized material are shown. The homogenized constants are computed with three different step sizes $h = \frac{1}{4}, \frac{1}{8}, \frac{1}{16}$,

		$C_{11}=C_{22}$	C_{12}	2G	ν
Fibers		3200000	800000	2400000	.25
Matrix		32000	8000	24000	.25
Bulk coefficient	h = 1/4	36637	8585	26613	.2343
	h = 1/8	36037	8675	26380	.2407
	h = 1/16	35822	8722	26271	.2434

Table 2.

2) Computation of the multiplication functions - (analogous to functions $\chi_i^{(k)}$).

The use of these functions is such that denoting $\Sigma_x, \Sigma_y, T_{xy}$ the (bulk) stresses of the homogenized material we have

$$\sigma_x^H = \Sigma_x \chi_{\Sigma_x}^{\sigma_x} + \Sigma_y \chi_{\Sigma_y}^{\sigma_x} + T_{xy} \chi_{T_{xy}}^{\sigma_x}$$

$$\sigma_y^H = \Sigma_x \chi_{\Sigma_x}^{\sigma_y} + \Sigma_y \chi_{\Sigma_y}^{\sigma_y} + T_{xy} \chi_{T_{xy}}^{\sigma_y}$$

$$\tau_{xy}^H = \Sigma_x \chi_{\Sigma_x}^{\tau_{xy}} + \Sigma_y \chi_{\Sigma_y}^{\tau_{xy}} + T_{xy} \chi_{T_{xy}}^{\tau_{xy}} .$$

Fig. 2 shows the graphs of $\chi_{\Sigma_x}^{\sigma_y}$ and $\chi_{\Sigma_x}^{\sigma_x}$ in sections A, B, C, in Fig. 1 respectively. Because of symmetry the graphs are drawn for $0 \leq x < \frac{1}{2}$ only. To compute σ^H we have to determine the bulk stresses $\Sigma_x \, \Sigma_y \, T_{xy}$ in the given point of interest and use (28) where multipliers χ_{Σ}^{σ} are evaluated in the relative point to the cell.

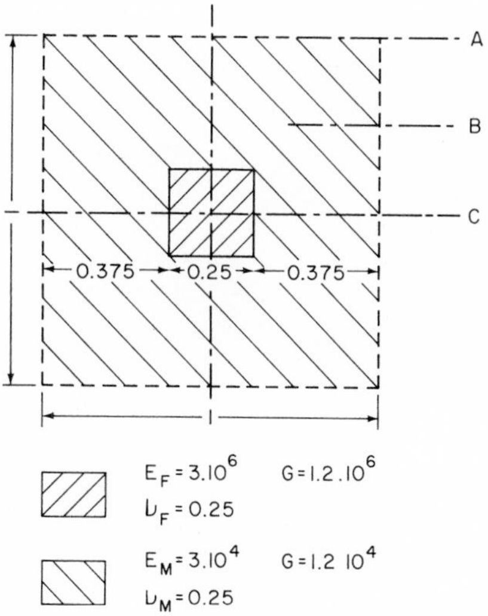

Figure 1

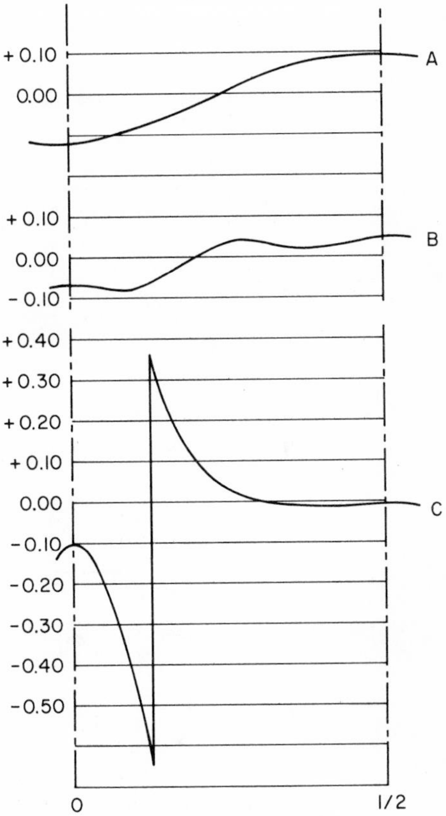

Figure 2a

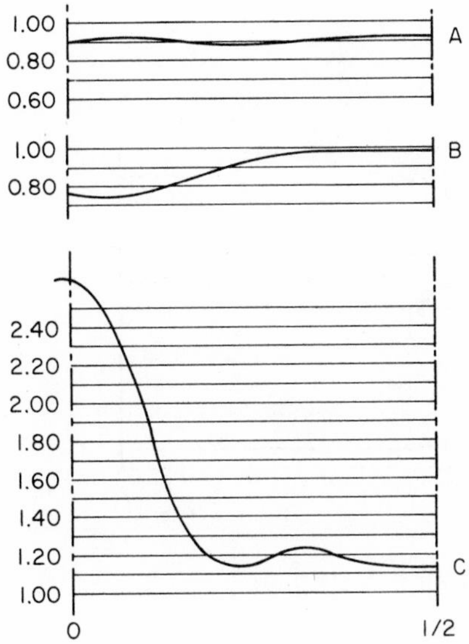

Figure 2b

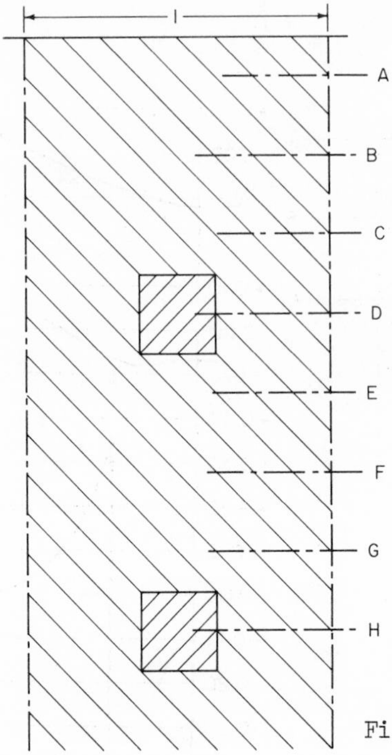

Figure 3

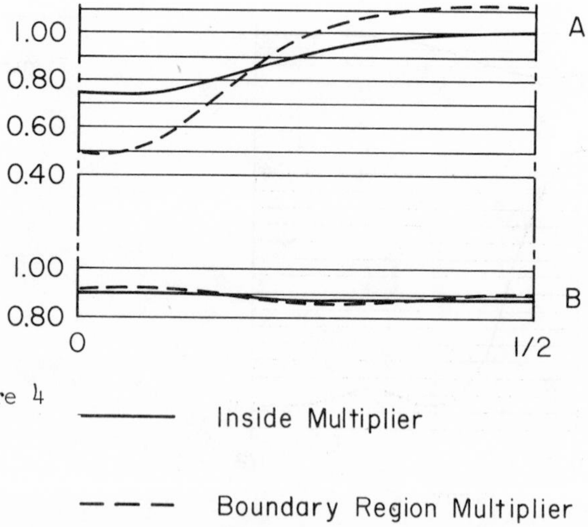

Figure 4

3) Computation of the multiplication functions for the halfplane.

As we said, we are interested in the case when the boundary is free of stresses. In this case when computing the bulk stresses we get on the boundary $\Sigma_y = T_{xy} = 0$ and $\Sigma_x \neq 0$. Now we have to modify functions

$$\chi_{\Sigma_x}^{\sigma_x}, \ \chi_{\Sigma_x}^{\sigma_y}, \ \chi_{\Sigma_x}^{\tau_{xy}} \ \text{to} \ \tilde{\chi}_{\Sigma_x}^{\sigma_x}, \ \text{etc.}$$

in the neighborhood of the boundary. The relative position of the boundary is of course now important. As an example we show in Fig. 4 the function $\chi_{\Sigma_x}^{\sigma_x}$ in cross sections A-B shown in Fig. 3. The boundary layer effect disappears rapidly so that in the cross sections C , ... , H the difference between $\chi_{\Sigma_x}^{\sigma_x}$ and $\tilde{\chi}_{\Sigma_x}^{\sigma_x}$ is negligible. Fig. 4 shows function $\chi_{\Sigma_x}^{\sigma_x}$ (full line) and $\tilde{\chi}_{\Sigma_x}^{\sigma_x}$ (dashed line).

REFERENCES

1. Poisson, S. D., "Second mêm. sur la théorie de magnetisme," Mem. de L'Acad. de France (1822), 5.
2. Mosotti, O. F., "Discussione analitica sul influenze che L'azione di in mezzo dielettrico ha sulla distribuzione dell' electtricita alla superficie di pin corpi electtici disseminati in esso," Mem. Di Math et Di Fisica in Modena 24, II (1850), 49.
3. Maxwell, C., Treatise on Electricity and Magnetismus, vol. 1, Oxford Univ. Press, 1873, 365.
4. Clausius, R., Die Mech. Wärme Theorie, Bd II, Vieweg, (1879), 62.
5. Rayleigh, J. W., "On the influence of obstacles in rectangular order upon the properties of a medium," Phil. Mag. (5), 34, (1892), 481.
6. Lichtenecker, K., "Die dielektrizitätskonstante natürlicher und künstlicher Mischkörper," Phys. Zeitschr. XXVII (1926), 115-158.

7. de Vries, "The thermal conductivity of granular materials," Bull. Inst. Inter. du Froid, Paris (1952).
8. Bruggeman, D. A. G., "Berechnung veischiedenen physikalischer Konstanten von heterogenen Subtanzen," Ann. der Phys. 5 Folge, Bd. 24 (1935), 636-679.
9. Bottcher, C. J. F., "The dielectric constant of crystalline powders," Recueil de Travaux Chimiques de Pays-Bas 64 (1945), 47-52.
10. Barrer, R. M., "Diffusion and permeation in heterogenous media," Diffusion in Polymers (J. Crank, G. S. Park, eds.), Academic Press, 1968.
11. Thornburgh, J. D., and Pears, C. D., "Prediction of the thermal conductivity of filled and reinforced plastics," No. 65/WA/HT-4, ASME Winter Annual Meeting, 1965.
12. Beran, M. J., Statistical Continuum Theories, Interscience Publishers, 1968.
13. Malvern, L. E., Introduction to the Mechanics of a Continuous Medium, Prentice-Hall, 1969.
14. Garg, S. K., Svalbonas, V., and Gurtman, G. A., Analysis of Structural Composite Materials, Marcel Dekker, Inc., New York, 1973.
15. Composite Materials, (L. J. Broutman, R. H. Krock, eds.), Vol. II Mechanics of Composite Materials, (G. P. Sendeckyj, ed.), Academic Press, 1974.
16. Hashin, Z., "Theory of fibers reinforced materials," (March 1972), NASA Report NASA CR-1974, 1-704.
17. Chams, C. C., and Sendeckyj, G. P., "Critique on theories predicting thermoelastic properties of fibrous composites," J. Comp. Mat. 2 (1968), 332-358.
18. Cox, H. L., "The elasticity and strength of paper and other fibrous materials," British J. Appl. Phys. 3 (1952), 72-79.
19. Gordon, J. E., "On the present and potential efficiency of structural plastics," J. Royal Aero. Soc. 56 (1952), 704-728.
20. Ekvall, J. C., "Elastic properties of orthotropic monofilament laminates," ASME paper No. 61-AV-56, presented at the ASME Aviation Conference, March 1961.
21. Whitney, J. M., and Riley, J. B., "Elastic stress - strain properties of fiber-reinforced composite materials," AIAAJ, 4 (1966), 1537-1542.
22. Paul, B., "Prediction of elastic constants of multiphase materials," Trans. M. Soc. AIME 218 (1960), 36-41.
23. Achenbach, J. E., and Sun, C. T., "The directional reinforced composite as a homogenous continuum with microstructure," Dynamics of Composite Materials, (E. H. Lee, ed.), The Am. Soc. of Mech. Eng., 1972, 48-70.

24. Hashin, Z., and Shtrikman, S., "A variational approach to the theory of the elastic behavior of multiphase materials," J. Mech. Phys. Solids 11 (1963), 127-140.
25. Rubenfeld, L. A., and Keller, J. B., "Bounds on elastic moduli of composite media," SIAM J. Appl. Math. 17 (1969), 495-510.
26. Pagano, N. J., and Rybicki, E. F., "On the significance of effective modules solutions for fibrous composites," J. Comp. Mat. 8 (1974), 214-229.
27. Tinsley, J. M., Williams, J. R., Tiner, R. L., and Malone, W. T., "Vertical fracture heigths. Its effect on steady state production increase," Soc. Petroleum Engr. Paper SPE 1900, (1967).
28. Cannon, J. R., and Mayer, G. H., "On diffusion in a fractured medium," SIAM J. Appl. Math. 20, No. 3 (1971).
29. Dynamics of Composite Material, (E. H. Lee, ed.), The Am. Soc. of Mech. Eng. 1972.
30. Spagnolo, S., "Sulla convergenza di soluzioni di equazioni paraboliche ed ellitiche," Ann. Scuola Norm. Sup Pisa 22 (1968), 571-597.
31. Spagnolo, S., "Convergence in energy for elliptic operators," this proceedings.
32. De Georgi, E., and Spagnolo, S., "Sulla convergenza degli integrali dell energia per operatozi ellittici del secondo ordine," Bolletiono UMI 48 (1973), 391-411.
33. Sanchez-Palencia, E., "Compartements local et macroscopique d'un type de milieux physiques heterogenes," Int. J. Eng. Sci. 12 (1974), 331-351.
34. Babuška, I., "Solution of interface problems by homogenization I," University of Maryland Tech. Note BN-782, January 1974.
35. Babuška, I., "Solution of interface problems by homogenization II," University of Maryland Tech. Note BN-787, March 1974.
36. Babuška, I., "Solution of interface problems by homogenization III," (to appear).
37. Babuška, I., and Kellogg, R. B., "Mathematical and computational problems in reactor calculations," Proceedings of Conference on Mathematical Models and Computational Techniques for Analysis of Nuclear Systems, Ann Arbor (1973), VII-67 - VII-94.
38. Babuška, I., "Numerical solution of partial differential equations," GAMM-Tagung, München (1973) ZAMM, Bd. 54 (1974) T1-T10.
39. Babuška, I., "Solution of problems with interfaces and singularities," Mathematical Aspects of Finite Elements in Partial Differential Equations (C. De Boor, ed.), Academic Press, 1974, 213-277.

40. Crank, J., "Diffusion in heterogenous media," (chapter 12), The Mathematics of Diffusion, Clarendon Press, Oxford, 1975, 266-286.
41. Burgers, M. J., "Separation between macro- and micro-features in fields determined by certain elliptic differential equations," manuscript, Institute for Fluid Dynamics, University of Maryland, June 1974.
42. Gegemier, G. A., "On a theory of interacting continua for wave propagation in composites," Dynamics of Composite Materials (E. H. Lee, ed.), The Am. Soc. of Mech. Eng., 1972, 70-121.
43. Pagano, N. T., "On the calculation of interlaminar normal stress in composite laminates," J. Comp. Mat. 8 (1974), 65-81.
44. Hruslov, E. Ja., "The method of orthogonal projections and the Dirichlet problem in domains with a fine-grained boundary," Math. USSR Sbornik, vol. 17, No. 1, (1972), 37-59.
45. Roitberg, Ja A., "Homeomorphism theorems and a Green formula for general elliptic boundary problems with non-normal boundary conditions," Math. Sb. USSR Sbornik 12, No. 2, (1970), 177-213.
46. Babuška, I., and Mayer, G., "Mathematical problems of intrusions," (to appear).
47. Marčenko, V. A., and Suzikov, G. V., "The second boundary value problem in domain with complicated boundary," Mat. Sb. 65 (1964), 458-472, and Mat. Sb. 69 (1966), 35-60.
48. Hruslov, E. Ja., "On the Neumann boundary problem in a domain with complicated boundary," Math. USSR Sbornik 12, No. 4, (1970), 553-571.
49. Opial, Zd., "Continuous dependence in linear system of differential equations," J. Diff. Eq. 3 (1967), 571-579.
50. Kurzweil, J., "Generalized ordinary differential equations and continuous dependence on a parameter," Czechoslovak Mat. J. 7 (1957), 418-479.
51. Senatorov, P. K., "The stability of the solution of an ordinary differential equation of second order and of a parabolic equation with respect to the coefficients of the equation," Diff. Urav. 7 (1971), 754-758.
52. Bathe, K. J., "An assessment of current finite element analysis of nonlinear problems in solid mechanics," these proceedings.

AN ASSESSMENT OF CURRENT FINITE ELEMENT ANALYSIS OF NONLINEAR PROBLEMS IN SOLID MECHANICS

Klaus-Jürgen Bathe*

1. Introduction

The finite element analysis of nonlinear problems in solid mechanics has acquired increasing attention. Presently, various large scale computer programs and smaller special purpose codes are used, that offer various analysis capabilities. However, although a large number of problems can already be solved effectively, all computer programs have serious limitations. The objective in this paper is to survey some important aspects of finite element solid mechanics nonlinear analysis, to point out the potential of the finite element method, and to identify some important present research areas with emphasis on numerical analysis procedures. In the presentation we consider what are believed to be the most effective finite element nonlinear analysis techniques presently in use.

The development of finite element nonlinear analysis procedures is based on knowledge in essentially three different disciplines; namely continuum mechanics, numerical analysis procedures and computer implementation. Figure 1 summarizes important areas in these three disciplines. Although the disciplines can be identified separately, an important point is that for the development of effective finite element nonlinear analysis techniques, it is necessary to take into account the

* Department of Mechanical Engineering, Massachusetts Institute of Technology.

interaction that exists between them. For example, the continuum mechanics formulation used in dynamic analysis must be amenable to the numerical time integration scheme that shall be employed, which in turn must be implemented in a cost effective manner on the computer. The objective in this paper is to discuss mainly the formulation of the finite element equations and the numerical procedures used for solution. However, when important, reference is briefly made to the computer implementation, and throughout the paper the important interaction between the areas summarized in Fig. 1 is emphasized.

The material presented in this paper is largely based on the experience gained during the development and use of the general purpose nonlinear static and dynamic finite element analysis computer program ADINA [1], which is a further development of program NONSAP [2] [3]. In order to enhance finite element nonlinear analysis procedures it is necessary, both, to develop new techniques, establish theoretical basis to the techniques used, and to test the procedures on actual difficult practical problems. In this way, the assumptions and limitations of the procedures are demonstrated and researchers are stimulated to improve and further develop the methods employed. It is for this reason that we shall discuss first the theory used and then present practical applications in order to emphasize the present capabilities and important research areas of nonlinear finite element analysis.

A. CONTINUUM MECHANICS	B. NUMERICAL METHODS	C. COMPUTER TECHNIQUES
1. FORMULATION OF NONLINEAR EQUATIONS OF MOTION	1. NUMERICAL INTEGRATION IN SPACE	1. PROGRAMMING METHODS
2. DEVELOPMENT OF FINITE ELEMENT EQUATIONS	2. TIME INTEGRATION	2. USAGE OF AVAILABLE HARDWARE AND SOFTWARE
3. DEVELOPMENT OF MATERIAL MODELS	3. SOLUTION OF EQUATIONS	3. EFFICIENT PROGRAM ORGANIZATION
	4. CALCULATION OF EIGENSYSTEM	4. FLEXIBILITY FOR MODIFICATIONS

FIGURE 1 IMPORTANT DISCIPLINES IN THE DEVELOPMENT OF FINITE ELEMENT ANALYSIS PROCEDURES

2. Finite Element Formulations

The finite element formulation of nonlinear problems comprises the kinematic formulation of the problems using continuum mechanics principles, the identification of the constitutive relations, and finally the discretization of the equations using finite element procedures. We deal in the following with these individual aspects and discuss the important relations between them.

2.1 Continuum Mechanics Formulation

For the continuum mechanics formulation of nonlinear problems, there exist, in general, two approaches, namely using a Lagrangian (material) formulation or a Eulerian (spatial) formulation [4] [5] [6]. The important difference between the two approaches is that in the Lagrangian formulations attention is focussed on the material itself as it moves, whereas in the Eulerian formulations attention is focussed on what happens at a specific spatial location. Eulerian formulations are used almost exclusively in fluid flow analysis, whereas Lagrangian descriptions are widely employed in finite element analysis of solid mechanics problems. In this context it may be noted that in some cases formulations have been termed Eulerian although material coordinates are used. Considering solid mechanics problems Lagrangian descriptions are generally employed, because the boundaries of the body under consideration change during the history of solution, and it is natural to use a description which follows the material of the body. In the following we summarize the important concepts of the Lagrangian formulation used in ADINA. The formulation has been presented in detail in references [7] and [8].

Consider a general body subjected to large displacements, large strains and linear or nonlinear constitutive relations. A schematic sketch of the body is shown in Fig. 2.

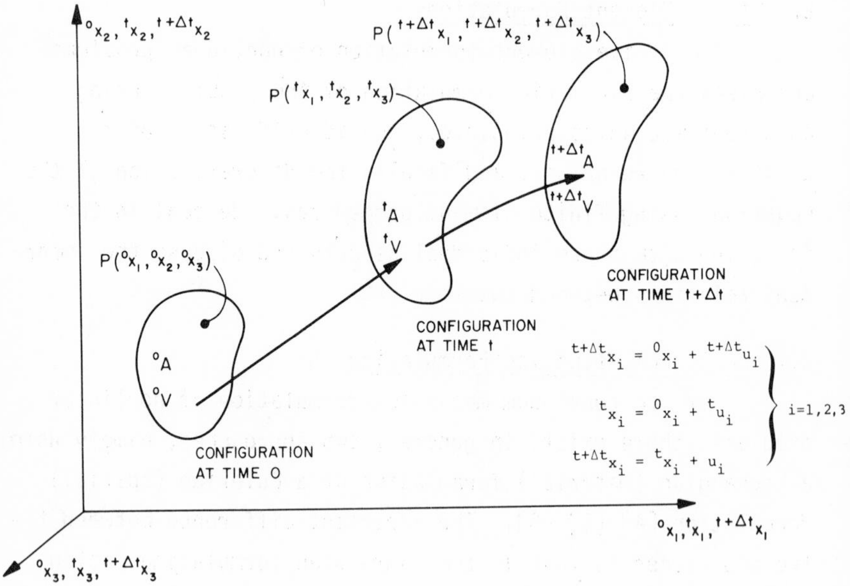

FIGURE 2 MOTION OF BODY IN CARTESIAN COORDINATE SYSTEM

To describe the motion of the body, we use rectangular Cartesian coordinates. The coordinates of a material particle are ${}^{0}x_i$ at time 0, ${}^{t}x_i$ at time t, and ${}^{t+\Delta t}x_i$ at time $t+\Delta t$, $i = 1,2,3$. Thus, we have

$$ {}^{t+\Delta t}x_i = {}^{0}x_i + {}^{t+\Delta t}u_i \tag{1} $$

$$ {}^{t}x_i = {}^{0}x_i + {}^{t}u_i \tag{2} $$

where ${}^{t}u_i$, ${}^{t+\Delta t}u_i$ are the Cartesian displacements of the particle at times t and $t+\Delta t$, respectively. We define the incremental displacements from times t to $t+\Delta t$ to be u_i, and hence

$$ u_i = {}^{t+\Delta t}u_i - {}^{t}u_i \quad . \tag{3} $$

The governing equilibrium equations can be derived using the principle of virtual displacements. This principle is used to formulate a displacement-based finite element solution, which at present is regarded to be most effective. Consider that the solution of the body under static or dynamic loading is required for times 0, Δt, $2\Delta t$, ..., t, $t+\Delta t$, ... and assume that the solution has been obtained up to time t. To calculate the solution for time $t+\Delta t$, in essence two different approaches can be followed that lead to explicit and implicit time integration. For explicit time integration the equilibrium equations are established at time t using a stress rate that is invariant with respect to body rotation [9] [10]. An important disadvantage of such formulations for general applications is that small time steps (load-steps in static analysis) must be used even though the constitutive relations may allow a much larger time step. The reason that small time steps must be employed in static or dynamic analysis lies in that simple forward integration is employed. In dynamic analysis it is further necessary to use a small enough time step for integration stability.

Although not for all types of problems most effective, it is believed that a formulation in which the equilibrium of the body is considered at time $t+\Delta t$ is most efficient for general static and dynamic analysis. In dynamic analysis such formulation leads to implicit time integration, which using an appropriate integration operator is unconditionally stable in linear analysis. In nonlinear analysis, a control on the accuracy of solution is established by satisfying the equilibrium equations at time $t+\Delta t$ within a specified tolerance.

Considering the equilibrium of the body at time $t+\Delta t$, the principle of virtual displacements states the

$$\int_{{}^{t+\Delta t}V} {}^{t+\Delta t}\tau_{ij}\ \delta_{t+\Delta t}e_{ij}\ {}^{t+\Delta t}dv = {}^{t+\Delta t}\mathcal{R} \tag{4}$$

where

$$ {}^{t+\Delta t}\mathcal{R} = \int_{{}^{0}A} {}^{t+\Delta t}_{0}t_k\ \delta u_k\ {}^{0}da + \int_{{}^{0}V} \rho\ {}^{t+\Delta t}_{0}f_k\ \delta u_k\ {}^{0}dv \tag{5}$$

and ${}^{t+\Delta t}\tau_{ij}$ are the Cartesian components of the Cauchy stress tensor at time $t+\Delta t$, ${}^{t+\Delta t}_{0}t_k$ and ${}^{t+\Delta t}_{0}f_k$ are the Cartesian components of the surface traction and body force vectors at time $t+\Delta t$ referred to time 0, ${}^{0}\rho$ is the specific mass of the material in the original configuration,

$$\delta_{t+\Delta t}e_{ij} = \delta\frac{1}{2}\left({}_{t+\Delta t}u_{i,j} + {}_{t+\Delta t}u_{j,i}\right) \tag{6}$$

with ${}^{t+\Delta t}u_{l,m} = \partial u_l / \partial\, {}^{t+\Delta t}u_m$ and δu_k is a variation in the current displacement components ${}^{t+\Delta t}u_k$. It should be noted that in Eq. (4) the internal virtual work is evaluated by integration over the volume at time $t+\Delta t$.

The virtual work principle given above is general and is used in linear, material and geometric nonlinear analysis. In small displacement and small strain analysis, the configuration of the body is assumed not to change and Eq. (4) can be used directly for the finite element solution. In this case Eq. (4) becomes

$$\int_{{}^0V} {}^{t+\Delta t}\sigma_{ij}\ \delta e_{ij}\ {}^0dv = {}^{t+\Delta t}\mathcal{R} \tag{7}$$

where ${}^{t+\Delta t}\sigma_{ij}$ are the Cartesian components of the small displacement stress tensor at time $t+\Delta t$, and e_{ij} are the Cartesian components of the infinitesimal strain increment tensor,

$$e_{ij} = \frac{1}{2}\,(u_{i,j}+u_{j,i}) \ . \tag{8}$$

The strain components are evaluated from the finite element displacement interpolations discussed in Section 2.3, and the stress components are obtained using

$$ {}^{t+\Delta t}\sigma_{ij} = {}^{t}\sigma_{ij} + C_{ijrs}\ e_{rs} \tag{9}$$

where C_{ijrs} is the tangent constitutive tensor at time t . Substituting for ${}^{t+\Delta t}\sigma_{ij}$ from Eq. (9) into Eq. (7), we obtain the equation that is discretized in finite element analysis,

$$\int_{{}^0V} C_{ijrs}\, e_{rs}\, \delta e_{ij}\, {}^0dv = {}^{t+\Delta t}\mathcal{R} - \int_{{}^0V} {}^t\sigma_{ij}\, \delta e_{ij}\, {}^0dv \,. \tag{10}$$

Considering now large displacement and large strain analysis, the configuration of the body changes continuously with the individual material particles being subjected to large rotations and stretches. Since the configuration at time $t+\Delta t$ is unknown, for solution the stress and strain variables are referred to a previously calculated known geometric configuration. A very general and effective formulation is the total Lagrangian formulation, in which the initial configuration of the body is used as reference. Using the relations

$$ {}^{t+\Delta t}\tau_{sr} = \frac{{}^{t+\Delta t}\rho}{{}^0\rho}\; {}^{t+\Delta t}_{\ \ 0}x_{s,i}\; {}^{t+\Delta t}_{\ \ 0}x_{r,j}\; {}^{t+\Delta t}_{\ \ 0}S_{ij} \tag{11}$$

where ${}^{t+\Delta t}_{\ \ 0}x_{s,i} = \partial\, {}^{t+\Delta t}x_s / \partial\, {}^0x_i$, and

$$ \delta_{t+\Delta t}\, e_{sr} = {}_{t+\Delta t}^{\ \ 0}x_{i,s}\; {}_{t+\Delta t}^{\ \ 0}x_{j,r}\; \delta\, {}^{t+\Delta t}_{\ \ 0}\varepsilon_{ij} \tag{12}$$

and

$$ {}^0\rho\; {}^0dv = {}^{t+\Delta t}\rho\; {}^{t+\Delta t}dv \tag{13}$$

where ${}^{t+\Delta t}_{\ \ 0}S_{ij}$ are the Cartesian components of the second Piola-Kirchhoff stress tensor and ${}^{t+\Delta t}_{\ \ 0}\varepsilon_{ij}$ are the Cartesian components of the Green-Lagrange strain tensor, we can express Eq. (4) as

(14) $$\int_{{}^{0}V} {}^{t+\Delta t}_{0}S_{ij} \; \delta \, {}^{t+\Delta t}_{0}\varepsilon_{ij} \; {}^{0}dv = {}^{t+\Delta t}\mathcal{R} \; .$$

The unknown stresses ${}^{t+\Delta t}_{0}S_{ij}$ can now be incrementally decomposed into known stresses ${}^{t}_{0}S_{ij}$ plus increments in stresses ${}_{0}S_{ij}$. Then Eq. (4) can be rewritten and linearized as given in Table 1, to obtain

(15) $$\int_{{}^{0}V} {}_{0}C_{ijrs} \; {}_{0}e_{rs} \; \delta_{0}e_{ij} \; {}^{0}dv + \int_{{}^{0}V} {}^{t}_{0}S_{ij} \; \delta_{0}\eta_{ij} \; {}^{0}dv$$

$$= {}^{t+\Delta t}\mathcal{R} - \int_{{}^{0}V} {}^{t}_{0}S_{ij} \; \delta_{0}e_{ij} \; {}^{0}dv \; .$$

Instead of using the initial configuration at time 0 , alternatively any other previously calculated configuration can be used. Specifically, if the configuration at time t is employed as reference the updated Lagrangian formulation also used in ADINA is obtained [1] [7] [8].

It must be noted that Eq. (14) is applicable to any magnitude of displacements and strains. However, Eq. (15) is the linearization of Eq. (14) about the configuration at time t and its solution only approximates the exact solution. If large nonlinearities are present within the individual time intervals, this linearization introduces large uncontrolled errors. In general, therefore, it is necessary to seek a more accurate solution within each interval of time, and such solution can frequently be obtained effectively using a modified Newton iteration. If we define

(16) $${}^{t+\Delta t}u_i^{(k)} = {}^{t+\Delta t}u_i^{(k-1)} + \Delta u_i^{(k)}$$

where ${}^{t+\Delta t}u_i^{(0)} = {}^{t}u_i$, the equation considered in the materially nonlinear only formulation is

$$(17)\quad \int_{{}^{0}V} C_{ijrs}\, e_{rs}^{(k)}\, \delta e_{ij}^{(k)}\, {}^{0}dv = {}^{t+\Delta t}\mathcal{R} - \int_{{}^{0}V} {}^{t+\Delta t}\sigma_{ij}^{(k-1)}\, \delta e_{ij}^{(k)}\, {}^{0}dv \qquad \Big/ k = 1,2,\ldots$$

and in the total Lagrangian formulation we use

$$(18)\quad \int_{{}^{0}V} {}_{0}C_{ijrs}\, {}_{0}e_{rs}^{(k)}\, \delta_0 e_{ij}^{(k)}\, {}^{0}dv + \int_{{}^{0}V} {}^{t}_{0}S_{ij}\, \delta_0\eta_{ij}^{(k)}\, {}^{0}dv$$

$$= {}^{t+\Delta t}\mathcal{R} - \int_{{}^{0}V} {}^{t+\Delta t}_{0}S_{ij}^{(k-1)}\, \delta\, {}^{t+\Delta t}_{0}\varepsilon_{ij}^{(k-1)}\, {}^{0}dv \qquad \Big/ k = 1,2,\ldots\ .$$

The important problem in using these equilibrium equations is the problem of convergence of the iteration in general analysis. The mathematical properties of the modified Newton iteration and experience show that convergence difficulties can generally be expected when the structure stiffens, which may be the result of a number of different physical phenomena. The result is that a very small time step may have to be used, and iteration may need to be dispensed with for economical reasons.

TABLE 1 TOTAL LAGRANGIAN FORMULATION

1. Equations of Motion

$$\int_{{}^{0}V} {}^{t+\Delta t}_{0}S_{ij}\,\delta\,{}^{t+\Delta t}_{0}\varepsilon_{ij}\,{}^{0}dv = {}^{t+\Delta t}\mathcal{R}$$

where

$${}^{t+\Delta t}_{0}S_{ij} = \frac{{}^{0}\rho}{{}^{t+\Delta t}\rho}\,{}^{0}_{t+\Delta t}x_{i,s}\,{}^{t+\Delta t}\tau_{sr}\,{}^{0}_{t+\Delta t}x_{j,r}\ ;$$

$$\delta\,{}^{t+\Delta t}_{0}\varepsilon_{ij} = \tfrac{1}{2}\delta\left({}^{t+\Delta t}_{0}u_{i,j} + {}^{t+\Delta t}_{0}u_{j,i} + {}^{t+\Delta t}_{0}u_{k,i}\,{}^{t+\Delta t}_{0}u_{k,j}\right).$$

2. Incremental Decompositions

a. stresses

$${}^{t+\Delta t}_{0}S_{ij} = {}^{t}_{0}S_{ij} + {}_{0}S_{ij}$$

b. strains

$${}^{t+\Delta t}_{0}\varepsilon_{ij} = {}^{t}_{0}\varepsilon_{ij} + {}_{0}\varepsilon_{ij}\ ;\quad {}_{0}\varepsilon_{ij} = {}_{0}e_{ij} + {}_{0}\eta_{ij}\ ;$$

$${}_{0}e_{ij} = \tfrac{1}{2}\left({}_{0}u_{i,j} + {}_{0}u_{j,i} + {}^{t}_{0}u_{k,i}\,{}_{0}u_{k,j} + {}_{0}u_{k,i}\,{}^{t}_{0}u_{k,j}\right);$$

$${}_{0}\eta_{ij} = \tfrac{1}{2}\,{}_{0}u_{k,i}\,{}_{0}u_{k,j}\ .$$

3. Equations of Motion with Incremental Decompositions

Noting that $\delta\,{}^{t+\Delta t}_{0}\varepsilon_{ij} = \delta\,{}_{0}\varepsilon_{ij}$ and ${}_{0}S_{ij} = {}_{0}C_{ijrs}\,{}_{0}\varepsilon_{rs}$

the equations of motion are

$$\int_{{}^{0}V} {}_{0}C_{ijrs}\,{}_{0}\varepsilon_{rs}\,\delta\,{}_{0}\varepsilon_{ij}\,{}^{0}dv + \int_{{}^{0}V} {}^{t}_{0}S_{ij}\,\delta\,{}_{0}\eta_{ij}\,{}^{0}dv = {}^{t+\Delta t}\mathcal{R}$$

$$- \int_{{}^{0}V} {}^{t}_{0}S_{ij}\,\delta\,{}_{0}e_{ij}\,{}^{0}dv\ .$$

(cont.)

TABLE 1 (cont.)

4. Linearization of Equations of Motion

Using the approximations ${}_0S_{ij} = {}_0C_{ijrs}\,{}_0e_{rs}$, $\delta_0\varepsilon_{ij} = \delta_0 e_{ij}$ we obtain as approximate equations of motion

$$\int_{{}^0V} {}_0C_{ijrs}\,{}_0e_{rs}\,\delta_0 e_{ij}\,{}^0dv + \int_{{}^0V} {}_0^tS_{ij}\,\delta_0\eta_{ij}\,{}^0dv = {}^{t+\Delta t}R - \int_{{}^0V} {}_0^tS_{ij}\,\delta_0 e_{ij}\,{}^0dv \,.$$

2.2 Constitutive Relations

In the previous section we considered the continuum mechanics equilibrium equations for small or large displacements. An important decision in nonlinear analysis is to use the appropriate constitutive relations in these equilibrium equations. The material behavior may be classed as linear or nonlinear elastic or viscoelastic, hyper-elastic, elastic-plastic, viscoplastic or other [5]. In the kinematic formulations discussed the stress increment from times t to $t+\Delta t$ was defined in terms of the constitutive tensor and the corresponding strain increments. It should be noted here that including creep and temperature strains the stress increment also depends on the thermal strains and creep strains, the effect of which is conveniently included as a contribution to the present stresses.

In material nonlinear only analysis, the stresses and strains to be used are the small displacement engineering quantities. However, in large displacement and large strain analysis, the stress-strain law may be defined corresponding to different stress and strain measures. If the constitutive tensor relates 2nd Piola-Kirchhoff stresses to Green-Lagrange strains as is the case in hyperelasticity, the stress-strain law can directly be used in the formulation [11]. However, if the material law is defined in terms of Cauchy stresses, it need be transformed as discussed in detail in references [7] [10] [12]. These transformations can add significantly to the total solution cost and for this reason more effective stress-strain relationships are sought.

Considering the difficulties that are encountered in the analysis of nonlinear material behavior, as probably expected, solution difficulties generally increase as the material to be described becomes more complex. One important practical problem is excessive solution cost, because in history dependent analysis the constitutive relations can depend on many previously calculated variables and may be expensive to evaluate. Solution difficulties are also frequently encountered when discontinuities are present in the material behavior. This is, for example, the case in elastic-plastic analysis under cyclic loading conditions and in crack propagation analysis. Convergence difficulties are encountered when the material suddenly stiffens. In order to deal with such situations, the material discontinuity has been smeared over a small time domain, or simply iterations have not been carried out, well realizing that this way errors are introduced in the solution. As another means to obtain a solution artificial viscosity is used as described in Section 3.2.

2.3 Finite Element Discretization

During the course of nonlinear finite element analysis various different finite elements for different stress and strain conditions have been developed. However, considering general nonlinear analysis the discretization of the continuum mechanics equations using isoparametric and related elements is considered to be most effective. In particular, the variable-number-nodes elements presented in [8] [9] [13] can be used very efficiently for the analysis of many problems.

In the isoparametric finite element discretization, the finite element equilibrium equations are obtained by interpolating the coordinates and displacements of the elements, using, respectively,

$$(19)\qquad \left.\begin{aligned} {}^0x_j = \sum_{k=1}^{N} h_k\, {}^0x_j^k \ ; \quad {}^tx_j = \sum_{k=1}^{N} h_k\, {}^tx_j^k \\ {}^{t+\Delta t}x_j = \sum_{k=1}^{N} h_k\, {}^{t+\Delta t}x_j^k \end{aligned}\right\} \begin{array}{l} j=1,2,3 \\ \text{as applicable} \end{array}$$

$$(20)\qquad {}^tu_j = \sum_{k=1}^{N} h_k\, {}^tu_j^k \ ; \quad \Delta u_j = \sum_{k=1}^{N} h_k \Delta u_j^k \qquad \begin{array}{l} j=1,2,3 \\ \text{as applicable} \end{array}$$

where ${}^0x_j^k$ is the coordinate of nodal point k corresponding to direction j at time 0, ${}^tx_j^k$, ${}^{t+\Delta t}x_j^k$, ${}^tu_j^k$ and Δu_j^k are defined similarly, and N is the total number of nodal points of the element. The function h_k is the interpolation function corresponding to nodal point k.

Figures 3 and 4 show the variable-number-nodes elements presently used in program ADINA. The effectiveness of these elements lies in that any number of nodes between the minimum and the maximum number can be chosen. This way it is possible to model adequately a variety of structural configurations and continua using basically one element. In particular, it is possible to change from a coarse to a fine mesh always preserving displacement compatibility between elements. An important extension of the variable-number-nodes isoparametric elements would be the possibility of degenerating the elements partially or completely to superparametric elements as shown in Fig. 5. With the options of using a variable number of nodes and degenerate forms of the basic element, in essence one element can be used to model two and three-dimensional continua and thick and thin shell structures with full compatibility between element boundaries.

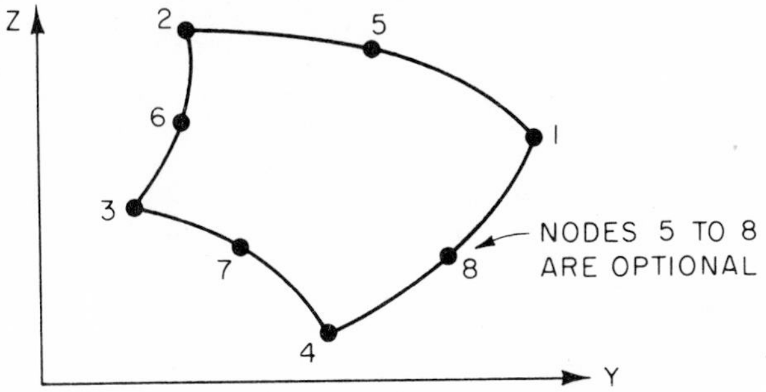

FIGURE 3 4 TO 8 VARIABLE-NUMBER-NODES TWO-DIMENSIONAL PLANE STRESS, PLANE STRAIN AND AXISYMMETRIC ELEMENTS

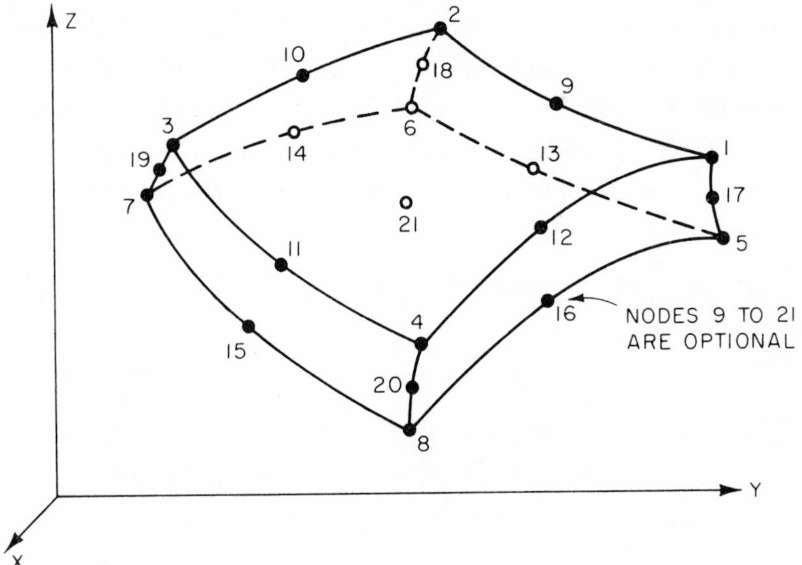

FIGURE 4 8 TO 21 VARIABLE-NUMBER-NODES THREE-DIMENSIONAL SOLID AND THICK SHELL ELEMENT

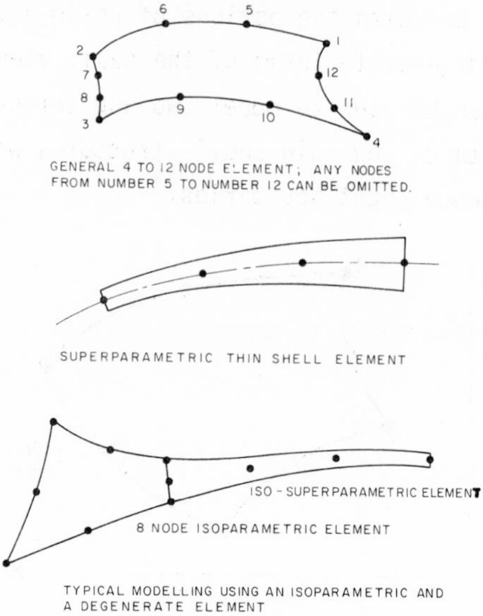

FIGURE 5 TWO - DIMENSIONAL SHELL ELEMENTS

Once the interpolation functions for an element domain have been defined, the expressions in Eqs. (19) and (20) are used in Eqs. (17) and (18) to evaluate the finite element equilibrium equations. Realizing that in dynamic analysis body force components include inertia forces and damping forces, the equilibrium equations for a single element corresponding to Eq. (17) are

$$(21) \quad {}^{t}K \, \Delta u^{(k)} = {}^{t+\Delta t}R - {}^{t+\Delta t}F^{(k-1)} - C \, {}^{t+\Delta t}\dot{u}^{(k)} - M \, {}^{t+\Delta t}\ddot{u}^{(k)}$$

$$k=1,2,\ldots \quad .$$

Similarly, including large displacement and large strain effects we obtain for a single element using the total Lagrangian formulation

$$(22)\quad ({}^{t}_{0}K_{L}+{}^{t}_{0}K_{NL})\ \Delta u^{(k)} = {}^{t+\Delta t}R - {}^{t+\Delta t}_{0}F^{(k-1)} - C\ {}^{t+\Delta t}\dot{u}^{(k)} - M\ {}^{t+\Delta t}\ddot{u}^{(k)}$$

$$k=1,2,\ldots\ .$$

where ${}^{t}K_{L}$ is the element linear strain stiffness matrix at time t when material nonlinear effects are considered only, ${}^{t}_{0}K_{L}$ and ${}^{t}_{0}K_{NL}$ are the linear and nonlinear strain stiffness matrices, ${}^{t+\Delta t}R$ is the vector of externally applied nodal point loads, ${}^{t+\Delta t}_{0}F^{(k-1)}$ is a vector of nodal point forces equivalent to the current element stresses, M and C are the mass and damping matrices and $\Delta u^{(k)}$ is a vector of nodal point displacement increments, with ${}^{t+\Delta t}u^{(k)} = {}^{t+\Delta t}u^{(k-1)} + \Delta u^{(k)}$. The vectors of nodal point accelerations and velocities are evaluated differently depending on the time integration scheme used (see Section 3.3). Table 2 summarizes the evaluation of the matrices. The calculation of the finite element matrices is given in more detail in references [7] and [8].

TABLE 2 FINITE ELEMENT MATRICES

ANALYSIS TYPE	INTEGRAL	MATRIX EVALUATION
IN ALL ANALYSES†	$\int_{{}^{0}V} {}^{0}\rho\ {}^{t+\Delta t}\ddot{u}_k\ \delta u_k\ {}^{0}dv$	$M\ {}^{t+\Delta t}\ddot{u} = {}^{0}\rho\left(\int_{{}^{0}V} H^T H\ {}^{0}dv\right){}^{t+\Delta t}\ddot{u}$
	${}^{t+\Delta t}R = \int_{{}^{0}A} {}^{t+\Delta t}_{0}t_k\ \delta u_k\ {}^{0}da + \int_{{}^{0}V} {}^{0}\rho\ {}^{t+\Delta t}_{0}f_k\ \delta u_k\ {}^{0}dv$	${}^{t+\Delta t}R = \int_{{}^{0}A} H_S^T\ {}^{t+\Delta t}_{0}t\ {}^{0}da + {}^{0}\rho \int_{{}^{0}V} H^T\ {}^{t+\Delta t}_{0}f\ {}^{0}dv$

(cont.)

TABLE 2 (cont.)

ANALYSIS TYPE	INTEGRAL	MATRIX EVALUATION
MATERIAL NONLINEARITY ONLY	$\int_{^0V} C_{ijrs}\, e_{rs}\, \delta e_{ij}\; {}^0dv$	${}^tK\, u = \left(\int_{^0V} B_L^T\, C\, B_L\; {}^0dv \right) u$
	$\int_{^0V} {}^t\sigma_{ij}\, \delta e_{ij}\; {}^0dv$	${}^tF = \int_{^0V} B_L^T\, {}^t\hat{\Sigma}\; {}^0dv$
TOTAL LAGRANGIAN FORMULATION	$\int_{^0V} {}_0C_{ijrs}\; {}_0e_{rs}\; \delta_0 e_{ij}\; {}^0dv$	${}_0^tK_L\, u = \left(\int_{^0V} {}_0^tB_L^T\; {}_0C\; {}_0^tB_L\; {}^0dv \right) u$
	$\int_{^0V} {}_0^tS_{ij}\; \delta_0\eta_{ij}\; {}^0dv$	${}_0^tK_{NL}\, u = \left(\int_{^0V} {}_0^tB_{NL}^T\; {}_0^tS\; {}_0^tB_{NL}\; {}^0dv \right) u$
	$\int_{^0V} {}_0^tS_{ij}\; \delta_0 e_{ij}\; {}^0dv$	${}_0^tF = \int_{^0V} {}_0^tB_L^T\; {}_0^t\hat{S}\; {}^0dv$

†The damping matrix is not considered; it may be established using various procedures [9] [14].

The equilibrium equations corresponding to the complete finite element assemblage which idealizes the continuum are obtained by assembling the individual finite element matrices in the usual manner [9] [15]. In this context, it should be noted that elements (or domains) representing the complete continuum may individually be described by different kinematic formulations. Thus, nonlinear elements with material nonlinearities only or with combined material and geometric nonlinearities may be used in one finite element mesh to represent the actual continuum.

3. Numerical Methods

A most important part of finite element analysis is the use of effective numerical methods, firstly, to evaluate the required finite element matrices and, secondly, to calculate the solution of the equilibrium equations. In this section, the important numerical techniques used in nonlinear finite element analysis are surveyed with the objective to point out where significant improvements are needed.

3.1 Numerical Integration in Space

The evaluation of the finite element matrices as summarized in Table 2 must be performed using numerical integration. The technique widely used in isoparametric finite element analysis is Gauss quadrature [9] [15]. For example, the linear strain stiffness matrix in the total Lagrangian formulation is evaluated using

$$\int_{{}^0V} {}^t_0B_L^T \; {}_0C \; {}^t_0B_L \; {}^0dv = \sum_i \alpha_i \; {}^t_0B_L^{iT} \; {}_0C^i \; {}^t_0B_L^i \tag{23}$$

where α_i is a weighting factor and the summation is carried out over all integration points. Since the numerical effort is directly proportional to the number of integration points used, it is desirable to employ integration methods that minimize the number of integration points required for a given accuracy. Various integration schemes have been devised specifically for finite element analysis. Some success has been reported using selective integration, in which different weighting factors are assigned to the different strain components. Although some excellent improvements in solution accuracy of some problems have been reported in linear analysis, the proposed techniques have not been evaluated to such an extent that they can be used with confidence in practical linear and nonlinear analysis.

It should be noted that in addition to ultimate convergence analyses, in particular, theoretical analyses are required that show under what conditions stable and for practical purposes sufficiently accurate solutions are obtained. Such guidelines are very difficult to establish when material and geometric nonlinearities are considered. However, for finite element analysis to be effective and reliable using a minimum number of integration points, rigorous guidelines for the choice of integration scheme and required order of integration are needed.

3.2 Solution of Static Equilibrium Equations

The equilibrium equations considered in static analysis are

$$ {}^{t}K\, \Delta u^{(k)} = {}^{t+\Delta t}R - {}^{t+\Delta t}F^{(k-1)} \qquad (24)$$

$$k=1,2,\ldots .$$

As discussed in Section 2.1, the above solution corresponds to a modified Newton iteration for the zero of the function f , where

$$ f = {}^{t+\Delta t}R - {}^{t+\Delta t}F^{(\infty)} \quad . \qquad (25)$$

The k'th iteration in essence requires the evaluation of the vector ${}^{t+\Delta t}F^{(k-1)}$ and then the solution of a set of linear algebraic equations. These equations could be solved using an iterative scheme or a direct method. Although probably not always most efficient, implementations of the basic Gauss elimination method are currently most effective for general applications [9]. An important advantage of Gauss elimination is that the number of numerical operations and hence the computer time required for the solution of the system of equations can be determined a priori, whereas using an iterative scheme such as the Gauss-Seidel method, the solution time can in general not be predicted accurately. It should also be noted that the direct solution solvers have been developed to be very efficient so that in many nonlinear analyses most of the computational effort is spent in the calculation of ${}^{t+\Delta t}F^{(k-1)}$ rather than in the solution of the equations.

The effectiveness of the direct equation solvers is largely a result of the specific storage scheme used and the specific implementation of the basic Gauss solution method. Namely, in the solution only the elements under the skyline of the matrix are considered as shown in Fig. 6 [1] [9] [16]. Although the number of operations required for the LDL^T or Cholesky factorization of a given matrix is predetermined, for a specific solution scheme, large reductions in the number of operations can frequently be obtained by rearranging the equations prior to the solution. It should be noted that for the compacted column storage reduction schemes, different algorithms than those developed for bandwidth minimization are required [17] [18].

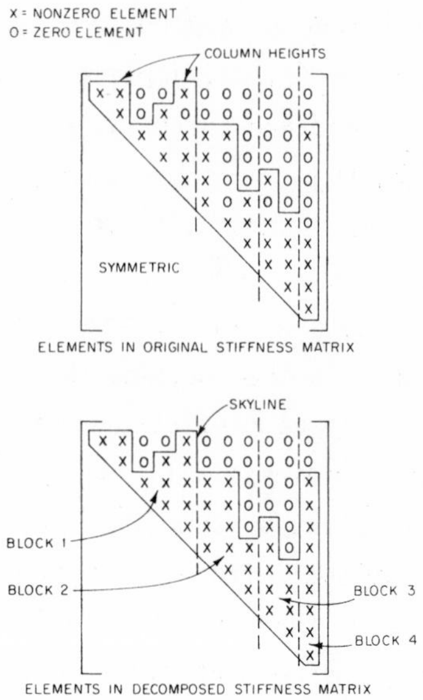

FIGURE 6 TYPICAL ELEMENT PATTERN IN A STIFFNESS MATRIX USING BLOCK STORAGE

The modified Newton iteration in Eq. (24) converges provided the correction $\Delta u^{(k)}$ is small enough, but relatively large corrections can be dealt with, if the first and second derivatives of f satisfy certain conditions. If a one degree of freedom system is considered, these conditions are that $f' < 0$ and $f'' \geq 0$. However, for multiple degree of freedom systems f represents a multiple dimension surface and the change in f corresponding to all components must be considered. In practice, those components that are subjected to the most significant changes will decide whether convergence within a specified tolerance will occur. The practical implications of these statements are that convergence difficulties can be expected if the finite element assemblage is stiffening under the applied load. The stiffening may be due to geometric nonlinear effects or changes in material properties.

If iteration divergence is encountered in practive, it frequently is effective to include in the analysis viscous effects in form of a damping matrix C, i.e. instead of Eq. (24) the following equilibrium relation is considered,

$$ {}^{t}K\, \Delta u^{(k)} = {}^{t+\Delta t}R - {}^{t+\Delta t}F^{(k-1)} - C\, {}^{t+\Delta t}\dot{u}^{(k)} \qquad (26)$$

$$ k=1,2,\ldots $$

where ${}^{t+\Delta t}\dot{u}^{(k)}$ is evaluated using a time integration scheme (see Section 3.3). The viscous nodal point forces prevent large sudden changes in displacements, and vanish once no more changes in displacements occur. Hence the static solution is obtained. In a number of static analyses with convergence difficulties, simply the Newmark time integration scheme was used with zero mass and small concentrated nodal damping values. These concentrated dampers were delected using engineering judgment.

Considering the difficulties in convergence, it is realized that a very important field is the development of solution strategies to speed up convergence in general analysis. The solution algorithms should detect convergence difficulties, and then choose appropriate measures to stabilize the solution. This can be achieved by means of automatic load step (time step) selection, over or under-relaxation, temporary modification of the equations or other measures. In addition to acceleration techniques for iteration convergence, also effective schemes for automatic load or time step selection are to be developed. It is in these areas that a great deal of research is still required in finite element analysis because very few generally applicable results have been obtained. Advances towards more effective solution strategies will only be possible by rigorous mathematical analysis, if the solution methods are to be generally applicable and self-adaptive.

3.3 Solution of Dynamic Equilibrium Equations

In dynamic analysis including mass and damping effects the equations to be solved are

$$(27)\quad {}^{t}K\ \Delta u^{(k)} = {}^{t+\Delta t}R - {}^{t+\Delta t}F^{(k-1)} - C\ {}^{t+\Delta t}\dot{u}^{(k)} - M\ {}^{t+\Delta t}\ddot{u}^{(k)} \qquad k=1,2,\ldots\ .$$

The solution to this system of equations is obtained using direct integration. As pointed out in Section 2.1, the equilibrium equations have been developed for use with an implicit integration method. The most effective implicit integration methods for general application are presently the Newmark method and Wilson θ-method [19] [20] [21] [22] [23]. The Houbolt method has similar characteristics of stability and accuracy as the Wilson method but the disadvantage that special starting strategies are required. Table 3 lists the assumptions used in these three methods. Although not included in the table, a great deal of potential appears to lie in the various stiffly stable methods [24] [25] [26].

TABLE 3 COMMON IMPLICIT DIRECT INTEGRATION METHODS

HOUBOLT METHOD:

$$ {}^{t+\Delta t}\ddot{u} = \frac{1}{\Delta t^2}\left\{2\,{}^{t+\Delta t}u - 5\,{}^{t}u + 4\,{}^{t-\Delta t}u - {}^{t-2\Delta t}u\right\} $$

$$ {}^{t+\Delta t}\dot{u} = \frac{1}{6\Delta t}\left\{11\,{}^{t+\Delta t}u - 18\,{}^{t}u + 9\,{}^{t-\Delta t}u - 2\,{}^{t-2\Delta t}u\right\} $$

NEWMARK METHOD:

$$ {}^{t+\Delta t}\dot{u} = {}^{t}\dot{u} + \left\{\left(1 - \delta\right){}^{t}\ddot{u} + \delta\,{}^{t+\Delta t}\ddot{u}\right\}\Delta t $$

$$ {}^{t+\Delta t}u = {}^{t}u + {}^{t}\dot{u}\,\Delta t + \left\{\left(\frac{1}{2} - \alpha\right){}^{t}\ddot{u} + \alpha\,{}^{t+\Delta t}\ddot{u}\right\}\Delta t^2 $$

WILSON θ - METHOD

$$ {}^{t+\theta\Delta t}\ddot{u} = \frac{6}{\theta^2\Delta t^2}\left({}^{t+\theta\Delta t}u - {}^{t}u\right) - \frac{6}{\theta\Delta t}\,{}^{t}\dot{u} - 2\,{}^{t}\ddot{u} $$

$$ {}^{t+\Delta\theta t}\dot{u} = \frac{3}{\theta\Delta t}\left({}^{t+\theta\Delta t}u - {}^{t}u\right) - 2\,{}^{t}\dot{u} - \frac{\theta\Delta t}{2}\,{}^{t}\ddot{u} $$

In the computer program ADINA the Newmark method and Wilson θ-method are presently used, but it is expected that additional integration schemes will be implemented. Considering the various classes of problems that shall be solved using a general purpose finite element program, it is likely that a library of different integration methods should be provided. The important task is then to identify clearly for which classes of nonlinear problems each of the available operators is most effective. Indeed, it might well be that substantial advantages can be obtained by changing during the step-by-step solution from one operator to another. At present, very little information other than engineering experience with specific nonlinear problems is available for the selection of an integration operator.

An important aspect is the generality of the integration algorithm used. Table 4 summarizes the dynamic step-by-step solution performed in program ADINA. Essentially one integration algorithm is used in which the constants are selected either to employ the Newmark method or the Wilson θ-method.

It should be noted from Table 4 that if no mass and damping effects are considered the step-by-step solution reduces to the solution of Eqs. (24) and (26) solved in static analysis. Therefore, the various solution aspects considering convergence and accuracy discussed in Section 3.2 are also applicable in dynamic analysis. However, it may be noted here that, as probably expected physically, the dynamic solution of a problem is more stable. This is mathematically demonstrated by the fact that the mass effects contribute to the effective stiffness of the system as shown in the calculation of ${}^{t}\hat{K}$ in Table 4. Convergence in the iteration is frequently assured provided the elements in the effective stiffness matrix are large enough, meaning in practical analysis that a small enough time step Δt has to be chosen (see Section 4).

It is interesting to note that until a few years ago almost all nonlinear dynamic finite element analysis has been performed without equilibrium iteration. Indeed, at present, it appears that the only general purpose computer programs that have specifically designed for equilibrium iteration in dynamic analysis are program ADINA and its forerunner NONSAP. This is the case, although the accurate solution of the equilibrium equations at each time step is even more important in dynamic analysis than in static analysis. Namely, any error that is introduced into the dynamic step-by-step solution will accumulate and cannot be compensated for later, as in the solution of many static geometrically nonlinear elastic analyses.

TABLE 4 STEP-BY-STEP INTEGRATION IN ADINA

- - INITIAL CALCULATIONS - -

1. Form linear stiffness matrix K, mass matrix M, and damping matrix C; initialize ${}^0u$, ${}^0\dot{u}$, ${}^0\ddot{u}$.

2. Calculate the following constants:

   tol ≤ 0.01 ; nitem ≥ 3 ; in static analysis $\theta = 1$ and go to 3.

   Wilson θ-method: $\theta \geq 1.37$, usually $\theta = 1.4$, $\tau = \theta\Delta t$

   $a_0 = 6/\tau^2$ $\quad a_1 = 3/\tau$ $\quad a_2 = 2a_1$ $\quad a_3 = 2$

   $a_4 = 2$ $\quad a_5 = \tau/2$ $\quad a_6 = a_0/\theta$ $\quad a_7 = -a_2/\theta$

   $a_8 = 1 - 3/\theta$ $\quad a_9 = \Delta t/2$ $\quad a_{10} = \Delta t^2/6$

   Newmark method: $\theta = 1.0$, $\delta \geq 0.50$, $\alpha \geq 0.25(0.5 + \delta)^2$, $\tau = \Delta t$

   $a_0 = 1/(\alpha\Delta t^2)$ $\quad a_1 = \delta/(\alpha\Delta t)$ $\quad a_2 = 1/(\alpha\Delta t)$ $\quad a_3 = 1/(2\alpha) - 1$

   $a_4 = \delta/\alpha - 1$ $\quad a_5 = \Delta t(\delta/\alpha - 2)/2$ $\quad a_6 = a_0$ $\quad a_7 = -a_2$

   $a_8 = -a_3$ $\quad a_9 = \Delta t(1 - \delta)$ $\quad a_{10} = \delta\Delta t$

3. Form effective linear stiffness matrix: $\hat{K} = K + a_0M + a_1C$.

4. In linear analysis triangularize $\hat{K}$.

- - FOR EACH TIME STEP --

A. IN LINEAR ANALYSIS

   (i) Form effective load vector

$$ {}^{t+\tau}\hat{R} = {}^{t}R + \theta({}^{t+\Delta t}R - {}^{t}R) + M(a_0\,{}^{t}u + a_2\,{}^{t}\dot{u} + a_3\,{}^{t}\ddot{u}) $$

$$ + C(a_1\,{}^{t}u + a_4\,{}^{t}\dot{u} + a_5\,{}^{t}\ddot{u}) \,. $$

(cont.)

TABLE 4 (cont.)

A. (ii) Solve for displacement increments:

$$\hat{K}\ {}^{t+\tau}u = {}^{t+\tau}\hat{R}\ ;\ u = {}^{t+\tau}u - {}^{t}u\ .$$

(iii) Go to C .

B. IN NONLINEAR ANALYSIS

(i) If a new stiffness matrix is to be formed, update $\hat{K}$ for nonlinear stiffness effects to obtain ${}^{t}\hat{K}$; triangularize ${}^{t}\hat{K}$:

$$ {}^{t}\hat{K} = LDL^{T}\ .$$

(ii) Form effective load vector:

$$ {}^{t+\tau}\hat{R} = {}^{t}R + \theta({}^{t+\Delta t}R - {}^{t}R) + M(a_2\ {}^{t}\dot{u} + a_3\ {}^{t}\ddot{u}) + C(a_4\ {}^{t}\dot{u} + a_5\ {}^{t}\ddot{u}) - {}^{t}F\ .$$

(iii) Solve for displacement increments using latest D , L factors:

$$LDL^{T}\ u = {}^{t+\tau}\hat{R}\ .$$

(iv) If required, iterate for dynamic equilibrium; then initialize $u^{(0)} = u$, $i = 0$

(a) $i = i + 1$.

(b) Calculate (i-1)st approximation to accelerations, velocities, and displacements:

$$ {}^{t+\tau}\ddot{u}^{(i-1)} = a_0\ u^{(i-1)} - a_2\ {}^{t}\dot{u} - a_3\ {}^{t}\ddot{u}\ ;$$

$$ {}^{t+\tau}\dot{u}^{(i-1)} = a_1\ u^{(i-1)} - a_4\ {}^{t}\dot{u} - a_5\ {}^{t}\ddot{u}\ ;$$

$$ {}^{t+\tau}u^{(i-1)} = u^{(i-1)} + {}^{t}u\ .$$

(cont.)

TABLE 4 (cont.)

B. (iv) (c) Calculate (i-1)st effective out-of-balance loads:

$$ {}^{t+\tau}\hat{R}^{(i-1)} = {}^{t}R + \theta({}^{t+\Delta t}R - {}^{t}R) - M\,{}^{t+\tau}\ddot{u}^{(i-1)} - C\,{}^{t+\tau}\dot{u}^{(i-1)} - {}^{t+\tau}F^{(i-1)} \quad . $$

(d) Solve for i'th correction to displacement increments:

$$ LDL^{T}\,\Delta u^{(i)} = {}^{t+\tau}\hat{R}^{(i-1)} \quad . $$

(e) Calculate new displacement increments:

$$ u^{(i)} = u^{(i-1)} + \Delta u^{(i)} \quad . $$

(f) Iteration convergence if $\|\Delta u^{(i)}\|_2 / \|u^{(i)} + {}^{t}u\|_2 < \text{tol}$

If convergence: $u = u^{(i)}$ and go to C ;

If no convergence and $i < \text{nitem}$: go to (a);

otherwise restart using new stiffness matrix and/or a smaller time step size.

C. CALCULATE NEW ACCELERATIONS, VELOCITIES, AND DISPLACEMENTS

Wilson θ-method:

$$ {}^{t+\Delta t}\ddot{u} = a_6\,u + a_7\,{}^{t}\dot{u} + a_8\,{}^{t}\ddot{u} $$

$$ {}^{t+\Delta t}\dot{u} = {}^{t}\dot{u} + a_9\,({}^{t+\Delta t}\ddot{u} + {}^{t}\ddot{u}) $$

$$ {}^{t+\Delta t}u = {}^{t}u + \Delta t\,{}^{t}\dot{u} + a_{10}\,({}^{t+\Delta t}\ddot{u} + 2\,{}^{t}\ddot{u}) $$

Newmark method:

$$ {}^{t+\Delta t}\ddot{u} = a_6\,u + a_7\,{}^{t}\dot{u} + a_8\,{}^{t}\ddot{u} $$

$$ {}^{t+\Delta t}\dot{u} = {}^{t}\dot{u} + a_9\,{}^{t}\ddot{u} + a_{10}\,{}^{t+\Delta t}\ddot{u} $$

$$ {}^{t+\Delta t}u = {}^{t}u + u \quad . $$

3.4 Solution of Eigenproblems

In nonlinear analysis of solid mechanics problems eigensolutions are required at various occasions. A most frequently considered eigenproblem is the calculation of the smallest eigenvalues and corresponding eigenvectors of the finite element system. In this case the generalized eigenproblem considered is

$$ {}^{t}K\,\phi = \omega^{2} M\,\phi \tag{28} $$

and the solution is sought for the p lowest eigenvalues $0 < \omega_1^2 \leq \omega_2^2 \leq \cdots \leq \omega_p^2$ and the corresponding eigenvectors $\phi_1, \phi_2, \cdots, \phi_p$. The ω_i and the ϕ_i represent the free vibration frequencies and mode shapes of the finite element system at time t. The frequencies are used to estimate an appropriate time step Δt for the direct integration solution, and they are usually calculated at some representative time in order to take into account that they change during the solution.

Another important eigenproblem arises in linearized buckling analysis. This problem can be written in the form of Eq. (28) by introducing a shift [9].

A large number of different solution methods are available for the problem in Eq. (28) [9] [27]. However, considering that in finite element analysis large systems shall be solved on a routine basis, only a relatively few number of solution methods can be used, and indeed more effective solution techniques are still required. At present, for most practical finite element analyses the determinant search method and subspace iteration technique are considered to be most effective [9] [28]. These two solution methods have been described in detail earlier, and the only objective is here briefly summarized the subspace iteration method in order to point out the large additional potential in the technique.

In essence, the subspace iteration method consists of the following steps:

(1) Establish q starting iteration vectors, $q > p$.

(2) Use simultaneous inverse iteration on the q vectors and Ritz analysis to extract the "best" eigenvalue and eigenvector approximations from the q iteration vectors.

(3) After iteration convergence use the Sturm sequence check to verify that the required eigenvalues and corresponding eigenvectors have been calculated.

The technique has been called subspace iteration method because the iteration is equivalent to iterating with a q-dimensional subspace, and should not be regarded as a simultaneous iteration with q individual iteration vectors. Let X_1 store the q linear independent starting iteration vectors, then the k'th iteration step can be summarized as follows:

Iterate from the subspace E_k to the subspace E_{k+1},

(29) $$^{t}K \, \overline{X}_{k+1} = M \, X_k \; .$$

Calculate the projections of ^{t}K and M onto E_{k+1},

(30) $$K_{k+1} = \overline{X}_{k+1}^{T} \; {}^{t}K \; \overline{X}_{k+1}$$

$$M_{k+1} = \overline{X}_{k+1}^{T} \; M \; \overline{X}_{k+1} \; .$$

Calculate the eigensystem of the projected stiffness and mass matrices,

(31) $$K_{k+1} Q_{k+1} = M_{k+1} \, Q_{k+1} \, \Omega_{k+1}^{2} \; .$$

Evaluate new eigenvector approximations,

(32) $$X_{k+1} = \overline{X}_{k+1} \, Q_{k+1} \; .$$

In Eqs. (31) and (32) the "best" eigenvalue and eigenvector approximations that can be obtained in the sense of a Ritz analysis are calculated, and provided that the starting vectors in X_1 are not orthogonal to the required eigenvectors convergence occurs to $\phi_1 , \ldots , \phi_p$ and $\omega_1^2 , \ldots , \omega_p^2$.

It should be noted that the selection of the starting iteration vectors and the Sturm sequence check are considered important parts of the subspace iteration method. Namely, the closer the starting subspace is to the p-dimensional least dominant required subspace, the smaller will be the number of iterations required for convergence. Indeed, if q starting vectors can be established that span the least dominant p-dimensional subspace, convergence will occur in the first step of the iteration. Hence, one way to improve the effectiveness of the iteration, is to find algorithms that establish more effective starting iteration vectors. At present, the starting vectors are selected using the magnitudes of the diagonal elements in ${}^{t}K$ and M and, depending on the problem and the required eigensolution accuracy, of the order 10 iterations are required. In addition to improving the starting subspace, also reliable shift strategies and other acceleration techniques should further be investigated.

The last phase of the subspace iteration method consists of the Sturm sequence check, which assures that the required eigenvalues and vectors have indeed been obtained. This check can be relatively costly when a large banded system is considered and it is certainly desirable to develop equivalent more effective techniques.

The important point to emphasize is that, in theory, any order of eigenproblem of the form in Eq. (28) can be solved. However, in practice, restrictions on the order of the eigenproblem that can be considered are reached surprisingly fast, mainly because of cost limitations. Therefore, additional research should be conducted to increase the effectiveness of the eigensolution of Eq. (28).

4. DEMONSTRATIVE SAMPLE SOLUTIONS

During the last years a large number of analysis results of nonlinear problems in solid mechanics have been obtained using various finite element computer programs [29]. But in almost all published material relatively little has been said about the practical difficulties encountered during the solutions. Apart from indicating the current capabilities and the potential of the finite element method for solving nonlinear problems, the primary objective in this section is to show some examples where the typical difficulties of nonlinear analysis described in the previous sections have been encountered. The analysis results used here for this purpose have been obtained using the computer programs ADINA [1] and SAP IV [13].

4.1 Static and Dynamic Large Displacement Analysis of a Cantilever Beam

A simple cantilever beam subjected to uniformly distributed loading was analyzed using the total Lagrangian formulation. The cantilever beam is shown in Fig. 7.

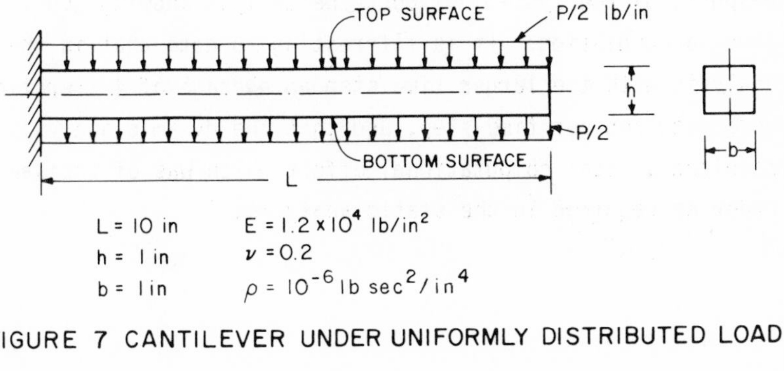

FIGURE 7 CANTILEVER UNDER UNIFORMLY DISTRIBUTED LOAD

For the finite element analysis the beam was idealized as an assemblage of 5 plane stress isoparametric elements.

This sample analysis demonstrates that excellent results can be obtained in some nonlinear static analyses, but a relatively large computational effort may be required. Figure 8 shows the calculated displacement response of the cantilever, when 100 equal load steps are used to apply the total load. The solution compares very well with the response predicted by Holden [30]. Although no attempt was made to optimize the solution cost, [8], a relatively large computational effort is required, in order to obtain accurate results, because the structure is stiffening with increasing displacements.

Considering next the dynamic behavior of the beam, the importance of equilibrium iterations in the dynamic analysis of this structure could be demonstrated. Figure 9 shows the displacement response predicted for an instantaneously applied pressure when two different values of solution time step Δt are used. It is noted that the response predicted with the three times larger time step is accurate provided equilibrium iterations are performed. If the step-by-step solution is used without equilibrium iterations errors accumulate, which in this case decrease the response significantly. In general, the calculated response would simply diverge from the actual response of the system, without the analyst knowing about the error accumulation. It is interesting to note that in the analysis with the larger time step an average of 4 iterations were required per time step, and that the dynamic analysis required a total computational effort which was of the same order as required in the static analysis.

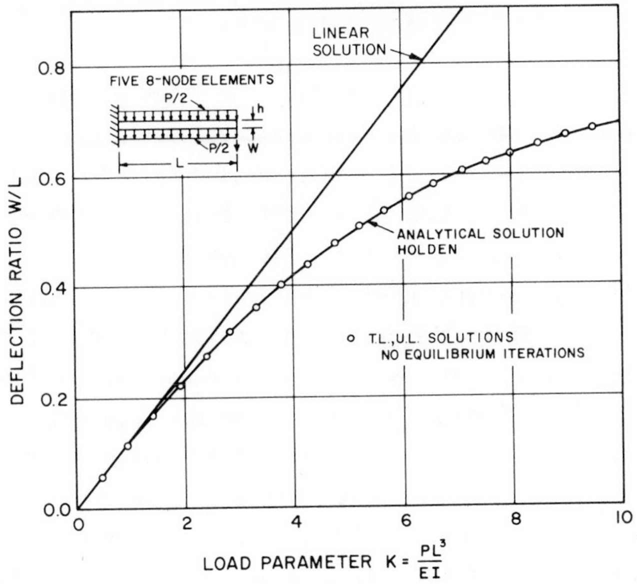

FIGURE 8 LARGE DISPLACEMENT ANALYSIS OF A CANTILEVER

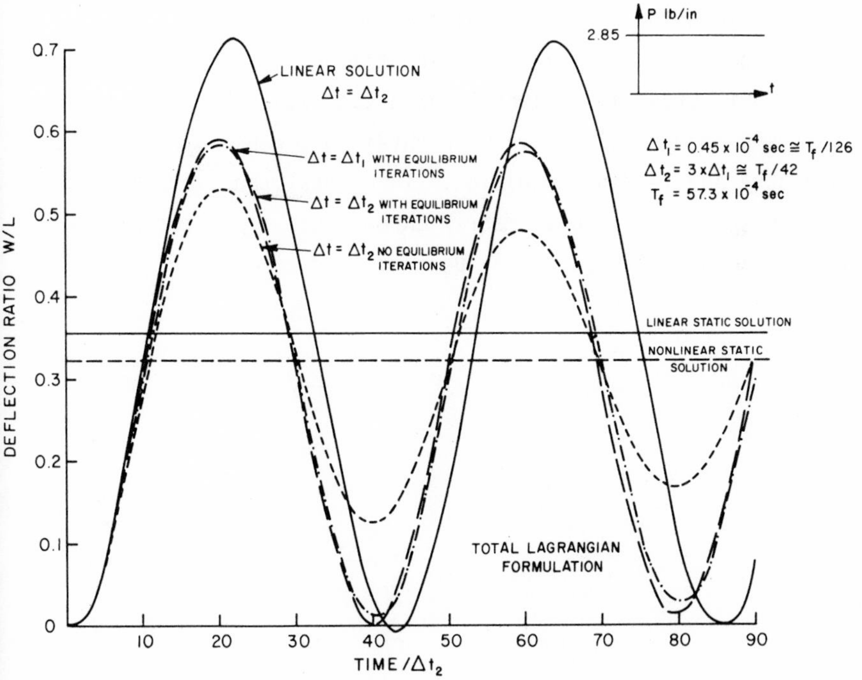

FIGURE 9 LARGE DISPLACEMENT DYNAMIC RESPONSE OF CANTILEVER UNDER UNIFORMLY DISTRIBUTED LOAD, NEWMARK METHOD δ=0.50, α=0.25

4.2 Elastic-Plastic Static Analysis of a Thick-Walled Cylinder

A very common computer program check-out problem is the elastic-plastic analysis of a thick-walled cylinder under internal pressure, because an accurate solution has been provided by Hodge and White [31]. To compare with their solution the finite element model shown in Fig. 10 was analyzed.

Figure 11 shows the predicted displacement response in the finite element analysis. It is observed that the finite element solution compares very well with the solution by Hodge and White. The reason for discussing this analysis here is to point out the difficulties that arise when unloading of the cylinder is considered. Namely, if at the unloading point the elastic-plastic stiffness matrix is used, the equilibrium iteration will not converge. In order to calculate in this finite element analysis the displacement response for loading and unloading the initial elastic stiffness matrix was used with equilibrium iterations in each load step. In this analysis good results have been obtained because of the special geometry and loading condition. In general, however, the use of the elastic stiffness matrix throughout an elastic-plastic analysis can introduce large uncontrolled errors.

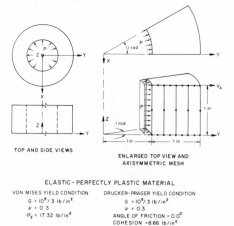

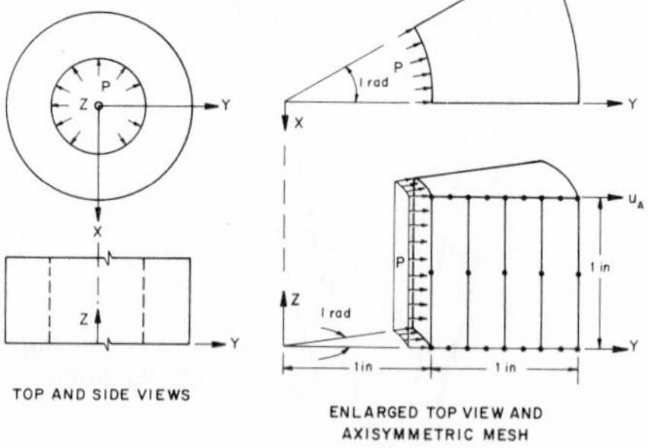

ELASTIC - PERFECTLY PLASTIC MATERIAL

VON MISES YIELD CONDITION	DRUCKER-PRAGER YIELD CONDITION
$G = 10^5/3$ lb/in^2	$G = 10^5/3$ lb/in^2
$\nu = 0.3$	$\nu = 0.3$
$\sigma_y = 17.32$ lb/in^2	ANGLE OF FRICTION = 0.0°
	COHESION = 8.66 lb/in^2

FIGURE 10 FINITE ELEMENT MESH OF THICK-WALLED CYLINDER

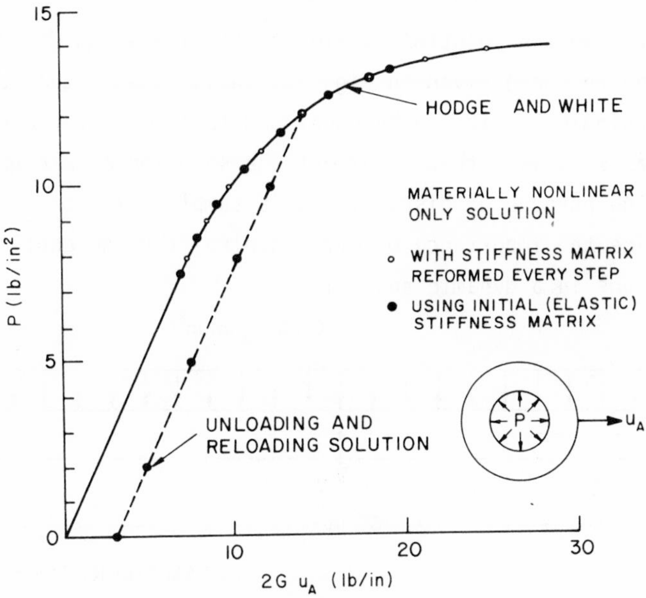

FIGURE 11 ELASTIC-PLASTIC DISPLACEMENT RESPONSE OF THICK-WALLED CYLINDER

4.3 Elastic-Plastic Dynamic Analysis of a Simply Supported Beam

The simply supported beam shown in Fig. 12 was analyzed for the step loading indicated. The figure also gives the finite element idealization used in the analysis. The material of the beam was assumed to be elastic-perfectly plastic.

Figure 13 shows the calculated displacement response. The initial calculated displacement response is shown again in Fig. 14, but this time normalized with respect to the static elastic deflection of the beam subjected to p_0. In the figure, also the displacement response predicted by Nagarajan and Popov is shown [32], who only presented the initial response and did not use equilibrium iterations. When the analysis was repeated with the time step used by Nagarajan and Popov and no equilibrium iterations close comparison with their results was obtained, as shown in Fig. 14. However, at a later time the solution started to oscillate and diverge from the response given in Fig. 13, until finally the solution was meaningless. It can be concluded that the error accumulation was so severe that in effect the solution became unstable. The purpose of discussing this sample solution is to underline the importance of error control through equilibrium iterations in a dynamic analysis.

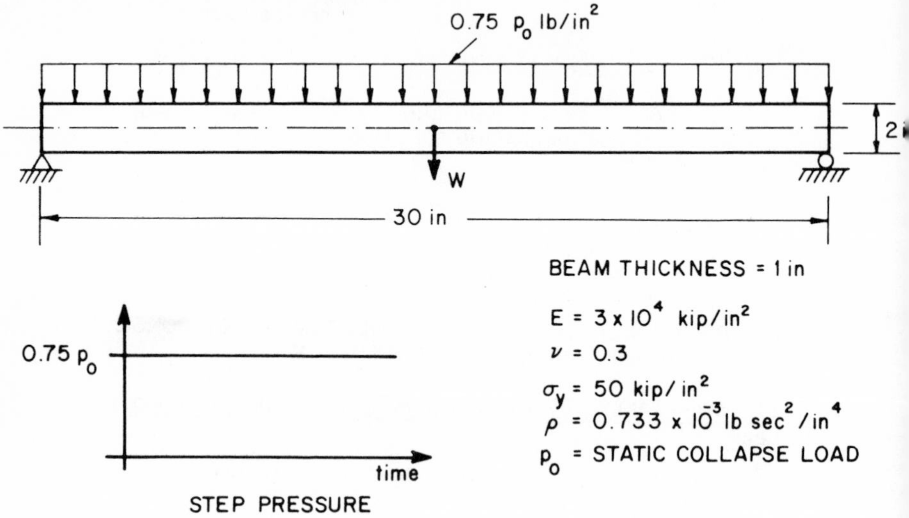

SIMPLY SUPPORTED BEAM AND APPLIED LOAD

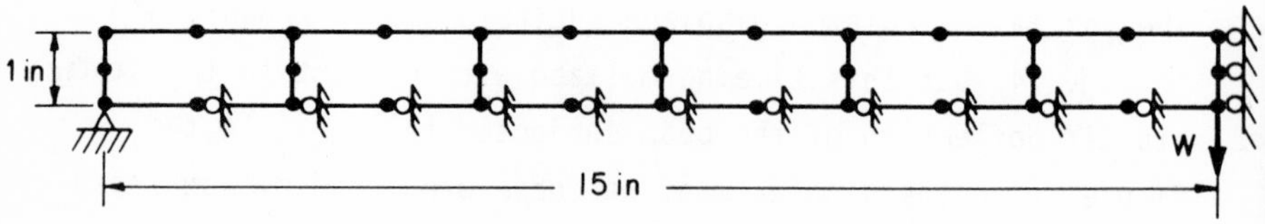

FINITE ELEMENT IDEALIZATION, SIX 8 NODE ELEMENTS

FIGURE 12 ELASTIC-PLASTIC DYNAMIC ANALYSIS OF SIMPLY-SUPPORTED BEAM

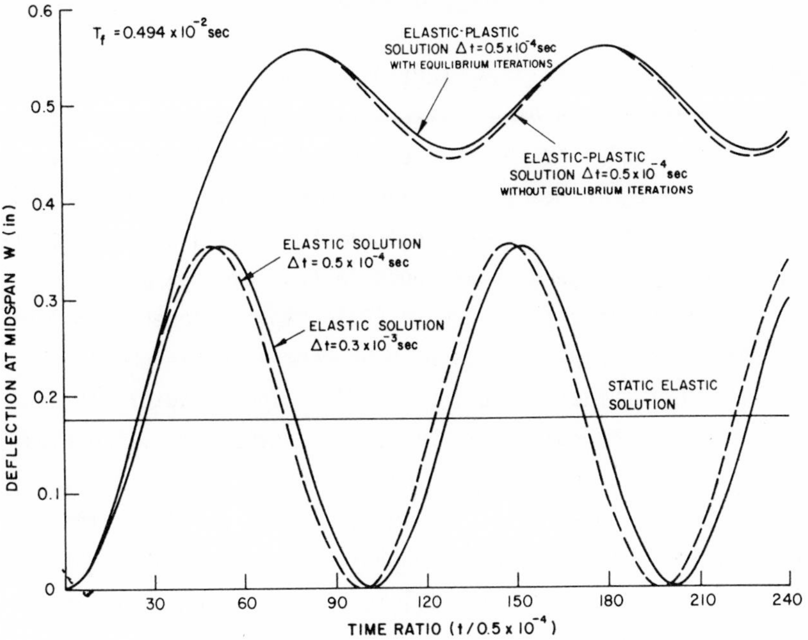

FIGURE 13 DYNAMIC SMALL DISPLACEMENT ELASTIC-PLASTIC RESPONSE OF SIMPLY SUPPORTED BEAM, NEWMARK METHOD, δ = 0.50, α = 0.25

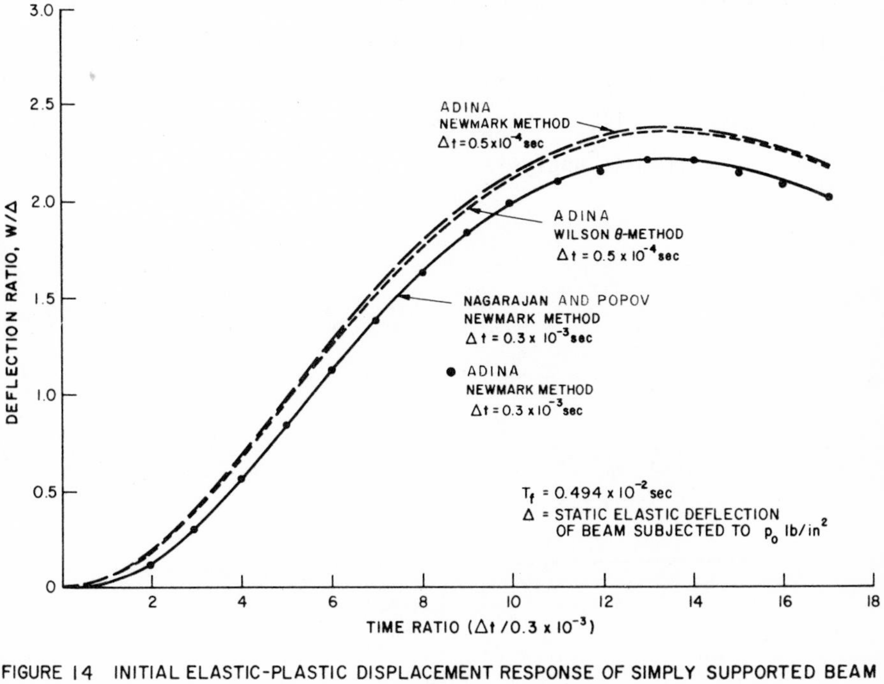

FIGURE 14 INITIAL ELASTIC-PLASTIC DISPLACEMENT RESPONSE OF SIMPLY SUPPORTED BEAM

4.4 Analysis of a Pipe Whip Problem

In the design of nuclear reactor piping equipment an important problem is the analysis of pipes that are subjected to high impact forces and impinge on displacement stops. The purpose of installing the displacement stops is to prevent the occurrence of large displacements in case of a pipe break.

Figures 15 and 16 show a simple finite element model of a cantilevered pipe and its displacement stop, which was analyzed using the program ADINA [33]. The behavior of the system is highly nonlinear, because the material of the pipe and the stop is assumed to be elastic-perfectly plastic and the stop introduces instantaneously a large stiffness into the system. The purpose of presenting this analysis is to indicate some present practical requirements in nonlinear analysis.

Figures 15 and 16 give the displacement and velocity response of the model. The response predicted for the pipe is compared with a solution obtained using the computer program HEMP using 720 zones [34], in which it was assumed that the stop is perfectly rigid. It is noted that the two solutions compare well, although only 6 elements have been used in the finite element analysis.

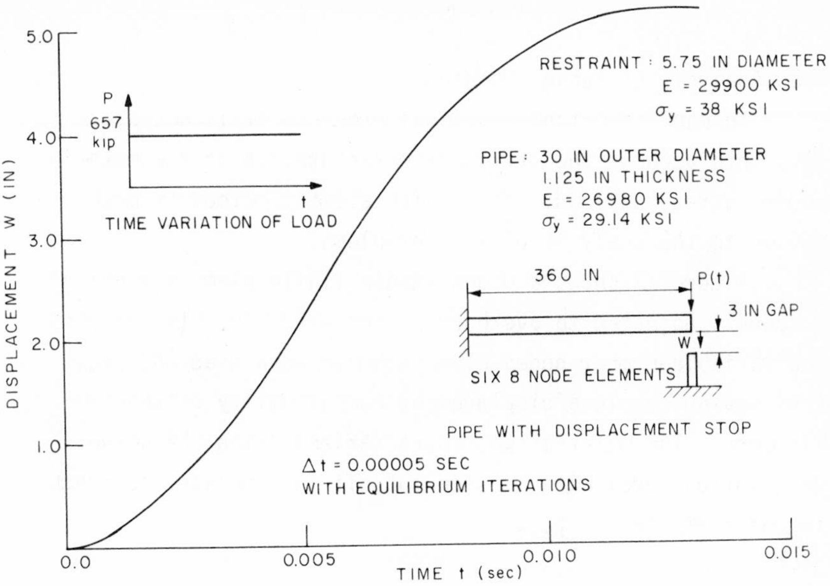

FIGURE 15 DISPLACEMENT RESPONSE OF PIPE WHIP MODEL USING ADINA, NEWMARK METHOD, $\delta = 0.50$, $\alpha = 0.25$

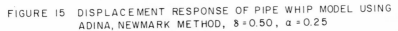

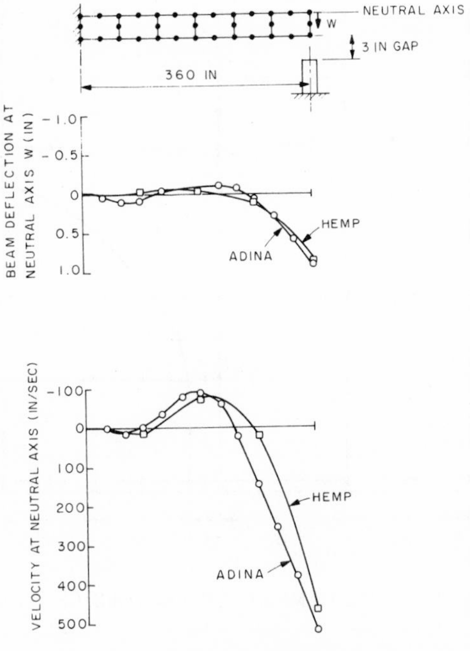

FIGURE 16 PIPE WHIP MODEL RESPONSE AT t = 0.003 SEC

4.5 Analysis of Tunnel Opening

In many important practical problems it is necessary to take into account the stiffness deterioration of the material under increasing load. The finite element method is much suited to the analysis of such problems.

Figure 17 shows a plane strain finite element model of a tunnel subjected to overburden pressure. In this analysis the variable-number-nodes elements have been used [8] [9], thus having complete displacement compatibility between elements. The material was characterized using the curve description model with tension cut-off to simulate the cracking of the material [8].

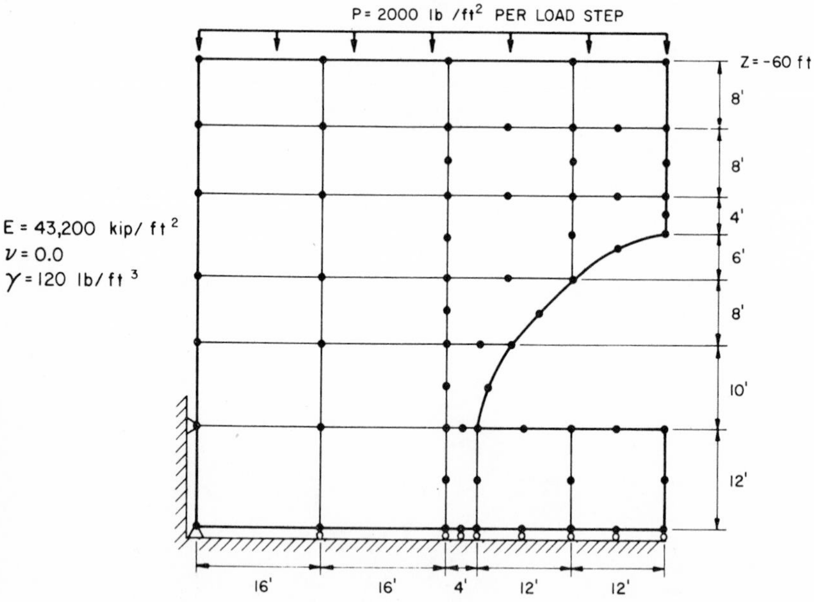

FIGURE 17 FINITE ELEMENT MESH FOR ANALYSIS OF UNDERGROUND OPENING

Figures 18 and 19 show the displacement response and crack distributions as a function of the overburden pressure. Accurate results can be obtained with this model provided small enough load steps are used and the material is loading. On the other hand, unloading conditions are difficult to analyze because the material is stiffening. Indeed, a great deal of additional research is still required to develop constitutive models that predict adequately stiffness deteriorating material response and are numerically stable and accurate under cyclic loading conditions.

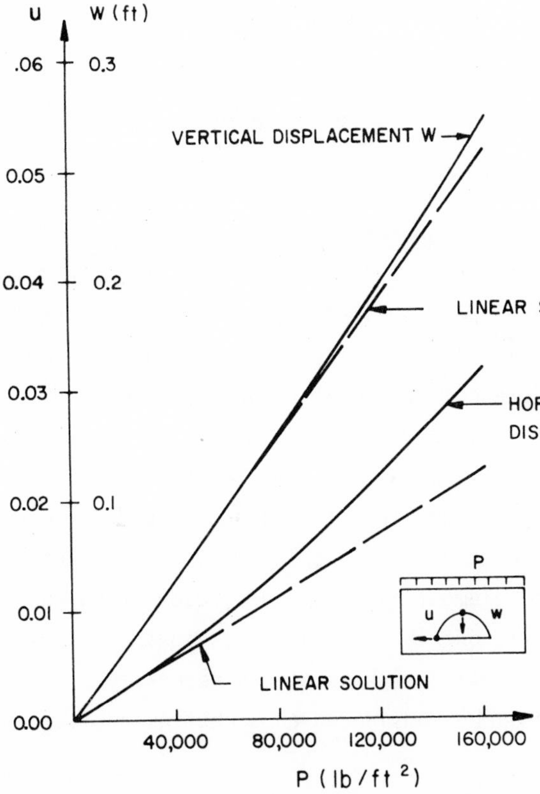

FIGURE 18 LOAD-DEFLECTION RESPONSE OF UNDERGROUND OPENING

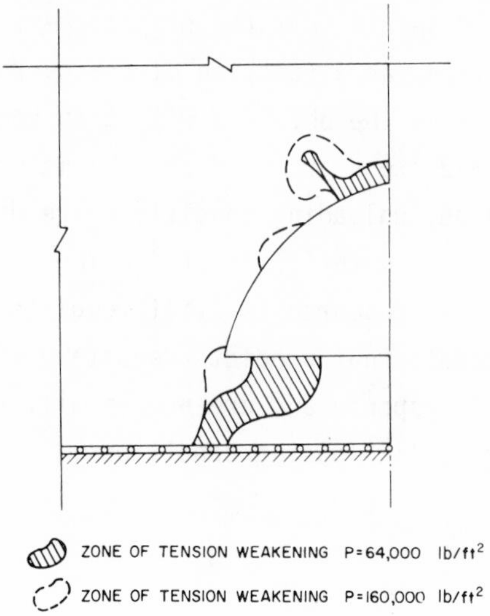

FIGURE 19 CRACKED REGIONS AROUND UNDERGROUND OPENING

4.6 Calculation of Frequencies and Mode Shapes

As was pointed out earlier a significant computational effort in finite element analysis is frequently expended in the calculation of eigenvalues and corresponding eigenvectors. To indicate the present capabilities and expense involved in the calculation of eigensystems, Table 5 summarizes the solution times for various problems using the subspace iteration method. In all cases the lowest frequencies and corresponding vibration mode shapes have been calculated. It is noted that very large finite element systems can already be solved without excessive cost, but that improvements in the efficiency of present eigensolution techniques are necessary, if very large systems shall be considered on a routine basis.

Table 5 CALCULATION OF FREQUENCIES AND MODE SHAPES USING SUBSPACE ITERATION METHOD

SYSTEM	SYSTEM ORDER n	MAXIMUM HALF BANDWIDTH	NUMBER OF EIGENPAIRS	COMPUTER USED	CENTRAL PROCESSOR SECONDS
PIPING SYSTEM	566	12	28	CDC 6600	142
REACTOR BUILDING WITH FOUNDA-TION	1174	138	45	CDC 6600	890
DAM	2916	492	4	CDC 7600	495
WIND-TUNNEL	5952	216	10	CDC 7600	1000
3-DIM. BUILDING FRAME	468	156	4	CDC 6400	160

5. Conclusions

The objective in this paper was to survey present nonlinear finite element analysis procedures with specific emphasis on the numerical techniques used. It is concluded that although already many practical problems can be solved effectively, there are a number of important research areas that require attention. The specific problems to be considered are, in essence, those of the stability, accuracy and cost of solution. Because of the inherent potential in the finite element method, major advances towards more effective solution strategies are anticipated, but it is concluded that an increasing amount of mathematical analysis will be required.

REFERENCES

1. Bathe, K. J., "ADINA - A Finite Element Program for Automatic Dynamic Incremental Nonlinear Analysis," Report No. 82448-1, Acoustics and Vibration Laboratory, Department of Mechanical Engineering, Massachusetts Institute of Technology, 1975.
2. Bathe, K. J., Wilson, E. L., and Iding R., "NONSAP - A Structural Analysis Program for Static and Dynamic Response of Nonlinear Systems," Report No. UCSESM 74-3, Department of Civil Engineering, University of California, Berkeley, February 1974.
3. Bathe, K. J., and Wilson, E. L., "NONSAP - A Nonlinear Structural Analysis Program," Nuclear Engineering and Design, Vol. 29, (1974).
4. Crandall, et al., Dynamics of Mechanical and Electromechanical Systems, McGraw-Hill, New York, 1967.
5. Malvern, L. E., Introduction to the Mechanics of a Continuous Medium, Prentice-Hall, Inc. Englewood Cliffs, New Jersey, 1969.
6. Washizu, K., Variational Methods in Elasticity and Plasticity, Pergamon Press, New York, 2nd edition, 1975.
7. Bathe, K. J., Ramm, E., and Wilson, E. L., "Finite Element Formulations for Large Deformation Dynamic Analysis," International Journal for Numerical Methods in Engineering, Vol. 9, 353-386.
8. Bathe, K. J., Ozdemir, H., and Wilson, E. L., "Static and Dynamic Geometric and Material Nonlinear Analysis," Report No.UCSESM 74-4, Department of Civil Engineering, University of California, Berkeley, February 1974.
9. Bathe, K. J., and Wilson, E. L., Numerical Methods in Finite Element Analysis, Prentice Hall, Englewood Cliffs, New Jersey, to appear 1976.
10. McMeeking, R. M., and Rice, J. R., "Finite Element Formulations for Problems of Large Elastic-Plastic Deformation," Technical Report, Nasa NGL 40-002-080/15, Division of Engineering, Brown University, Providence, R. I., May 1974.
11. Oden, J. T., Finite Elements of Nonlinear Continua, McGraw-Hill, 1972.
12. Bathe, K. J., and Ozdemir, H., "Elastic-Plastic Large Deformation Static and Dynamic Analysis," J. Computers and Structures, in press.
13. Bathe, K. J., Wilson, E. L., and Peterson, F. E., "SAP IV - A Structural Analysis Program for Static and Dynamic Response of Linear Systems," Report No. EERC 73-11, University of California, Berkeley, 1973.

14. Clough, R. W., and Bathe, K. J., "Finite Element Analysis of Dynamic Response," 2nd U.S.-Japan Symposium, Berkeley, August 1972; Advances in Computational Methods in Structural Mechanics and Design, University of Alabama Press, Huntsville, Alabama, U.S.A.
15. Zienkiewicz, O. C., The Finite Element Method in Engineering Science, McGraw-Hill, London, 1971.
16. Mondkar, D. P. and Powell, G. H., "Large Capacity Equation Solver for Structural Analysis," Computers and Structures, Vol. 4 (1974), 699-728.
17. Rosen, R., "Matrix Bandwidth Minimization," Proceedings National Conference A.C.M. (1968), 585-595.
18. Cuthill, E. H., and McKee, J. M., "Reducing the Bandwidth of Sparse Symmetric Matrices," Proceedings, National Conference A.C.M. (1969), 151-172.
19. Newmark, N. M., "A Method of Computation for Structural Dynamics," ASCE, Journal of Engineering Mechanics Division, Vol. 85 (1959), 67-94.
20. Wilson, E. L., Farhoomand, I., and Bathe, K. J., "Nonlinear Dynamic Analysis of Complex Structures," International Journal of Earthquake Engineering and Structural Dynamics, Vol. 1 (1973), 241-252.
21. Bathe, K. J., and Wilson, E. L., "Stability and Accuracy Analysis of Direct Integration Methods," Earthquake Engineering and Structural Dynamics, Vol. 1 (1973), 283-291.
22. Belytschko, T., and Schoeberle, D. F., "On the Unconditional Stability of an Implicit Algorithm for Nonlinear Structural Dynamics," to be published.
23. Houbolt, J. C., "A Recurrence Matrix Solution for the Dynamic Response of Elastic Aircraft," Journal of Aeronautical Science, Vol. 17 (1950), 540-550.
24. Gear, C. W., "The Automatic Integration of Stiff Ordinary Differential Equations," Information Processing, Vol. 68 (1969), 187-193.
25. Jain, M. K., and Srivastava, V. K., "Optimal Stiffly Stable Methods for Ordinary Differential Equations," Report 402, Department of Computer Sciences, University of Illinois, Urbana, June 1970.
26. Jensen, P. S., "Transient Analysis of Structures by Stiffly Stable Methods," J. Computers and Structures, Vol. 4 (May 1974), 615-626.
27. Wilkinson, J. H., The Algebraic Eigenvalue Problem, Clarendon Press, Oxford, 1965.
28. Bathe, K. J., and Wilson, E. L., "Solution Methods for Eigenvalue Problems in Structural Mechanics," International Journal for Numerical Methods in Engineering, Vol. 6 (1973), 213-226.

29. Marcal, P. V., "Survey of General Purpose Programs for Finite Element Analysis," 2nd U.S.-Japan Symposium, Berkeley, August 1972; Advances in Computational Methods in Structural Mechanics and Design, University of Alabama Press, Huntsville, Alabama, U.S.A.
30. Holden, J. T., "On the Finite Deflections of Thin Beams," Int. J. Solids Struct., Vol. 8 (1972), 1051-1055.
31. Hodge, P. G., and White, G. H., "A Quantitative Comparison of Flow and Deformation Theories of Plasticity," J. Appl. Mech., Vol. 17 (1950), 180-184.
32. Nagarajan, S., and Popov. E. P., "Elastic-Plastic Dynamic Analysis of Axisymmetric Solids," Report No. UC SESM 73-9, Department of Civil Engineering, University of California, Berkeley, 1973.
33. Dunder, V. and Ma, S. M., Private Communication, Bechtel Int. Corp., San Franciso, California, 1975.
34. Wilkins, M. L., "Calculation of Elastic-Plastic Flow," Report UCRL-7322, Rev. I, Lawrence Radiation Laboratory, University of California, Livermore, Ca. 1969.

CRUDE NUMERICAL APPROXIMATION OF TURBULENT FLOW*

Alexandre Joel Chorin**

Introduction

The numerical solution of the Navier-Stokes equations, with given sure (= non random) initial data and boundary conditions, when the Reynolds number R is very large, is extremely difficult. The difficulty stems mainly from the fact that at large R the solution has components of widely differing scales, whose interaction must be taken into account, but which are very difficult to represent simultaneously by the usual approximate procedures, such as finite difference methods, finite element and Galerkin methods. The experimental fact is that perturbations whose order of magnitude is small compared with any truncation error one may encounter can alter significantly the nature of the flow at high R, and there is no realistic hope that such perturbations will be resolvable on a computer any time soon. Analyses of this difficulty, all essentially equivalent, are available, e.g., Chorin [1] gives error estimates; Dorodnicyn [11] discusses the limitations imposed by the thinness of the boundary layers; Keller and Takami [17] discuss the limitation due to mesh Reynolds number restrictions; Fox, Herring, Kraichnan and Orszag [13] discuss the problem of obtaining an adequate representation in terms of Fourier components.

---

*This work was partially supported by the Office of Naval Research under contract No. N00014-69-A-200-1052.

**University of California

To study turbulence one should construct, it would seem, not only single flows but families of flows, i.e., do the impossible many times. Fortunately, it is almost a general principle that it is easier to construct averages of solutions of difficult problems than construct individual solutions of these problems (see e.g. Kac [16]). In this talk I would like to present two hopeful approaches to the construction of such averages.

Physical background

Statistical mechanics and kinetic theory provide prototypes of what one would like to achieve in turbulence theory. Consider a gas of elastic spheres contained in a box. The gas soon reaches a statistical equilibrium independent of the initial data. If, at equilibrium, one considers a very small volume ω contained in the box, very small compared to the volume of the box, but large enough to contain very many spheres, one sees that the distribution of velocities in the box is very nearly invariant, although the individual spheres are moving about quickly and unsteadily. If one is interested in non-equilibrium phenomena, one may proceed by considering perturbations from equilibrium (although this is more difficult than once thought).

If one wanted to adapt the methods of statistical mechanics to turbulence theory, one could be encouraged by their recent success in field theory, a subject formally similar to turbulence theory, and by their applicability to formal models of turbulent flow in which the mean energy at every point is allowed to be infinite, (Hopf [14]),or in which the equations are suitably modified by a cut-off or a renormalization (Kraichnan [20]). The difficulty in using such ideas in real turbulence are, however, very substantial (Hopf [14], Chorin [8]). The main stumbling blocks are problems of coherency and scale: an examination of a turbulent flow typically reveals a number of large scale structures, containing many scales of motion locked into almost fixed phase relations; these structures are slowly changing, irregularly spaced, and produced by mechanisms which do not seem to be independent of the history of the flow (discussions can be found, e.g., in Saffman's excellent lecture notes [29], in Corrsin [10], Chorin [3]; the whole subject is surveyed in Chorin [7]). The flow is intermittent, and it is not generally true that the distribution of velocities in a small volume ω is almost independent of ω. Thus, the methods of statistical mechanics must be deeply modified before they can be used.

In this talk, I shall discuss two distinct approaches to the application of statistical methods in the analysis of turbulent flow:

i) One can argue that specific solutions of the Navier-Stokes equations, corresponding to well-defined initial and boundary conditions, are not only difficult to obtain, but also meaningless, since real flows are determined by uncontrollable small perturbations; one should therefore concentrate on constructing numerically flows which are in some sense typical, and whose averages should converge to the averages of the solutions of the Navier-Stokes equations. To do this, one must identify turbulent structures which are in some sense generic. Such structures can be identified in two-dimensional flow, when circular compound vortices are known to play a major role in the dynamics of the flow. There is a simple relation between the structure of these vortices and the energy spectrum of the flow (see e.g. Chorin [7]). In three dimensional flow the problem is much more difficult.

(ii) Alternatively, one can give up any attempt to obtain detailed information about the flow, and consider only averages of the flow over regions large enough so that they contain typically a number of basic structures, and are thus capable of statistical analysis. Because of the problem of scale, these regions are not very small, and the resulting analysis must perforce be crude.

These two ways will be described in the next two sections.

The random vortex method

We begin by discussing a model of two dimensional random flow (not everybody agrees that such flow can be viewed as turbulent). Consider flow in a region D of the $(x,y) \equiv \underline{r}$ plane, with boundary ∂D. The Navier-Stokes equations can be written in the form

$$(1) \quad \frac{D\xi}{Dt} - R^{-1}\Delta\xi = 0$$

$$\Delta\psi = - \xi$$

$$u = -\partial_y\psi \ , \ v = \partial_x\psi \ , \ (u,v) \equiv \underline{u} \ , \ \xi = -\partial_x v + \partial_y u \ ,$$

where D/Dt denotes a Lagrangian derivative, ψ is the stream function, ξ is the vorticity, and $\underline{u}$ is the velocity. The boundary conditions are

$$\underline{u} = 0 \quad \text{on} \quad \partial D \ .$$

Assume ξ can be considered to be a sum of circular vortices,

$$\xi = \sum_{j=1}^{n} \xi_j \xi_0(|\underline{r}-\underline{r}_j|) \ ,$$

where ξ_0 is a fixed function of $|\underline{r}|$ and the ξ_j are constants. In the terminology introduced by Onsager [28] (see also Joyce and Montgomery [15], Edwards and Taylor [12]) the fluid is in a continuum analogue of a "negative temperature state". Assume ξ_0 has small support. It is important to notice that ξ_0 of small support does not preclude the creation of larger compound vortices, and that shear layers, which may play an important role in two dimensional turbulence, (Saffman [30]), can be approximated by a sum of vortices of small support (Chorin and Bernard [9]). The Leith spectrum can be matched with the following choice of ξ_0 :

$$\xi_0 = \begin{cases} 0 & |\underline{r}| \geq \sigma \\ |\underline{r}|^{-1} & |\underline{r}| < \sigma \end{cases}$$

(see Chorin [7]). If $R^{-1} = 0$, and only the normal component of $\underline{u}$ is made to vanish on ∂D , equations (1) lead to

$$(2) \quad \frac{d\underline{r}_j}{dt} = \sum_{i \neq j} \underline{u}_i + \underline{u}_0$$

$$u = -\partial_y \psi_i \ , \ v_i = \partial_x \psi_i \ , \ i = 0,\ldots,n$$

$$\Delta\psi_i = -\xi_i \xi_0(|r - r_i|) \ , \ \psi_i \to 0 \ \text{ at } \ |\underline{r}| \to \infty \ , \ i = 1,\ldots,n$$

$$\Delta\psi_0 = 0 \ , \ \sum_{i=0}^{n} \psi_i = 0 \ \text{ on } \ \partial D \ .$$

ψ_i , $i \neq 0$, is readily found, and ψ_0 can be determined by a variety of means (e.g. by solving an integral equation on ∂D). The first of these equations can be approximated by

$$(3) \qquad r_j^{n+1} - r_j^n = k \sum_{i=0,\ i \neq j}^{n} \underline{u}_i^n \ ,$$

where k is a time step, and $\underline{r}_j^n \equiv \underline{r}_j(nk)$, $\underline{u}_i^n \equiv \underline{u}_i(nk)$. If $R^{-1} \neq 0$, and the boundary is temporarily forgotten, (3) can be replaced by

(4) $$\underline{r}_j^{n+1} - \underline{r}_j^n = k\Sigma \underline{u}_i^n + \underline{\eta}$$

where $\underline{\eta} = (\eta_1, \eta_2)$, and η_1, η_2 are Gaussian random variables with mean 0 and variance $2k/R$. To satisfy the condition $\underline{u} = 0$ on ∂D one can borrow a device from boundary layer theory: at each time step one can imagine a thin layer at the boundary in which the tangential velocity tapers down from its finite value away from the boundary; the vorticity in this layer can be readily estimated, distributed among finite area vortices, and allowed to participate in the motion (4). Details can be found in Chorin [6]. The algorithm mimics the physics of vorticity production and dispersal.

When the flow is laminar, the method just described should approximate that flow. In this guise, it has been applied to a number of problems (see e.g. Chorin [4], Shestakov [32]). Some effort at rigorous justification has been carried out (Marsden [24], [25]). Shestakov [32] has written a program in which this method is used near a boundary, and a classical finite difference method is used away from the boundary. Extension to three dimensional flow has been attempted by A. Leonard [23].

Vortex methods for inviscid flow had been developed i.a. by Westwater [33], Sarpkaya [31], Laird [22]. A random vortex approximation has independently been attempted by D. W. Moore [27]. The need for a finite vortex structure has been realized i.a. by Kuwahara and Takami [21]. An interesting paper on the relation between turbulence theory and vortex motion is due to Koga [18].

Coarse grained approximation

We now turn to the second possibility outlined earlier, and attempt to apply statistical methods to averages evaluated over regions which are not microscopical. Consider, for the sake of simplicity, a flow in a two dimensional region D, with boundary ∂D; divide it into subregions $\omega_1,\ldots,\omega_M$ of equal area, and consider the variables

$$\xi_i = \int_{\omega_i} \xi dxdy$$

(ξ is the vorticity). Assume the flow is turbulent (assumption which may be meaningful only in the three dimensional case, mentioned below), and that vorticity is continuously produced at the boundaries and dispersed by the flow (which may exclude such phenomena as a turbulent mixing layer between two fluids). Consider an ω_i away from a boundary. Assume that the ω_i can be picked so that its area is small compared to the area of D, yet large enough to contain typically more than one structure such as a vortex. Make the following assumption: ξ_i is a Gaussian variable, with mean $\overline{\xi_i}$ and variance $\overline{\xi_i^2}$. In picturesque language, the fluid is viewed as a gas of vortices in local thermal equilibrium. Furthermore, assume that if ω_i,ω_j are away from ∂D, $i \neq j$, then ξ_i,ξ_j, are independent. (One could refine these assumptions by developing ξ_i in a coarse grained Wiener-Hermite series (Chorin [5]), but we shall refrain from such considerations here.) We now wish to construct a procedure for determining $\overline{\xi_i},\overline{\xi_i^2}$ from the Navier-Stokes equation.

Consider the problem of flow in a channel,

$$D = \{x,y \mid -\infty < x < +\infty , 0 < y < 1\} ,$$

with a pressure gradient which ensures that $u(u,\frac{1}{2}) = 1$. The regions ω_i are taken to be squares of side h; $h = 1/N$, N integer. Assume the velocity and vorticity fields are statistically translation invariant in the x direction, and write

$$\int_{\substack{ih \le x \le (i+1)h \\ jh \le y \le (j+1)h}} \xi dx\, dy = \xi_{j+1/2} = \overline{\xi}_{j+1/2} + \xi'_{j+1/2}/h$$

(the normalizing factor h^{-1} is convenient). One can readily find the velocity field $\underline{U}$ induced at the point (ih,ph) by a vortex by unit strength placed at the point $(0,(q + 1/2)h)$, taking into account only the condition $v = 0$ on $y = 0$, $y = 1$. Let

$$\underline{U} = (U(i,p,q + 1/2) , V(i,p,q + 1/2)) ,$$

Let

$$S_u(p,q + 1/2) = \sum_i U^2(i,p,q + 1/2)$$

$$S_v(p,q + 1/2) = \sum_i V^2(i,p,q + 1/2) .$$

Let u_j be the x component of $\underline{u}$ at $y = jh$ (we omit the x dependence of the variables because of the assumed statistical invariance in x). One can readily verify that if

$$u_u = \overline{u_j} + u_j'$$

then $\overline{u_j'^2} = h^2 \sum_q \overline{\xi_{q+1/2}'^2}\, S_u(j,q + 1/2)$ with a similar expression for $\overline{v_j'^2}$ in terms of S_v. Furthermore,

$$- \overline{\xi} = \frac{\partial \overline{u}}{\partial y} .$$

The vorticity equation can be written in the form

$$\partial_t \xi + \operatorname{div}(\xi \underline{u}) = R^{-1} \Delta \xi . \tag{5}$$

A discrete form of this equation is

(6)
$$\xi^{n+1}_{j+1/2} - \xi^{n}_{j+1/2} = -\frac{k}{h}\left((\xi v)^{n}_{j+1} - (\xi v)^{n}_{j}\right) + \frac{k}{Rh^2}\left(\xi^{n}_{j+3/2} + \xi^{n}_{j-1/2} - 2\xi^{n}_{j+1/2}\right) + \cdots$$

where the three dots denote terms approximating x derivatives, k is a time step, $\xi^n \equiv \xi(nk)$, etc. Equation (6) does not approximate equation (5) for any reasonable h (if it did, there would be no difficulty in numerical turbulence theory). However, suppose equation (6) were averaged, and suppose the averages of ξ and $\underline{u}$ were smooth enough; then the resulting equation would be meaningful. Of course, as usual, nonlinear terms such as $(\xi v)^n_j$ will appear. The assumption above allows one to express them in terms of $\overline{\xi'^2}$; an equation for $\overline{\xi'^2}$ can be obtained, and a closure thus achieved. This closure is not applied to the flow itself, but to a discrete collection of its functionals; once it is invoked, the limit $h \to 0$ is forever illegitimate, since the main assumption (Gaussianity and independence of the ξ_i) does not survive such a limit. The appropriate values of h and k remain undetermined, and one must make sure that the result depends little on the choices one makes. It is clear that $h \geq O(R^{-1/2})$, $k \geq O(R^{-1/2})$, since viscosity creates correlations over distances at that scale. Thus, no accuracy better than $O(R^{-1/2})$ can be hoped for. Furthermore, not only one cannot resolve boundary layers, but one must resolutely refrain from doing so. The tangential boundary condition $u = 0$ on $y = 0$ and $y = 1$ can be satisfied by a creation device analogous to the one used in the previous section. The modifications required for the three dimensional case are minor.

The details of the derivation, the complete set of equations, and the numerical results obtained with them, are given elsewhere [6], [7]. The accuracy of the calculation, estimated by comparison with experimental results, is of the order of 5 to 10%, which, although not magnificent, is quite respectable for a calculation based on so crude an analysis; the range of Reynolds numbers for which such results are available is rather modest.

One obvious way of improving these results is to use values of h and k extrapolated from experimental data. This may ultimately be necessary; we have so far refrained from doing so in order to demonstrate that the theory does provide a substantial insight into the mechanics of turbulence.

Some further remarks may be appropriate. The system of difference equations one obtains from the present analysis is formally rather similar to the difference equations which approximate the Navier-Stokes equations when the flow is laminar and smooth, except for the fact that h is large (too large ever to resolve a boundary layer or shear layer). This may be the germ of an explanation for the well-known but paradoxical fact that finite difference calculations have sometimes led to successful predictions in flow problems where, on the basis of non-statistical considerations, they should have been totally meaningless.

The method of this section relates to the method of the previous section; it can be viewed as a way of averaging the random vortex calculation. There is a curious analogy between this method and the work of Kraichnan [19]. Kraichnan introduced a stochastic model, which he, however, did not use directly, but did use as a point of departure for an averaging procedure. This stochastic model is made up of coupled oscillators, which are less physically meaningful then vortices in fluid mechanics; the averaging procedures are similar in a number of details, in particular, the functions S_u , and S_v above have obvious counterparts in Kraichnan's work. One major difference lies in the fact that we are working here in the energy range, while Kraichnan's work was mostly applied to the analysis of the inertial range.

The idea of coarse graining had apparently occurred to Hopf [14], last footnote.

Conclusions

We have briefly discussed two numerical methods for studying random flow. They are clearly not the last work on the subject, and the practical experience with their use has so far been limited. Their ingredients are borrowed from numerical analysis and probability theory; the numerical part and the probablistic part are inseparable; these methods have encountered sufficient success for one to believe that further progress in this direction will lead to a substantial understanding of turbulence and of the appropriate way to deal with it numerically.

REFERENCES

1. A. J. Chorin, Math. Comp., 23, 341 (1969).
2. A. J. Chorin, "Inertial Range Flow and Turbulence Cascades," AEC Research and Development Report NYO-1480-135, New York University (1969).
3. A. J. Chorin, "Computational Aspects of the Turbulence Problem," Proc. 2nd. Int. Conf. Num. Meth. Fluid Dynamics, Springer, 1970.
4. A. J. Chorin, J. Fluid Mech., 57, 785 (1973).
5. A. J. Chorin, J. Fluid Mech., 63, 21 (1974).
6. A. J. Chorin, "Analysis of Turbulent Flow with Shear," to appear.
7. A. J. Chorin, "Lectures on Turbulence Theory," Publish or Perish, Boston, 1975.
8. A. J. Chorin, "Numerical Experiments with Truncated Spectral Representations of Random Flow," to appear.
9. A. J. Chorin and P. Benard, J. Comp. Phys., 13, 423 (1973).
10. S. Corrsin, Phys. Fluids, 5, 1301 (1962).

11. A. A. Dorodnicyn, Proc. 3rd. Int. Conf. Num. Meth. Fluid Mechanics, Springer, 1972.
12. S. F. Edwards and J. B. Taylor, Proc. Roy. Soc. London, A, 336, 257 (1974).
13. J. R. Herring, S. A. Orszag, R. H. Kraichnan, and D. G. Fox, J. Fluid Mech., 66, 417 (1974).
14. E. Hopf, J. Rat. Mech. Anal., 1, 87 (1952).
15. G. Joyce and D. Montgomery, J. Plasma Physics, 10, 107 (1973).
16. M. Kac, Probability and its Application to the Physical Sciences, Interscience (1959).
17. H. B. Keller and H. Takami, in Numerical Solution of Partial Differential Equations (D. Greenspan, editor), 1966.
18. T. Koga, "A Kinetic Theory of Turbulence in Incompressible Fluids," PIBAL Report 68-5, Dept. of Aerospace Eng., Brooklyn Pol. Inst. (1969).
19. R. H. Kraichnan, J. Fluid Mech., 5, 497 (1959).
20. R. H. Kraichnan, "Remarks on Turbulence Theory," to appear in Advances in Mathematics.
21. K. Kuwahara and H. Takami, J. Phys. Soc. Japan, 34, 747 (1973).
22. A. D. K. Laird, J. of Hydraulics Div., Proc. ASCE, 763 (1971).
23. A. Leonard, Proc. 4th Int. Conf. Num. Meth. Fluid Mechanics, Springer, 1974.
24. J. Marsden, Bull. Am. Math. Soc., 80, 154 (1974).
25. J. Marsden, "Applications of Global Analysis to Mathematical Physics," Publish or Perish, Boston, 1974.
26. D. Montgomery and G. Joyce, Physics of Fluids, 1974.
27. D. W. Moore, Personal Communication, 1974.
28. L. Onsager, Nuovo Cimento, 6 (Suppl), 229 (1949).
29. P. G. Saffman, in Topics in Nonlinear Physics, (N.Zabusky, editor), Springer, 1968.
30. P. G. Saffman, Studies in Applied Math., L, 377 (1971).
31. T. Sarpkaya, J. Fluid. Mech., 68, 109 (1975).
32. A. I. Shestakov, Ph.D. Thesis, Berkeley (1975).
33. F. L. Westwater, ARC Reprot and Memo, No. 1692 (1936).

HETEROGENEITY EFFECTS IN NEUTRON TRANSPORT COMPUTATIONS*

Ely M. Gelbard**

A nuclear reactor is, generally, an intricate heterogenous structure whose adjacent components may differ radically in their neutronic properties. The heterogeneities in the structure of the reactor complicate the work of the reactor analyst and tend to degrade the efficiency of the numerical methods used in reactor computations. Two types of heterogeneity effects will be considered. First, certain singularities in the solution of the neutron transport equation, induced by heterogeneities, will be briefly described. Second, the effect of heterogeneities on neutron leakage rates, and consequently on effective diffusion coefficients, will be discussed.

1. Nature of Problem

One of the fundamental numerical problems involved in the design of nuclear reactors is the computation of the neutron distribution, given a distribution of neutron sources. If we postulate that all neutrons move at the same speed the neutron distribution is governed by the one-group neutron transport equation, Eq. (1):

---

*Work supported by the U.S. Energy Research and Development Administration.

**Argonne National Laboratory

(1)
$$\hat{\Omega} \cdot \nabla F(\underline{r},\hat{\Omega}) + \Sigma_t(\underline{r})F(\underline{r},\hat{\Omega}) = \int \Sigma_s(\hat{\Omega}' \to \hat{\Omega},\underline{r})F(\underline{r},\hat{\Omega}')d\hat{\Omega}' + S(\underline{r},\hat{\Omega}) \ ,$$
$$F(\underline{r},\hat{\Omega}) = vN(\underline{r},\hat{\Omega}) \ .$$

We have assumed, here, that each neutron moves in a straight line until it collides with one of the nuclei in its environment. The density of neutrons is always so small that collisions between neutrons can be ignored. As a result the neutron transport equations is linear. In Eq. (1) $\hat{\Omega}$ is a unit vector parallel to the neutron's velocity; v is the neutron's speed, and $N\,d\underline{r}\,d\hat{\Omega}$ is the number of neutrons, in the volume element $d\underline{r}$, with velocities in the solid angle $d\hat{\Omega}$. The function F is generally referred to as the "angular flux".

When a collision occurs many different processes can ensue. The neutron might simply be absorbed and disappear; it may be scattered, and exit from the collision with a new velocity: in some cases it will enter the nucleus and trigger a fission event in which other neutrons will be emitted. If one thinks of the neutrons as spheres, then the Σ's in Eq. (1) might be regarded as the total effective cross sections subtended by all the spheres in a unit volume. Since these are only effective cross sections, the nuclei will subtend different cross sections for different processes. Thus Σ_s, in Eq. (1), is the cross section for scattering from $\hat{\Omega}'$ to $\hat{\Omega}$ at point $\underline{r}$. Finally S is a neutron source density, considered here as a known function of $\underline{r}$ and $\hat{\Omega}$. Note that Eq. (1) is grossly oversimplified in that many processes have been neglected. In particular, fission has been neglected. If S is, in fact, the rate at which fission neutrons are born, per unit volume, then S will be proportional to the neutron density and the transport equation will become homogeneous. The neutron flux F is, then, the solution of an eigenvalue problem.

Of course neutrons do not all have the same energy. Neutrons born in fission have a broad spectrum of energies, and lose energy in each scattering collision. Thus the transport equation, in its most general form, is an integro-differential equation in six variables. The neutron position is characterized by three variables, and the velocity by three more.

Now, it is perfectly feasible to solve the general transport equation by Monte Carlo methods, and such methods play a very important role in reactor analysis. But Monte Carlo calculations are costly, and they do not provide us with the detailed information which the reactor designer and analyst needs. Therefore, Monte Carlo is not a wholly satisfactory substitute for deterministic methods. On the other hand, to solve the multienergy transport equation, in all its complexity, by deterministic methods alone is still prohibitively expensive. In practice, then, it is necessary to make some drastic approximations in the transport equation before it can be solved routinely by the nuclear designer. These approximations, which may all be invoked simultaneously or introduced separately, fall into three main classes. First, the cross sections, which are complicated and jagged functions of energy, are replaced by relatively smooth energy averages of some sort. This is a very important approximation, but one which involves a great deal of reactor physics which seems inappropriate here. I will assume, therefore, that the energy-averaging process has been carried out, somehow, and say no more about it.

Secondly, the cross sections, which may be very complicated functions of position, are often replaced by cross sections which are averaged over volumes whose dimensions vary considerably from case to case. I will discuss various aspects of this averaging process in some detail later.

Finally, the transport equation may be replaced by an equation which is much easier to solve, namely the diffusion equation. To bring out the relation between the transport and diffusion equations, I will derive the diffusion equation from the transport equation in a simple one-dimensional geometry.

Suppose a reactor is composed of plates oriented perpendicular to the x-axis. Imagine that the source and the cross sections depend only on x, and not on y or z. In this case the one-group transport equation takes the form

(2)

$$\mu \frac{\partial F(x,\mu)}{\partial x} + \Sigma_t(x)F(x,\mu) = \frac{1}{2}\int \Sigma_s(\hat{\Omega}' \to \hat{\Omega},x)F(x,\mu')d\mu' + \frac{1}{2}S(x) \ .$$

Here μ is the cosine of the angle between $\hat{\Omega}$ and the x axis. For simplicity I have postulated that the source is isotropic, i.e. that it is independent of $\hat{\Omega}$. Now we expand the flux, F, in Legendre polynomials, retaining only the first two polynomials $P_0(\mu)$ and $P_1(\mu)$, and integrate over μ. After some manipulation we get Eq. (3):

(3a) $$\frac{\partial j(x)}{\partial x} + \Sigma_a(x)\phi(x) = S(x) \ ,$$

(3b) $$\frac{1}{3}\frac{\partial \phi(x)}{\partial x} + \Sigma_1(x)j(x) = 0 \ ,$$

where

$$\phi(x) \equiv \int_{-1}^{1} F(x,\mu)d\mu \ ,$$

$$j(x) \equiv \int_{-1}^{1} F(x,\mu)\mu \ d\mu \ ,$$

$$\Sigma_{s0} \equiv \frac{1}{4\pi}\int \Sigma_s(\hat{\Omega} \to \hat{\Omega}',x)d\hat{\Omega}' \ ,$$

and

$$\Sigma_{s1} = \frac{1}{4\pi} \int \Sigma_s(\hat{\Omega} \rightarrow \hat{\Omega}', x)\hat{\Omega} \cdot \hat{\Omega}' \, d\hat{\Omega} \quad .$$

The quantities ϕ and j are generally referred to as the scalar flux and the current. From the definition of j it is easy to show that this quantity is, in fact, the rate at which neutrons cross a unit area whose surface is normal to the x axis. Elimination of j from Eqs. (3) leads us, finally to Eq. (4),

$$(4) \quad -\frac{\partial}{\partial x} D \frac{\partial \phi}{\partial x} + \Sigma_a \phi = S \ , \ D \equiv 1/3 \ \Sigma_1 \ , \ -D \frac{\partial \phi}{\partial x} = j \ ,$$

which is the neutron diffusion equation in slab geometry.

It has been assumed, implicitly, in our derivation that the diffusion coefficient, D, is a differentiable function of x. In practice, however, this will generally not be true. At interfaces between different materials the cross sections (and, consequently, the diffusion coefficient) will generally be discontinuous. It is necessary, therefore, that Eq. (4) be supplemented by appropriate auxiliary conditions at such interfaces. The conditions normally used can be obtained by the following argument. Suppose that the cross sections are not actually discontinuous, but that they vary rapidly over a thin boundary layer. One can easily show that, as the thickness of the layer goes to zero,

$$(5) \qquad D^- \frac{\partial \phi}{\partial x}\bigg|_{x_0-\varepsilon} = D^+ \frac{\partial \phi}{\partial x}\bigg|_{x_0+\varepsilon} \ , \ \phi\bigg|_{x_0-\varepsilon} = \phi\bigg|_{x_0+\varepsilon} \ .$$

In Eq. (5) D^+ and D^- are, respectively, the diffusion coefficients immediately to the left and right of the interface. Equation (5) is the interface condition normally used in neutron diffusion compuations.

A straightforward generalization of the procedure just described yields the neutron diffusion equation in three dimensions, Eq. (6):

$$-\nabla \cdot D\nabla\phi + \Sigma_a\phi = S \quad , \quad -D\nabla\phi = \underline{J} \quad ,$$

$$\hat{n}_0 \cdot D^-\nabla\phi|_- = \hat{n}_0 \cdot D^+\nabla\phi|_+ \quad , \quad \phi_- = \phi_+ \quad . \tag{6}$$

In Eq. (6) $\hat{n}_0$ is a unit vector normal to the interface: $\underline{J}$ is, again, the neutron current, in the sense that $\underline{J} \cdot \hat{n}$ is the rate at which neutrons cross a unit area whose unit normal vector is $\hat{n}$.

Since the diffusion equation is a good deal easier to solve than the neutron transport equation it is often used in place of the transport equation in situations where the implied approximations seem to be valid. When the diffusion equation is used, quite frequently a two-step approximation procedure is involved. In one step the transport equation is replaced by the diffusion equation: in another step (generally referred to as "homogenization") the complicated position-dependent reactor parameters are replaced by spatially smooth averages. It is important to note that these two steps do not necessarily commute: when they don't, the spatial averaging must be carried out before the diffusion approximation is introduced. Otherwise the diffusion approximation may obliterate important transport effects which are due to heterogeneities in the original problem configuration.

The heterogeneity effects I will deal with here fall into two broad classes. First, cross-section discontinuities induce singularities in the solution of the transport equation, as in the solution of the diffusion equation. Singularities in the solution of the diffusion equation have been studied intensively for some time, and a good deal is now known about them. Singularities in the solution of the transport equation are much more complicated, and not nearly as well understood. It must be expected that the presence of singularities will degrade the accuracy of numerical approximation methods and complicate the formulation of higher-order difference equations designed for use in transport computations.

Secondly, heterogeneity effects which are large enough to influence reactor analysis and design must, somehow, be taken into account when the cross sections are homogenized. Since the detailed structure of the reactor cannot be represented explicitly in most reactor computations, all significant heterogeneity effects must be incorporated into a simplified, homogenized, model reactor, a model which is computationally tractable, yet realistic enough to be useful.

2. Singularities in the Solution of the Transport Equation

The only independent variable which appears in the diffusion equation is the scalar flux. Singularities in the solution of the diffusion equation are, by definition, singularities in the scalar flux and its derivatives. On the other hand, the character of singularities in the solution of the transport equation is often most easily understood by examining the properties of the angular flux. Consider, for example, the behavior of the angular flux in the neighborhood of a vacuum boundary. If there are no neutron sources in the vacuum, and if the boundary of the diffusing medium is convex, no neutrons will enter the medium from the vacuum. But neutrons may leak from the medium into the vacuum. It is easy to show that, if $\hat{n}$ is normal to the surface of the diffusing medium the angular flux will generally be discontinuous, as a function of $\hat{\Omega}$, at $\hat{\Omega} \cdot \hat{n} = 0$.

The effect of this discontinuity on the scalar flux can be treated analytically in a simple model problem usually called the "Milne problem".[1] The Milne problem configuration is sketched in Fig. 1. In the simplest form of the Milne problem neutrons are assumed to be monoenergetic, and the medium in which they move is a pure scatterer. That is to say that all collisions are scattering collisions: no neutrons are absorbed. An infinite plane source of neutrons is located at an infinite distance to the left of the planar boundary. It can be shown that as one moves to the left the scalar flux becomes, asymptotically, a linear function of x. Near the boundary there is, however, a small zone in which the flux is not linear. In fact the scalar flux drops rapidly near the boundary, and its derivatives at the boundary are all infinite. It is to be expected that, unless special steps are taken to deal with this singularity, its presence will slow the convergence of any finite element or difference approximation. The nature of this singularity is discussed in a paper by Abu-Shumays and Bareiss,[2] who have developed a finite element approximation which incorporates the singularity into one of the basis functions. Numerical experiments show that the inclusion of such a basis function improves the performance of the finite element method, in this case, quite effectively.

But it should be noted that from a practical point of view, the Milne problem is not very important, not, at least, in reactor computations. And, as one might expect, practical problems are considerably more complicated. There is, however, a problem of some practical interest which is similar to the Milne problem. Suppose an infinite absorbing slab is embedded in an infinite medium, as in Fig. 2. In the slab, $\Sigma_s = 0$, i.e. all neutrons which collide in the slab are absorbed. You see that there are neutron sources in Regions 1 and 3 on either side of the slab, but not in the slab itself. Now consider the fluxes, at point P, in the two indicated directions. Most of the neutrons with velocities parallel to $\hat{\Omega}_1$ have travelled a long distance through the absorbing slab. The angular flux $F(\hat{\Omega}_1)$ will therefore be small. On the other hand, neutrons moving along $\hat{\Omega}_2$ have not passed through the slab, and therefore $F(\hat{\Omega}_2)$ may be relatively large. As in the Milne problem the angular flux will, generally, be discontinuous at $\hat{\Omega} \cdot \hat{n} = 0$.

As soon as we turn our attention from slab geometry to other geometries the situation becomes still more complicated. It seems pointless, here, to try to enumerate all the singularities which might be encountered in transport computations. Instead I will consider, as an illustrative example, the singular behavior of the angular flux at a corner point. It is easy to show that, in a simple model problem, the angular flux cannot have well-defined spatial derivatives at corners.

In Fig. 3 you see an infinite, purely absorbing, medium with constant cross sections, divided into two regions. If Ω_x and Ω_y are positive the angular flux, $F(\hat{\Omega},\underline{r})$, is continuous across the x and y interfaces. It follows that $\partial F/\partial x$ is continuous across the horizontal interface while $\partial F/\partial y$ is continuous across the vertical interface. At P_1 the flux satisfies the transport equation

(7) $$\Omega_x \frac{\partial F}{\partial x} + \Omega_y \frac{\partial F}{\partial y} + \Sigma_T F = \frac{1}{4\pi} \quad ,$$

while at P_2

(8) $$\Omega_x \frac{\partial F}{\partial x} + \Omega_y \frac{\partial F}{\partial y} + \Sigma_T F = 0 \quad .$$

Now let P_1 approach P_2 . If the partial derivatives exist at the corner point then, because of their continuity properties, it must be true that

(9) $$\left.\frac{\partial F}{\partial x}\right|_{P_1} \rightarrow \left.\frac{\partial F}{\partial x}\right|_{P_2} \; , \; \left.\frac{\partial F}{\partial y}\right|_{P_1} \rightarrow \left.\frac{\partial F}{\partial y}\right|_{P_2} \quad .$$

But it is impossible for the flux to satisfy Eqs. (7), (8), and (9) simultaneously. It is clear, then, that the spatial derivatives of the angular flux at the corner must be undefined in at least one of the regions.

Figure 4 represents a more complicated configuration, where four regions meet at a point. The dotted line in Fig. 4 is drawn from the corner point, in a direction parallel to some arbitary unti vector, $\hat{\Omega}$. Here $\hat{n}$ is another unit vector, perpendicular to $\hat{\Omega}$. It can be shown that, generally, the directional derivative $\hat{n} \cdot \nabla F(\underline{r},\hat{\Omega})$ is discontinuous at all points on the dotted line. This pathological behavior of the angular flux was first noted by Aruszewski, Kulokowski and Mika,[3] who refer to the dotted line as a "singular characteristic". Of course each direction, $\hat{\Omega}$, defines its own singular characteristic, along which the normal derivative of $F(\hat{\Omega})$ is discontinuous. The effects of corner points on the derivatives of the angular flux are closely examined in a recent paper by Kellogg.[4]

Usually, in practice, the neutron transport equation is treated by what is called the "S_N method". Aruszewski, Kulokowski and Mika have shown[3] that the indicated singularities in the angular flux have a deleterious effect on the accuracy of the difference equations normally used in the S_N method. These authors have proposed an alternative differencing scheme which takes into account the presence of the singular characteristics, and their test results using this scheme are encouraging. On the other hand their difference equations are rather complicated and have never been tested in any full-scale, production-oriented, transport code.

Apparently the discontinuities along the singular characteristics also have a noticeable deleterious effect in finite element calculations. Reed, at Los Alamos Scientific Laboratory, has studied the accuracy of finite element methods as applied, in various forms, to the transport equation.[5] He finds that in a series of simple test problems the errors in scalar fluxes seem to be $O(h)$, whatever the order of his basis polynomials. Apparently theoretical results on the accuracy of the finite element method near corners (in transport computations) are totally lacking.

3. Homogenization

So far I have simply assumed that the cross sections used in the flux computation are given functions of position, and have said nothing about their physical significance. In fact, these cross sections may be local cross sections in individual fuel pins, plates, or other reactor constituents, or they may be cross sections which have been averaged, somehow, over many different constituents. Individual reactor components are often represented explicitly in computations covering small subregions of a reactor, but it is generally impossible to incorporate fine details of the reactor's structure in computations which treat the reactor as a whole. In computations involving the whole reactor the cross sections and diffusion coefficients are usually average parameters which, in some sense, embody properties of all the materials in a fairly large, heterogeneous, subregions of the reactor. Many different processes have been used to obtain such homogenized parameters, and none are rigorously justifiable. All rely, to some extent, on engineering judgment. For the sake of simplicity I will not attempt to describe any homogenization procedure in full detail. I will, however, describe those features of a typical homogenization process which seem to me to be most interesting and important. Figure 5 represents schematically the core of a hypothetical reactor, viewed from above. Each small square in the core is a subassembly which might contain a fuel pin, as in A, or fuel plates as in B. The core is, of course, finite in height, and surrounded by a blanket or reflector region which is not shown. In this grossly oversimplified reactor each subassembly would probably be homogenized, so that the core would be represented as a uniform, homogeneous medium, surrounded by a homogeneous reflector or blanket.

The homogenization procedure would take a particularly simple form if the core were infinite in all directions so that the homogenized reactor would consist, simply, of an infinite homogeneous medium. I think it is instructive to examine a homogenization procedure which might be used in such a case.

Suppose, for example, that the subassembly is made up of plates, as in Fig. 5B. Suppose, further, that the whole reactor is critical, i.e. that the number of neutrons produced, per second, by fission is exactly equal to the number captured. For simplicity we again assume that all neutrons have the same energy. In this case the neutron transport equation takes the form

$$(10) \qquad \hat{\Omega} \cdot \nabla F + \Sigma_t F = \frac{1}{4\pi} \Sigma_s \phi + \frac{1}{4\pi} \nu \Sigma_f \phi \quad , \quad \phi \equiv \int F \, d\hat{\Omega} \quad .$$

In Eq. (10) I have taken the scattering to be isotropic, i.e. I have assumed that the probability that a neutron will be scattered into the solid angle $d\hat{\Omega}$ is equal to $\frac{1}{4\pi} d\hat{\Omega}$: Σ_s is the scattering cross section, Σ_f is the fission cross section and ν is the average number of neutrons produced in a single fission event.

An operating reactor is held critical through the action of control rods and other control mechanisms. On the other hand we cannot expect that the model reactor represented in our computations will also be critical. Therefore, it is necessary to introduce into our computations a parameter which artifically maintains criticality, and appears in the transport equations as an eigenvalue. For computational purposes, then, we modify the transport equation as in Eq. (11):

$$\hat{\Omega} \cdot \nabla F + \Sigma_t F = \frac{1}{4\pi} \Sigma_s \phi + \frac{1}{4\pi\lambda} \nu\Sigma_f \phi \; . \tag{11}$$

Equation (11), with its boundary conditions, determines a set of permissible values for λ . If our model is reasonably realistic, the maximum λ will be close to one (perhaps within one or two percent of one) and the corresponding flux, F , will be reasonably close to the true flux in the reactor.

Suppose that at some point on a subassembly boundary, $\hat{n}$ is a unit vector normal to the boundary. For every vector, $\hat{\Omega}$, at the subassembly boundary one can define a mirror image vector, $\hat{\Omega}'$, as in Eq. (12):

$$\hat{\Omega}' = \hat{\Omega} - 2\hat{n}(\hat{\Omega} \cdot \hat{n}) \; . \tag{12}$$

Because of the symmetry properties of our hypothetical subassembly (properties often found in real subassemblies), one can show that, at each boundary point, $\underline{r}_B$, the angular flux must satisfy Eq. (13):

$$F(\hat{\Omega}, \underline{r}_B) = F(\hat{\Omega}', \underline{r}_B) \; . \tag{13}$$

Given this condition it is not necessary to solve the transport equation over the whole reactor configuration. Equation (13) can be regarded as a boundary condition imposed on the flux in a single subassembly (or "cell"), and Eq. (11) can be solved in a single cell with this boundary condition.

If we integrate Eq. (11) over a single cell we get Eq. (14):

$$(14) \qquad \bar{\Sigma}_a\bar{\phi} = \frac{\overline{\nu\Sigma_f}}{\lambda}\bar{\phi} \quad , \quad \bar{\Sigma}_a = \bar{\Sigma}_t - \bar{\Sigma}_s \quad ,$$

$$\bar{\phi} \equiv \left(1/V_{cell}\right)\int_{cell} \phi \, dV \quad ,$$

$$\bar{\Sigma}_t = \left(1/\bar{\phi}V_{cell}\right)\int_{cell} \Sigma_t\phi \, dV \quad ,$$

$$\bar{\Sigma}_s = \left(1/\bar{\phi}V_{cell}\right)\int_{cell} \Sigma_s\phi \, dV \quad ,$$

$$\overline{\nu\Sigma_f} = \left(1/\bar{\phi}V_{cell}\right)\int \nu\Sigma_f\phi \, dV \quad .$$

It will be seen that Eq. (14) is simply the neutron transport equation for an infinite homogeneous medium with absorption cross section $\bar{\Sigma}_a$, and fission production cross section $\overline{\nu\Sigma_f}$. Thus $\overline{\Sigma_a}$ and $\overline{\nu\Sigma_f}$ are uniform cross sections for an infinite homogeneous medium which has the same eigenvalue as the original infinite lattice. If the core is very large (though <u>not</u> infinite) one might expect that the same $\overline{\Sigma_a}$ and $\overline{\nu\Sigma_f}$ could still be used as homogenized core cross sections. But the homogenization prescription defined by Eqs. (14) has a serious, and perhaps obvious, defect. In a real reactor, neutrons will leak from the core into the adjoining reflector or blanket. You will note, first, that we have not taken this leakage into account in any way in our very primitive homogenization procedure. Secondly, in any diffusion computation of the reactor eigenvalue, and of flux shapes in the reactor, we will need diffusion coefficients for the core, blanket, and reflector. I have, so far, suggested no procedure for the computation of such diffusion coefficients. In fact, the problems involved in computing homogenized diffusion coefficients are among the most difficult problems of reactor physics.

When one examines the prescriptions which I suggested earlier for the homogenization of absorption and fission cross sections, two ad hoc prescriptions for the computation of homogenized diffusion coefficients come to mind, both equally plausible. Recall that in a homogeneous medium, the diffusion coefficient is given by Eq. (15):

(15) $$D \equiv 1/3\,\Sigma_1 \quad , \quad \Sigma_1 \equiv \Sigma_t - \Sigma_{s1} \quad ,$$

$$\Sigma_{s1} \equiv \frac{1}{4\pi} \int \Sigma_s(\hat{\Omega} \rightarrow \hat{\Omega}')\hat{\Omega} \cdot \hat{\Omega}'\, d\hat{\Omega}' \; .$$

One might think it reasonable to define the homogenized diffusion coefficient as a flux-weighted average of the local diffusion coefficients as in Eq. (16):

(16) $$\bar{D} = (1/\bar{\phi}V_{cell}) \int_{cell} D\phi \; dV \quad .$$

On the other hand it may seem just as reasonable to compute a homogenized diffusion coefficient from flux-weighted average cross sections, as in Eq. (17):

(17) $$\bar{D} = (1/3\,\bar{\Sigma}_1) \; , \; \bar{\Sigma}_1 = (1/\bar{\phi}V_{cell}) \int_{cell} \Sigma_1\phi \; .$$

As a matter of fact both definitions have been used in practice, as well as other more complicated ad hoc prescriptions. If Σ_1 does not change radically from region to region all the various ad hoc prescriptions tend to give much the same homogenized diffusion coefficients. Further, if the leakage rate from the core is small (as it is in most large power reactors) small differences in $\bar{D}$ have little effect on computed eigenvalues or flux shapes. Unfortunately there are situations in which Σ_1 varies sharply as a function of position, and in which leakage from the core is quite important. This is true, for example, in a gas-cooled fast reactor. In such a reactor the coolant, which is generally helium, flows through many channels which are otherwise empty, and are interspersed throughout the system. These channels constitute excellent escape routes through which neutrons stream out of the core. Heterogeneities are not normally so important in the LMFBR, the liquid metal fast breeder reactor, which is cooled by liquid sodium. However, in the analysis of hypothetical accidents it is necessary to deal with situations in which some or all of the sodium has actually boiled out of the core. If this were to occur we would again be left with nearly empty coolant channels, severe heterogeneities, and important leakage effects. In such situations ad hoc prescriptions for homogenized diffusion coefficients are not always adequate, and more sophisticated homogenization techniques may be required. Before we can discuss such techniques it will be necessary to introduce the "buckling" concept, a fundamental concept of reactor physics. For the sake of simplicity, I will continue to confine my attention, here, to a one-group model reactor, though everything I have to say applies, as well, to a multi-energy reactor.

It is generally true that, near the center of a large homogeneous reactor core, the scalar flux (approximately) satisfies the Helmholz equation, $\nabla^2\phi = -B^2\phi$, for a value of B determined only by the core cross sections.* The constant B^2 is usually referred to as the "buckling". Any real solution of the Helmholz equation may be written in the form of an integral over the unit sphere, as in Eq. (18):

---

*This "folk theorem" plays a peculiar role in reactor physics. It is assumed to be true, is often observed to be true in reactor experiments, and has been verified in innumerable transport and diffusion computations. Yet this "fundamental theorem of reactor physics" is discussed very little in the reactor physics literature. For more insight into the physics underlying the buckling concept see, for example, Ref. 6.

(18) $$\phi(\underline{r}) = \int d\hat{b}\ g(\hat{b})\ e^{iB\hat{b}\cdot\underline{r}} \quad .$$

Correspondingly one can Fourier analyze the angular fluxes as in Eq. (19):

(19) $$F(\underline{r},\hat{\Omega}) = \int d\hat{b}\ f(\hat{\Omega},\hat{b})\ e^{iB\hat{b}\cdot\underline{r}} \quad ,$$

Here $\hat{b}$ is a unit vector, while f and g are arbitrary functions subject only to the constraints $g(-\hat{b}) = g^*(\hat{b})$, $f(\hat{\Omega},-\hat{b}) = f^*(\hat{\Omega},\hat{b})$. If the reactor core is roughly rectangular the fluxes may be expected to take on the particularly simple form

(20) $$\phi(\underline{r}) = c_1 \cos(\underline{B} \cdot \underline{r} + c_2) = R\{g\ e^{i\underline{B}\cdot\underline{r}}\} \quad ,$$

(21) $$F(\underline{r},\hat{\Omega}) = R\{f(\hat{\Omega})\ e^{i\underline{B}\cdot\underline{r}}\} \quad ,$$

where f and g are generally complex.

Almost invariably derivations of homogenized diffusion coefficients start from a generalization of Eqs. (20) and (21). Most methods for computing homogenized diffusion coefficients can be regarded as variants of the Benoist method,[7] so I will discuss this method first.

Suppose that an infinite lattice of some sort contains a distributed neutron source whose density has the following separable form,

(22) $$S(\underline{r},\hat{\Omega}) = s(\underline{r}) \cos(\underline{B} \cdot \underline{r}) = R\{s(r)\ e^{i\underline{B}\cdot\underline{r}}\} \quad ,$$

where the (real) function, $s(r)$ has the periodicity of the lattice. The flux in this lattice is governed by the transport equation, Eq. (23):

(23) $$\hat{\Omega} \cdot \nabla F + \Sigma_t F = \frac{1}{4\pi} \Sigma_s \phi + \frac{1}{4\pi} S \ , \ \phi = \int F\ d\hat{\Omega} \quad .$$

To solve Eq. (23) we make the substitution shown in Eq. (24):

$$F = R\{f(\underline{r},\hat{\Omega})\, e^{i\underline{B}\cdot\underline{r}}\} \,, \quad f \equiv R + iI \tag{24}$$

Here R and I are the real and imaginary parts of the complex function, f . It is easy to show that R and I satisfy Eqs. (25) and (26):

$$\Omega \cdot \nabla R + \Sigma_t R = \frac{1}{4\pi}\Sigma_s \psi + \hat{\Omega} \cdot \underline{B}\, I + \frac{1}{4\pi} s \,, \quad \psi \equiv \int R\, d\hat{\Omega} \,, \tag{25}$$

$$\Omega \cdot \nabla I + \Sigma_t I = \frac{1}{4\pi}\Sigma_s \chi - \hat{\Omega} \cdot \underline{B}\, R \,, \qquad \chi \equiv \int I\, d\hat{\Omega} \,. \tag{26}$$

Now we will assume that R and I can be expanded in Taylor series in the components of the vector $\underline{B}$. It should be noted that this is by no means an innocuous assumption. There are very important situations in which such an expansion is not possible,[7,8] so that alternative methods of attack are required.[8,9] But if a series expansion is possible, then, to leading order in B, Eqs. (25) and (26) are equivalent to Eqs. (27) and (28):

$$\Omega \cdot \nabla R + \Sigma_t R = \frac{1}{4\pi}\Sigma_s \psi + \frac{1}{4\pi} s \,, \tag{27}$$

$$\hat{\Omega} \,.\, \tilde{I} + \Sigma_t \tilde{I} = \frac{1}{4\pi}\Sigma_s \tilde{\chi} - \hat{\Omega} \cdot \hat{b}R \,, \; I = B\,\tilde{I} \,, \; \tilde{\chi} = B\tilde{\chi} \,. \tag{28}$$

It will be seen that Eqs. (27) and (28) are formally identical with the conventional transport equation. Further R and I have the periodicity of the lattice, so that they satisfy simple boundary conditions on each cell boundary. Therefore it is possible to solve these equations via transport computations which are confined to a single cell, and need not extend over the entire lattice.

Now suppose, for simplicity, that the lattice cells are rectangular, and that their bounding surfaces are parallel to the coordinate planes. Let L_x be the total number of neutrons leaking out of a cell, per second, across the two boundaries normal to the x axis. It can be shown that, to leading order in B, L_x is given by Eq. (29):

$$L_x = \cos(\underline{B} \cdot \underline{r}_0) B_x^2 \left[\int j_{xx} \, dV + \int_{\text{cell}} (x - x_0) \frac{\partial}{\partial x} j_{xx} \, dV \right], \tag{29}$$

$$j_{xx} \equiv - \int \Omega_x \tilde{I} \, d\hat{\Omega} ,$$

where $\underline{r}_0$ lies at the center of the cell.

Now suppose we are to replace the lattice by a homogeneous medium. Suppose that, in this medium, the scalar flux has the simple form

$$\phi(r) = c \cos \underline{B} \cdot \underline{r} . \tag{30}$$

Imagine a rectangular region, in this infinite medium, which is coextensive with one of the cells of the lattice. Suppose we want the average flux in this region to be equal to the average flux in the cell, and the leakage, L_x, from the region to be equal to the leakage, L_x from the cell. Further the leakage rates in the homogeneous medium are to be computed in a <u>diffusion</u> approximation. Now, in diffusion theory, the leakage rate in the x direction, in a homogeneous medium, is given by Eq. (31)

$$\frac{\partial J_x}{\partial x} = - \frac{\partial}{\partial x} D \frac{\partial \phi}{\partial x} = -D \frac{\partial^2 \phi}{\partial x^2} , \tag{31}$$

and the leakage out of the rectangular region is given by Eq. (32):

$$(32) \qquad L_{x(HOM)} = -\int_R D \frac{\partial^2 \phi_{(HOM)}}{\partial x^2} dV \quad .$$

Here R is the volume of the rectangular region coextensive with the given cell. Note that the flux and leakage rates, here, are fluxes and leakage rates in the homogeneous medium. Since we have assumed that the flux is separable we may write

$$(33) \qquad L_{x(HOM)} = D\, B^2 V_{cell}\, \bar{\phi}_{(HOM)} = D\, B^2 V_{cell}\, \bar{\phi} \quad ,$$

where $\bar{\phi}$, on the extreme right-hand side of Eq. (33) is now the average flux in the lattice cell. Equating the leakage rates in the lattice and in the homogeneous medium we are led to define the homogenized diffusion coefficient, to leading order in B, as in Eq. (34):

$$(34) \quad D_x = \left[\int_{cell} j_{xx} dV + \int_{cell} (x-x_0) \frac{\partial}{\partial x} j_{xx} dV\right] \Big/ \int_{cell} \psi \, dV \quad ,$$

$$\psi \equiv \int d\hat{\Omega}\, R \quad .$$

Thus the x diffusion coefficient is now determined entirely by the two coupled transport equations for R and I.

It should be noted, however, that the Benoist method gives us different diffusion coefficients in different directions. If, for example, a reactor contains empty channels parallel to the z-axis we must expect that neutrons will stream preferentially in the z-direction, so that D_z will be greater than D_x or D_y. When the diffusion coefficients are direction-dependent (or "anisotropic") the diffusion equation takes the form shown in Eq. (35):

(35) $$-\frac{\partial}{\partial x} D_x \frac{\partial \phi}{\partial x} - \frac{\partial}{\partial y} D_y \frac{\partial \phi}{\partial y} - \frac{\partial}{\partial z} D_z \frac{\partial \phi}{\partial z} + \Sigma_a \phi = S \quad .$$

Average absorption cross sections and fission production rates may still be defined as flux-weighted averages, just as in a infinite lattice with $\underline{B} = 0$, although slightly more refined procedures are often used.

As you may recall, I have already pointed out the homogenization process and the introduction of the diffusion approximation are operations which do not necessarily compute. You will note that the Benoist method involves the solution of the transport equation in the heterogeneous lattice cell. The homogenized diffusion coefficients, which are designed for use in the diffusion equation, are derived from solutions of the transport equation. It is easy to think of situations where the diffusion approximation must not be introduced until the homogenization process has been completed. Suppose, again, that empty channels run through the core in the z-direction. In these channels the cross sections vanish, so that the local diffusion coefficient is infinite. Now, in a diffusion approximation, the z-leakage rate at each point would be given by the expression $L_z(r) = DB_z^2\phi(\underline{r})$, so that the total z-leakage rate from the cell would be infinite. On the other hand, the Benoist method, based on a transport computation in the lattice, will give a finite leakage rate except in those pathological cases where the Benoist Taylor series expansion is invalid. Many other peculiar anomalies occur when the diffusion approximation is applied directly to heterogeneous lattice cells. On the other hand, the homogenized lattice can often be treated by diffusion theory, even when the heterogeneous lattice cannot.

So far I have discussed homogenization only in an infinite homogeneous medium. In a multiregion reactor homogenized diffusion coefficients and cross sections would be computed separately over several large subregions in which the lattice structure is fairly uniform. The resulting average parameters would then be used in a multiregion diffusion calculation in which the reactor would be represented as a collection of large homogeneous regions. Such a computational procedure is certainly not beyond reproach. After all, if a well-defined buckling exists anythere in a reactor it exists only near the center of a large core. Yet the homogenized diffusion coefficient must be used everywhere in the reactor, even near the boundary of the core where, strictly speaking, there is no buckling. Even if one assumes that a well-defined buckling exists, some interesting objections to the Benoist method remain.[10] In discussing the Benoist method I referred repeatedly to a unit cell. But in an infinite lattice there is no uniquely defined unit cell. The same lattice can generally be constructed from many different unit cells. Unfortunately different definitions of the unit cell imply different values of the Benoist diffusion coefficients. Given the usual prescriptions for homogenizing absorption and fission cross sections, it turns out that the different diffusion coefficients lead to slightly different lattice eigenvalues for any specified buckling. Under such circumstances it seems most important that the Benoist method (as well as other similar homogenization prescriptions) be checked extensively against experiment, and against Monte Carlo computations. There is some experimental evidence that the Benoist method is adequate for use in nuclear design and analysis[7,11] but in my opinion, the range of validity of the method has still not been thoroughly investigated.

Now, before closing, I want to turn briefly to a method which is related to Benoist's, but look totally different. In earlier sections of this paper I have written the transport equation for a reactor in differential form, as in Eq. (36):

(36) $$\hat{\Omega} \cdot \nabla F + \Sigma_t F = \frac{1}{4\pi} \Sigma_s \phi + \frac{1}{4\pi\lambda} \nu \Sigma_f \phi \quad .$$

The same equation can also be written in integral form as in Eq. (37):

(37) $$\lambda S_f(\underline{r}) = \int K(\underline{r}' \rightarrow \underline{r}) S_f(\underline{r}') d\underline{r}' \quad .$$

In Eq. (37) $S_f(\underline{r})$ is the fission source density at $\underline{r}$, i.e. $S_f(\underline{r})$ is equal to $\nu\Sigma_f(\underline{r})\phi(\underline{r})$. Further, the kernel $K(\underline{r}' \rightarrow \underline{r})$ is a Green's function which has the following significance. Suppose there is an isotropic δ function source of fission neutrons at $\underline{r}'$. Each of the fission neutrons born at $\underline{r}'$ will diffuse through the reactor until it is absorbed. At the point where it is absorbed it may trigger a fission, producing offspring which I will call "second-generation neutrons": $K(\underline{r}' \rightarrow \underline{r})$ is the birth rate of second-generation neutrons produced at $\underline{r}$ by the δ-function source at $\underline{r}'$. Of course second-generation neutrons will produce third-generation neutrons, and so on, but only the production rate of second-generation neutrons is to be included in the kernel, K .

Now suppose that

(38) $$S_f(\underline{r}) = s(\underline{r})\, e^{i\underline{B}\cdot\underline{r}} \quad ,$$

and that $s(\underline{r})$ has the periodicity of the lattice. Substituting from Eq. (38) into Eq. (37), we get Eq. (39):

(39) $$\lambda s(\underline{r}) = \int K(\underline{r}' \rightarrow \underline{r})\, e^{i\underline{B}\cdot(\underline{r}'-\underline{r})} s(\underline{r}') d\underline{r}' \quad .$$

If, in Eq. (39), we set $\underline{B} = 0$ we get Eq. (40):

(40) $$\lambda_0 s_0(\underline{r}) = \int K(\underline{r}' \rightarrow \underline{r}) s_0(\underline{r}') d\underline{r}' \quad .$$

Now Eqs. (39) and (40) are integral equations with slightly different kernels and, as $B \to 0$, we ought to be able to estimate the eigenvalue, λ, by perturbation theory. Perturbation theory leads us, in fact, to Eq. (41):

(41)

$$\lambda - \lambda_0 \equiv \Delta = -(1/S_T)\int d\underline{r}\ s_0^*(\underline{r})\int K(\underline{r}' \to \underline{r})[1-e^{iB\cdot(r'-r)}]s_0(\underline{r}')d\underline{r}',$$

$$S_T = \int d\underline{r}\ s_0^*(\underline{r})s_0(\underline{r}) \quad .$$

Here $s_0^*(\underline{r})$ is the adjoint fission source which is the solution of the adjoint integral equation, Eq. (42):

(42) $$\lambda_0 s_0^*(\underline{r}) = \int K(\underline{r} \to \underline{r}')s_0^*(\underline{r}')d\underline{r}' \quad .$$

It is now easy to show that the fractional change in λ is given by Eq. (43):

(43) $$(\Delta/\lambda_0) = \frac{\int d\underline{r}\ s_0^*(\underline{r}) \int K(\underline{r}' \to \underline{r})[1-e^{iB\cdot(\underline{r}'-r)}]s_0(\underline{r}')dr'}{\int d\underline{r}\ s_0^*(\underline{r}) \int K(\underline{r}' \to \underline{r})s_0(\underline{r}')d\underline{r}'} \quad .$$

At this point I am going to assume that the lattice cell has three planes of symmetry, each symmetry plane being normal to the x, y, or z axis. I will assume, further, that the numerator on the rhs of Eq. (43) can be expanded in a Taylor series in B. It turns out, then, that the perturbed eigenvalue is given, to leading order in B, by Eq. (44):

(44) $$(\Delta/\lambda_0) = -\frac{1}{2}\left(B_x^2\overline{\ell_x^2} + B_y^2\overline{\ell_y^2} + B_z^2\overline{\ell_z^2}\right),$$

$$\overline{\ell^2_{x,y,z}} = \frac{\int d\underline{r}\ s_0^*(\underline{r}) \int K(\underline{r}' \to \underline{r})(r'_{x,y,z}-r_{x,y,z})^2 s_0(\underline{r}')d\underline{r}'}{\int d\underline{r}\ s_0^*(\underline{r}) \int K(\underline{r}' \to \underline{r})s_0(\underline{r}')d\underline{r}'} \quad .$$

The quantities $\overline{\ell_x^2}$, $\overline{\ell_y^2}$, and $\overline{\ell_z^2}$ are mean-square distances from birth to fission, weighted by the value of the adjoint source at the point where fission occurs.

Equation (44) is interested for three reasons. First, it establishes a conceptually simple relation between the leakage rate, for a given $\underline{B}$, and the geometry of the lattice. Secondly, it is convenient for Monte Carlo computations since estimates of the mean-square distances in Eq. (44) yield a direct estimate of the leakage rate as a function of buckling. I know of no other simple way to impose a specified buckling in a Monte Carlo eigenvalue calculation. Finally, Eq. (44) clarifies the relation between leakage rates and mean-square path lengths. Some such relation seems to have been assumed in important early work by Behrens,[12] but never stated previously. Theoretical clarification of Behrens' work still seems to be of interest, since his method is still useful. It should be noted that Eq. (44) generalizes to lattices a relation long known to be valid for infinite homogeneous media.

I think it is natural to ask what can be deduced via perturbation theory if the cell does not have the symmetry properties I have postulated. You see that the right-hand side of Eq. (41) will generally be complex and, if the cell is not symmetric, one cannot show that Δ will be real. Moreover, one can construct an asymmetric cell such that if a core no matter how large is composed of such cells, a well-defined buckling will not exist anywhere in the core. Now all homogenization methods, so far as I know, rely on the existence of a buckling. Thus it is not at all certain, in principle, that any method available today can be applied to lattices of non-symmetric cells.

It seems very clear that the homogenization problems of reactor physics still urgently need a good deal of attention.

Perhaps there are, in the interdisciplinary group attending this conference, specialists from other disciplines who can bring, to these reactor physics problems, a fresh point of view and a new approach.

REFERENCES

1. K. M. Case and P. F. Zweifel, "Linear Transport Theory," Addison-Wesley (1967).
2. I. I. Abu-Shumays and E. H. Bareiss, "Adjoining Appropriate Singular Elements to Transport Theory Computations," J. Math. Analy. Applic., 48, 200 (1974).
3. J. Arkuszewski, T. Kulikowska and T. Mika, "Effect of Singularities on Approximation in S_N Methods," Nucl. Sci. Eng., 49, 20 (1972).
4. R. B. Kellogg, "First Derivatives of Solution of the Plane Neutron Transport Equation," Technical Note BN-783, Institute for Fluid Dynamics and Applied Mathematics, Univ. Maryland (1974).
5. W. H. Reed and T. R. Hill, "Triangular Mesh Methods for the Neutron Transport Equation," Proc. Conf. Mathematical Models and Computational Techniques for Analysis of Nuclear Systems, USAEC-CONF-730414-P1, Vol. I, p. 10 (1973).
6. A. M. Weinberg and E. P. Wigner, "The Physical Theory of Neutron Chain Reactors," University of Chicago Press (1958).
7. P. Benoist, "Théorie du Coefficient de Diffusion des Neutrons dans un Réseau Comportant des Cavités," CEA-R-2278, Centre d'Etudes Nucleaires de Saclay (1964).
8. P. Köhler and J. Ligou, "Axial Neutron Streaming in Gas-Cooled Fast Reactors," Nucl. Sci. Eng., 54, 357 (1974).
9. E. M. Gelbard and R. Lell, "Role of the Mean Chord Length in Lattice Reactivity Calculations," Trans. Am. Nucl. Soc., 20, (1975).
10. E. M. Gelbard, "Anisotropic Neutron Diffusion in Lattices of the Zero-Power Plutonium Reactor Experiments," Nucl. Sci. Eng., 54, 327 (1974).
11. E. Fischer et al, "An Investigation of the Heterogeneity Effect in Solium Void Reactivity Measurements," Intern. Symp. Physics of Fast Reactors, Tokyo, October 16-19, 1973, Committee for the Intern. Symp. Physics of Fast Reactors (1973), Vol. II, p. 945.
12. D. J. Behrens, "The Effect of Holes in a Reacting Material on the Passage of Neutrons," J. Phys. Soc. London, A, 62, 607 (1949).

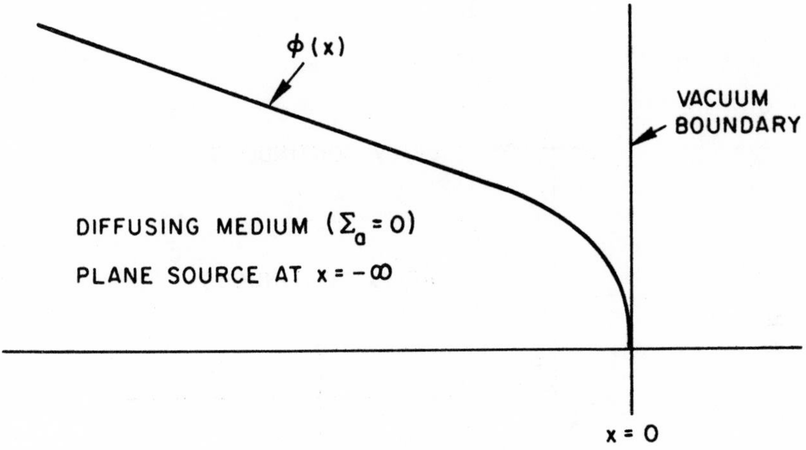

Fig. 1 Simple Milne problem configuration.

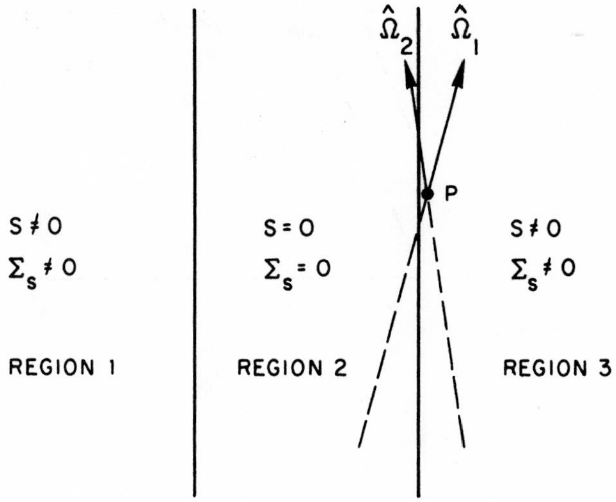

Fig. 2 Source-free absorbing slab imbedded in a scattering medium.

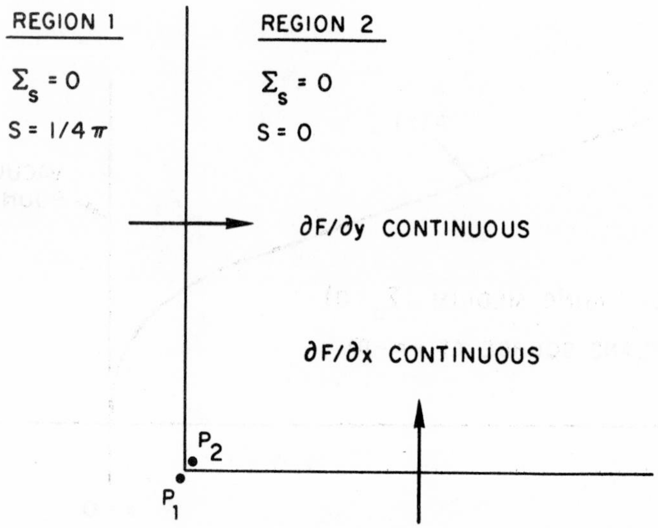

Fig. 3 Infinite absorbing medium with a source in Region 1 and no source in Region 2.

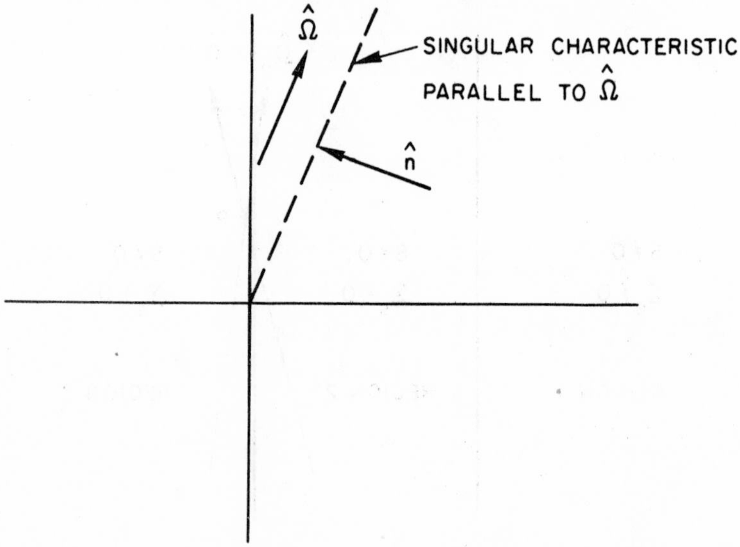

Fig. 4 Singular characteristic extending from a corner point: $\hat{n} \cdot \nabla F$ is discontinuous along dotted line.

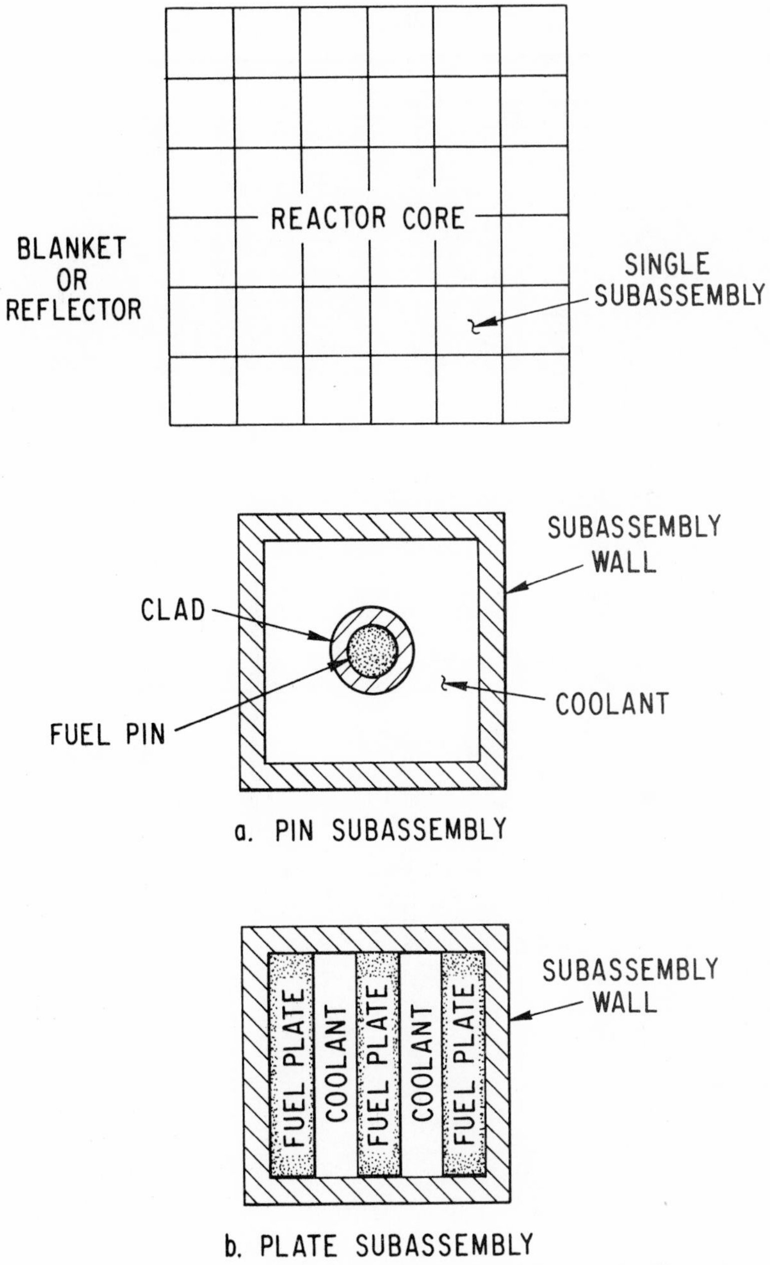

Fig. 5 Core lattice structure viewed from above.

BEHAVIOR OF THE SOLUTIONS OF AN ELLIPTIC BOUNDARY VALUE PROBLEM IN A POLYGONAL OR POLYHEDRAL DOMAIN

Pierre Grisvard*

Summary: The aim of this talk is to describe the main results about the regularity (as well as the singularities) of the solutions of an elliptic boundary value problem in domains with a non smooth boundary. Up to now, results concerning general elliptic boundary value problems, are available only if the domain is two-dimensional, mainly with a polygonal boundary. On the contrary few results are known in higher dimension when the domain presents singularities more complicated than conical points. For instance for a domain with edges but with no vertices, only boundary value problems for elliptic operator of the second order have been investigated. I will give a brief survey of recent results on this subject with precise bibliographical references together with some new contributions to the study of a second order elliptic boundary value problem in a three dimensional polyhedron. Except for new results the material is presented without proofs. The topics mentioned above are closely related to the study of mixed boundary value problems for an elliptic equation and of problems with interface conditions (conjugacy problems). For these problems, I will only mention briefly the results which are similar to those concerning our main subject, with bibliographical references.

---

* University of Nice, France

1. Introduction

1.1. For the sake of simplicity, I will describe the known results only for the Laplace equation with Dirichlet, Neuman or mixed boundary conditions. Nevertheless I will briefly explain after each result, its actual range of validity. Let us first consider the two-dimensional case: We consider the Laplace equation

$$\Delta u = f \tag{1.1}$$

in a bounded, connected and open set Ω. We assume the boundary Γ of Ω is a polygon, i.e. it is the union of a finite number of linear segments denoted by Γ_j, $1 \le j \le J$ (Γ_{j+1} follows Γ_j in the positive sense). The measure of the angle (interior to Ω) at the point σ_j where Γ_j meets Γ_{j+1} is denoted by ω_j [1]. On each of the Γ_j we impose either a Dirichlet condition or a Neuman condition: Let D and N be the subsets of $\{j;\ 1 \le j \le J\}$ for which

$$u|_{\Gamma_j} = g_j\ ,\ j \in D \tag{1.2}$$

$$\frac{\partial u}{\partial \nu_j}|_{\Gamma_j} = g_j\ ,\ j \in N\ , \tag{1.3}$$

here ν_j is the inner normal vector on Γ_j. In order to include the case of the usual mixed problem, we admit the value π for ω_j when the boundary condition is not of the same type on Γ_j and Γ_{j+1}. Of course in the opposite case we assume that $\omega_j \neq \pi$. Now let m be a non negative integer and p a real number > 1. $W^{m,p}(\Omega)$ denotes the usual Sobolev space of order m related to $L^p(\Omega)$; it is well known that the restrictions to Γ_j of the functions of $W^{m,p}(\Omega)$ span the space

---

1) It is in intended that $\Gamma_{N+1} = \Gamma_1$.

$W^{m-1/p,p}(\Gamma_j)$ [1]. Now given $f \in W^{m,p}(\Omega)$, $g_j \in W^{m+2-1/p,p}(\Gamma_j)$, $j \in D$ and $g_j \in W^{m+1-1/p,p}(\Gamma_j)$, $j \in N$, let u be any variational solution of (1,1)(1,2)(1,3); our problem is to decide whether u belongs to $W^{m+2,p}(\Omega)$. Of course, we assume that the given functions satisfy the natural consistency condition at the points σ_j for which j and $j+1 \in D$, i.e.

$$g_j(\sigma_j) = g_{j+1}(\sigma_j) \ . \tag{1.4}$$

In other words we consider the operator

$$T_{m,p} = u \to \{\Delta u; u/\Gamma_j, j \in D; \frac{\partial u}{\partial \nu_j}\Big|_{\Gamma_j}, j \in N\}$$

from $W^{m+2,p}(\Omega)$ into the subspace of

$$W^{m,p}(\Omega) \times \prod_{j\in D} W^{m+2-1/p,p}(\Gamma_j) \times \prod_{j\in N} W^{m+1-1/p,p}(\Gamma_j)$$

defined by (1.4). The main result is that $T_{m,p}$ is an index operator if no ω_j belongs to a certain sequence of exceptional values. Moreover we can calculate the index of $T_{m,p}$. In order to state this result in a more precise way, let us first introduce two integer valued functions defined for all real ω and real s :

$$X_S(\omega;s) = \text{Card } \{k \in N;\ 1 \le k < \frac{\omega}{\pi}(s+2)\}$$

$$X_M(\omega;s) = \text{Card } \{k \in N;\ 1 \le k < \frac{\omega}{\pi}(s+2) + \frac{1}{2}\}$$

then let us denote by S and M the subsets of $\{j;\ 1 \le j \le J\}$ for which the boundary conditions are the same on Γ_j and Γ_{j+1} , for $j \in S$ and differ from each other when $j \in M$.

---

1) A detailed description of the function spaces used is given in §2.

THEOREM 1. We assume that $\frac{\omega_j}{\pi}(s+2)$ for $j \in S$ and $\frac{\omega_j}{\pi}(s+2) + \frac{1}{2}$ for $j \in M$ are non integer for $s = m - 2/p$. Then there exist X continuous linear forms ℓ_k, $1 \leq k \leq X$ on

$$W^{m,p}(\Omega) \times \prod_{j\in D} W^{m+2-1/p,p}(\Gamma_j) \times \prod_{j\in N} W^{m+1-1/p,p}(\Gamma_j)$$

such that: given

$$f \in W^{m,p}(\Omega), g_j \in W^{m+2-1/p,p}(\Gamma_j),\ j\in D, g_j \in W^{m+1-1/p,p}(\Gamma_j),\ j\in N,$$

the variational solutions u of (1.1)(1.2)(1.3) are in $W^{m+2,p}(\Omega)$ iff ℓ_k vanishes at $\{f;\ g_j,\ 1 \leq j \leq J\}$ for all k. Furthermore

$$X = \prod_{j\in M} X_M(\omega_j;\ m-2/p) + \prod_{j\in S} X_S(\omega_j;\ m-2/p)\ .$$

In other words the index of the operator $T_{m,p}$ defined earlier is $-X$; moreover when D is non empty, $T_{m,p}$ is one to one and its range is closed and of codimension X; when $D = \emptyset$ (a pure Neuman problem), the dimension of the kernel is one, its range is closed and of codimension $x + 1$. It must be emphasized that the index depends only on the measure of the angles (i.e. the ω_js, $1 \leq j \leq J$) and on the Sobolev exponent $m - 2/p$.

The first proof of this result is in [8] when $p = 2$, $m = 0$, $N = \emptyset$, $g_j = 0$, $\forall_j$; as a particular case it is proved that Δ is an isomorphism from $\{u \in W^{2,2}(\Omega); u_{|\Gamma_j} = 0, 1 \leq j \leq J\}$ onto $L^2(\Omega)$ iff $\omega_j < \pi$ for all j, which means that Ω is convex.

The extension of this particular result to a general convex bounded open set (i.e even non polygonal) was achieved in [24] and later in [41] for the Neuman problem and for an oblique derivative boundary condition such that the angle between the oblique vector along which the derivative is calculated and the tangent to the boundary is constant (this makes sense since the boundary of a convex set is always locally Lipshitzian in the sense of [43] for instance, thus the normal vector is defined a.e. along the boundary). In the case of a Dirichlet problem the same result has been proven in [23] for a curvilinear polygon, the boundary of which is a finite union of C^3 curves meeting at corners or at points with a contact of finite order. Then in [29] th. 1 is proved for $p = 2$ in the general case when $X = 0$, but the method used there covers all the cases where $p = 2$. A complete proof of th. 1 when $p = 2$ is given in [19] for the Dirichlet problem and in [41] for the Neuman problem. Only recently the general case with an arbitrary p, was solved in [38], [39] (even with non integer orders of differentiation m). Th. 1 remains true when $p = +\infty$, but it must be modified as follows: m is a real non integer positive number ($m = k+\sigma$ with k integer and $0 < \sigma < 1$) and $W^{m,p}(\Omega)$ must be replaced by the usual $C^{k,\sigma}(\bar{\Omega})$ space, i.e. the space of the k times continuously differentiable functions for which the derivatives of order k satisfy a uniform Hölder condition of order σ in $\bar{\Omega}$. The number X is then

$$X = \sum_{j\in M} X_M(\omega_j;m) + \sum_{j\in S} X_S(\omega_j;m) .$$

These results are proved in [54] for right angles only and in [55] for general angles.

A statement analogous to th. 1 in the framework of the Nikolski spaces was proved earlier in [44] for right angles and in [14] for the general case. Finally, weighted Sobolev spaces are used in [5].

The meaning of the condition on the angles stated in th. 1 is that the values of ω_j for which $X_s(\omega_j,s)$ $(j\in S)$ or

$X_M(\omega_j,s)$ $(j\in M)$ change, must be avoided. Actually, if the condition is violated then the conclusion of th. 1 is false: for $p = 2$ and $N = \emptyset$ at least, it is proved in [19] that the range of $T_{m,2}$ is not closed, thus it cannot have an index. The method used in [8] [24] [19] starts with a direct a priori estimate for the solution fulfilling homogeneous boundary conditions; this estimate is of the same kind as the so-called "second fundamental inequality" of [31]; it only deals with second order problem with $p = 2$. However the constants occuring in the a priori estimates may be calculated exactly and do not depend on Ω; that is why the results in [24] [41] [23] may be obtained by considering Ω as a limit of an increasing sequence of polygonal domains. By another limiting procedure, namely the elliptic regularisation, one can obtain results for the heat equation in a polygon (with respect to the time and one space variable) as is done in [48]. On the other hand the papers [29] [27] [39] deal with a general elliptic boundary value problem of arbitrary order the solution of which is studied locally, referring to the case where Ω is an infinite sector; in this case the solution is built explicitly by means of a Mellin transformation. This second method has the further interest of giving an explicit description of the singularity of the solution occuring near the vertices when it is not in $W^{m+2,p}(\Omega)$.

Actually following this idea we will give an alternative statement for th. 1 where for simplicity we restrict ourselves to the case where $m = 0$ and the boundary conditions are homogeneous. We introduce local polar coordinates related to each of the vertices σ_j (see figure 1): let ρ_j be the distance from x to σ_j and θ_j be the angle between the vector $\sigma_j x$ and Γ_{j+1} ($\theta_j > 0$ when $x \in \Omega$ in a neighborhood of σ_j); then we define the following set of exceptional functions

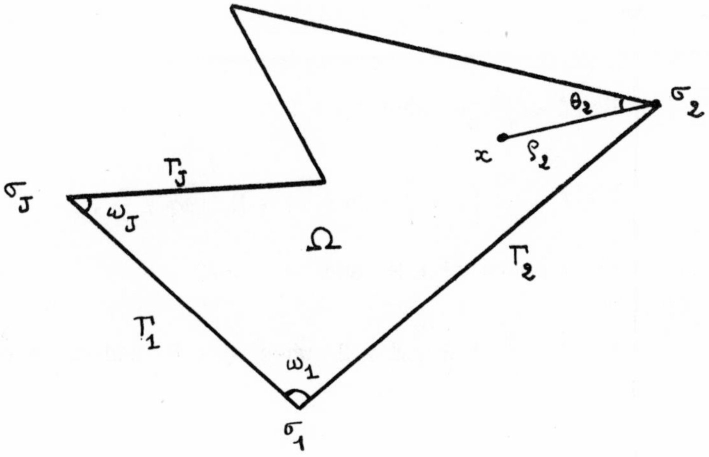

Figure 1

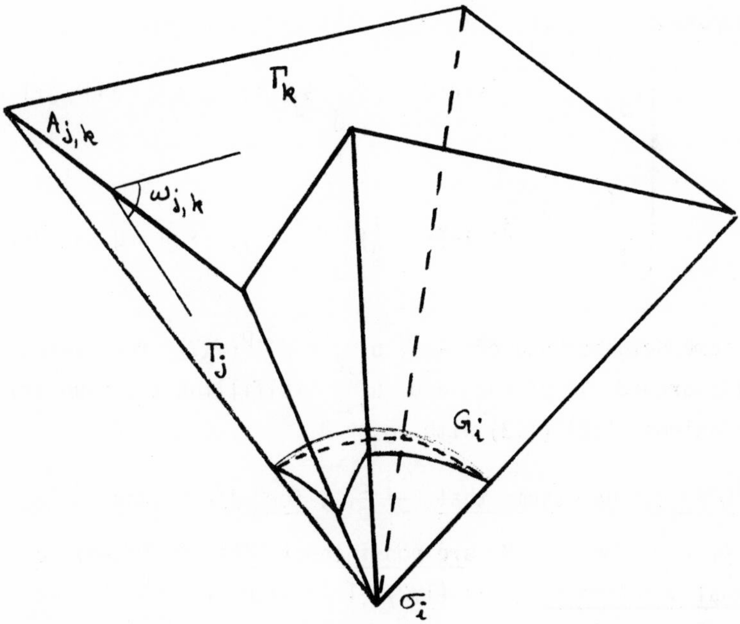

Figure 2

$u_{j,\ell}$: for

$$(1.5)\quad \begin{cases} 1 \le \ell < \frac{\omega_j}{\pi}\frac{2}{p'} \text{ when } j \in S\ ; \\ 1 \le \ell < \frac{\omega_j}{\pi}\frac{2}{p'} + \frac{1}{2} \text{ when } j \in M\ ,\ \omega_j \ne \frac{\pi}{2}\ ,\ \frac{3\pi}{2}\ , \\ \text{no } \ell \text{ when } j \in M \text{ and } \omega_j = \pi/2\ ; \\ 1 \le \ell < \frac{3}{p'} + \frac{1}{2}\ ,\ \ell \ne 2 \text{ when } j \in M \text{ and } \omega_j = \frac{3\pi}{2}\ ; \end{cases}$$

$$u_{j,\ell}(x) = \begin{cases} \rho_j^{\ell\pi/\omega_j} \sin \ell\pi \frac{\theta_j}{\omega_j}\ , \text{ if } j \text{ and } j+1 \in D\ , \\ \rho_j^{\ell\pi/\omega_j} \cos \ell\pi \frac{\theta_j}{\omega_j}\ , \text{ if } j \text{ and } j+1 \in N \\ \rho_j^{(\ell - \frac{1}{2})\frac{\pi}{\omega_j}} \sin(\ell - \frac{1}{2})\frac{\pi}{\omega_j}\,\theta_j, \text{ if } j \in D \text{ and } j+1 \in N \\ \rho_j^{(\ell - \frac{1}{2})\frac{\pi}{\omega_j}} \sin(\ell - \frac{1}{2})\frac{\pi}{\omega_j}(\omega_j - \theta_j), \text{ if } j \in N \text{ and } J+1 \in D \end{cases}$$

in some neighborhood of σ_j , $u_{j,\ell} \in W^{2,p}(\Omega\setminus V)$ for every neighborhood V of σ_j and $u_{j,\ell}$ fulfilling the boundary conditions (1.2) (1.3) with $g_j = 0$.

THEOREM 2. *We assume that* $\frac{\omega_j}{\pi}\frac{2}{p'}$, *for* $j \in S$ *and* $\frac{\omega_j}{\pi}\frac{2}{p'} + \frac{1}{2}$ *for* $j \in M$ *are non-integer; let* u *be any variational solution of* (1.1) (1.2) (1.3) *with* $f \in L^p(\Omega)$ *and* $g_j = 0$, $1 \le j \le J$; *then there exist numbers* α_j^ℓ *such that*

$$u - \Sigma \alpha_j^{\ell} u_{j,\ell} \in W^{2,p}(\Omega)$$

where the sum is over all j's with $1 \le j \le J$ and all ℓ integers fulfilling (1.5).

The interest of such a statement is clear in numerical analysis: in order to compute u with a Ritz-Galerkin method, we must include the exceptional functions $u_{j,\ell}$ in the space in which we seek the approximate solution, in order to be allowed to use the usual error estimate for a solution in $W^{2,p}(\Omega)$, according to [52]. A rough description of the corresponding exceptional functions for an elliptic boundary value problem of arbitarty order near a corner point is given in [29] [27].

Of course, the classical procedure of Korn allows us to extend the results of th. 1 to the case of variable coefficients in a curvilinear polygon owing to the invariance of the index under small perturbation; this is done in [29] when $p = 2$. This may be also performed by means of pseudo-differential methods as in [27]. Unfortunately these procedures cannot be used to build the exceptional functions for elliptic problems with variable coefficients, because those functions are hard to follow under perturbation: the difficulty occurring is of the same kind as that of following eigen-functions of an operator under perturbation.

To conclude this review of results in dimension two let us indicate that the result of th. 1 in the case of a mixed problem for a flat angle, agrees with those in [50] [46] where the existence of an index for a higher order mixed elliptic boundary value problem is proved. Analogous results for the conjugacy problem for a second order operator are proved in [26] for $m = 0$ and $p = 2$ and [32] for arbitrary m and $p = 2$, by means of a priori methods; the Mellin transformation has been used recently in [6] to obtain results with $p \neq 2$ (conjugacy problems for higher order operators of a very general kind are studied in [13] but with less accuracy in the regularity results).

1.2. Now let us turn to the three-dimensional case. From now on, Ω is a bounded, connected open set in $\mathbb{R}^3$ with a polyhedral boundary Γ. This means that Γ is the union of a finite number of plane faces Γ_j, $1 \le j \le J$. We shall denote by $A_{j,k}$ the edge where Γ_j and Γ_k meet (for those j and those k for which they meet) and by $\omega_{j,k}$ the measure of the corresponding angle inside Ω. Let σ_i, $1 \le i \le I$ be an enumeration of the vertices of Ω; each of the σ_i's is the vertex of a cone (determined by the faces which meet at σ_i) which corresponds to a solid angle G_i, G_i is a subset of the unit sphere S^2 (see figure 2). We must emphasize the fact that G_i does not have regular boundary ∂G_i since ∂G_i is the finite union of a certain number of arcs of great circles, that we denote by $S_{i,j}$ for those j for which Γ_j meets σ_i.

We will only consider the case of equation (1.1) under the homogeneous boundary conditions

(1.6) $$u_{|\Gamma_j} = 0 \ , \ j \in D$$

(1.7) $$\frac{\partial u}{\partial \nu_j}{}_{|\Gamma_j} = 0 \ , \ j \in N$$

where D and N are two disjoint subsets of $\{j; 1 \le j \le J\}$ such that $D \cup N = \{j; 1 \le j \le J\}$.

Let L be the Laplace Betrami operator on S^2; we consider it as a self-adjoint operator in $L^2(G_i)$ under homogeneous Dirichlet conditions on $S_{i,j}$ for $j \in D$ and homogeneous Neuman conditions on $S_{i,j}$ for $j \in N$. This realization of

L has a discrete sequence of eigenvalues calculated, for instance, by using the minimax principle; we denote them by

$$-\lambda_i^{(\ell)} \ , \ \ell = 1, \ , \cdots\cdot$$

with $0 \leq \lambda_i^{(1)} \leq \lambda_i^{(2)} \leq \cdots\cdot$, repeating each eigenvalue according to its multiplicity (see [11]). Now we are able to state a result similar to th. 1; $W_0^{m,p}(\Omega)$ will be the closure of $C_0^\infty(\Omega)$ in $W^{m,p}(\Omega)$ under the natural topology (of course when $m = 0$ all these spaces are reduced to $L^p(\Omega)$).

THEOREM 3. *We assume that* $\frac{\omega_{j,k}}{\pi}(m + 2 - \frac{2}{p}) < 1$ *for all* (j,k) *such that the boundary condition is the same on* Γ_j *and* Γ_k *and that* $\frac{\omega_{j,k}}{\pi}(m + 2 - \frac{2}{p}) < \frac{1}{2}$ *for all* (j,k) *such that the boundary conditions on* Γ_j *and* Γ_k *differ from each other; furthermore we assume that* $\lambda_i^{(\ell)} \neq (m + 2 - {}^3/_p)(m + 3 - {}^3/_p)$ *for every* i *and every* ℓ *. Then there exist* X *continuous linear forms* ℓ_k , $1 \leq k \leq X$ *on* $W_0^{m,p}(\Omega)$ *such that: given* $f \in W_0^{m,p}(\Omega)$, *the variational solutions* u *of* (1.1)(1.6)(1.7) *are in* $W^{m+2,p}(\Omega)$ *iff* $\ell_k(f) = 0$, $1 \leq k \leq X$, *and*

$$X \leq \sum_{i=1}^{I} \text{Card}\,\{\ell \in \mathbb{N};\ \lambda_i^{(\ell)} < (m+2-{}^3/_p)(m+3-{}^3/_p)\}\ .$$

As in the bidimensional case X depends only on the solid-angles G_i and on the Sobolev exponent $m - {}^3/_p$. The first restriction on the edges is exactly the condition on the angles ensuring that $X = 0$ in th. 1. If this condition is violated the range of Δ as an operator from

$$\{u \in W^{m+2,p}(\Omega);\ \ u \ \text{ fulfills } (1.6)(1.7),\ \Delta u \in W_0^{m,p}(\Omega)\}$$

into $W_0^{m,p}(\Omega)$, is of infinite codimension (and sometimes not closed). Roughly speaking this means that every point of an edge gives a contribution to the index of the problem equal to the contribution of a corner point in the plane with the same angle; since the edges are infinite sets, the index will be always infinite except if all the corresponding corners give regularity. If the last restriction in the statement (that $\lambda_i^{(\ell)}$ differ from $(m+2 - {}^3/_p)(m+3 - {}^3/_p))$, is violated then the range of Δ is non closed.

It is easy to see that the linear forms which must vanish at f in th. 3, are exactly given by any base of the space of the weak solutions of the homogeneous adjoint problem, namely

$$\{v \in W^{-m,p'}(\Omega);\ \Delta v = 0 \text{ in } \Omega,\ v_{|\Gamma_j} = 0\ \forall\ j \in D,\ \frac{\partial v}{\partial \nu_j}_{|\Gamma_j} = 0, \forall\ j \in N\}$$

where $W^{-m,p'}(\Omega)$ is the dual space of $W_0^{m,p}(\Omega)$ and the boundary conditions are intended in the weak sense defined in [34] [35]. In the general case where $m > 0$, we assume in th. 3 that f fulfills some homogeneous boundary conditions which are superfluous in our problem. However it is possible to extend the statement of th. 3 to a general $f \in W^{m,p}(\Omega)$ by using the traces th. (stated in §2).

Th. 3 is proved in [24] when $p = 2$, $m = 0$ with Dirichlet conditions and Ω is convex. It is then extended to an arbitrary convex bounded open set of $\mathbb{R}^3$ (and even $\mathbb{R}^n$, $n \geq 3$). For the Dirichlet problem with $p = 2$, $m = 0$, th. 3 is proved in [22] and for $m > 0$ in [20]. In these three papers the method starts with a priori estimates as in [31]. The part of the results of th. 3 concerning regularity near the edges but far from the vertices is proved when $p = 2$ in [30] in the case of Dirichlet condition and extended to problems with variable coefficients in domains with curvilinear edges. The oblique derivative problem in the same kind of domains is studied in [42] and very general boundary conditions are considered in [28]. The three last papers mentioned above deal also with edges in higher dimension. In the case of a mixed problem along a flat dihedral angle (i.e. with angle π) the results of th. 3 agree with those in [51] [57]. As far as I know the results of th. 3 when $p \neq 2$ are new and the remaining paragraphs of this paper are devoted to a sketch of the proof.

Before going further in technical considerations concerning the proof of th. 1, let us give a non formal statement of a result similar to th. 2. Let us assume that all the hypothesis of th. 3 are fulfilled with $m = 0$; then the exceptional functions which we must add to $W^{2,p}(\Omega)$ in order to get all the solutions of (1.1) (1.6) (1.7) with $f \in L^p(\Omega)$ are among the

$$w_{i,\ell}(x) = \rho_i^{\frac{1}{2} + \sqrt{\frac{1}{4} + \lambda_i^{(\ell)}}} \varphi_i^{(\ell)}(\omega_i) \quad \text{(near } \sigma_i \text{) ;}$$

here ρ_i is the distance between x and σ_i , ω_i the variable on $\sigma_i + S^2$ and $\varphi_i^{(\ell)}$ an eigenfunction of L related to the eigenvalue $\lambda_i^{(\ell)}$ defined earlier, and the indices ℓ to be considered are bounded by the condition

$$\lambda_i^{(\ell)} < (2 - \frac{3}{p}) \frac{3}{p'} .$$

The difficulty which arises here is that due to the general form of the $\varphi_i^{(\ell)}$, we are not able to decide whether or not $w_{i,\ell}$ is of class $W^{2,p}$ near σ_i , when the power of ρ_i is an integer (i.e. $\lambda_i^{(\ell)} = 2$) . However this question is not as important as it may appear at once since it may happen that the condition on the angles $\omega_{j,k}$ implies that $\lambda_i^{(\ell)} > (m + 2 - \frac{3}{p})(m + 3 - \frac{3}{p})$ for all i and ℓ : this is the case for instance when $p = 2$, $N = \emptyset$ and for each vertex σ_i there exists a neighborhood V of σ_i such that $V \cap \Omega$ is contained in one of the dihedral angles which meet σ_i (for instance if Ω is convex); see §5.

Let us now make some comments on the proof of the regularity near one vertex in th. 3. Obviously it will be enough to consider an infinite cone Ω with vertex at the origin and with solid angle $G \subset S^2$. In spherical coordinates equation (1.1) is

$$\frac{\partial^2 u}{\partial r^2} + \frac{2}{r}\frac{\partial u}{\partial r} + \frac{1}{r^2} Lu = f ;$$

and after the Euler substitution $r = e^t$, equation (1.1) becomes

$$\frac{\partial^2 u}{\partial t^2} + \frac{\partial u}{\partial t} + Lu = e^{2t} f$$

in a cylinder $Q = R \times G$. Then after Fourier-Laplace transformation with respect to t or equivalently Mellin transformation with respect to r , we get the equation

$$(1.8) \qquad (-z^2+iz)\tilde{u} + L\tilde{u} = (e^{2t}f)^{\sim}$$

in G , which depends on the complex parameter z , with boundary conditions which remain the same from the beginning. Thus we are led to an elliptic problem with parameter in G which is a non regular subset of S^2 ; this problem is equivalent after a stereographic projection to an elliptic boundary value problem with parameter in a curvilinear polygon of R^2 . Unfortunately nothing equivalent to [2] and [4] is known for non regular domains in R^2 , and this is certainly one of the main problems for further development of the theory.

In order to avoid the difficulty mentioned above there are two possibilities : the first one consists in restricting oneself to the case where G has a regular boundary and using the estimates of [4]; this leads to the consideration of domains having only conical points as in [29] [27]. The second possibility is the one that we follow here, i.e. we consider only second order problems and use a variant of the maximum principle to get estimates for the solution of (1.8).

Finally let us explain further the difficulties which arise in solving (1.8) rewritten simply as

$$-\lambda\varphi + \Delta\varphi = \psi$$

with say, Dirichlet conditions in a plane sector. Using once more the Euler substitution and partial Fourier transform we get equation

$$-\lambda\tilde{\varphi}(\tau+2i) + L\tilde{\varphi}(\tau) - \tau^2\tilde{\varphi}(\tau) = \tilde{\psi}(\tau+2i)$$

which is no longer a differential equation with respect to $\tilde{\varphi}$ because of the shift of the variable in $-\lambda\tilde{\varphi}$.

Let us now turn to the proof of th. 3; the techniques used here are a mixture of functional calculus for unbounded operators, pseudo-differential calculus and interpolation using Besov spaces, I am convinced that the proof may be simplified; however, the same kind of proof could be used for solving a second order conjugacy problem in three variables. I shall not do this here due to lack of space. The plan is as follows

§2 Functional spaces and trace-theorem

§3 Regularity in weighted Sobolev spaces of fractional order

§4 Regularity in weighted Sobolev spaces of integer order

§5 Regularity in usual Sobolev spaces

Appendices 1 and 2

Bibliography

2. Functional Spaces and Trace Theorems

2.1. We will use Sobolev spaces and Besov spaces defined as follows in an open set Ω of $\mathbb{R}^n$, where $1 < p < +\infty$, m is a non negative integer, $s = m + \sigma$ with $0 < \sigma < 1$ and $1/p + 1/p' = 1$:

(a) $W^{m,p}(\Omega)$ is the space of all functions u defined in Ω and such that(1)

$$\|u\|_{W^{m,p}(\Omega)} = \left\{ \sum_{|\alpha| \leq m} \int_\Omega |D^\alpha u(x)|^p dx \right\}^{1/p} < +\infty .$$

(b) $W^{s,p}(\Omega)$ is the space of all functions u defined in Ω and such that(1)

$$\|u\|_{W^{s,p}(\Omega)} = \|u\|_{W^{m,p}(\Omega)} + \left\{ \sum_{|\alpha|=m} \iint_{\Omega\times\Omega} |D^\alpha u(x) - D^\alpha u(y)|^p \frac{dx\,dy}{|x-y|^{n+\sigma p}} \right\}^{1/p} < +\infty .$$

(c) $W_0^{m,p}(\Omega)$ (resp. $W_0^{s,p}(\Omega)$) is the closure of $C_0^\infty(\Omega)$ in $W^{m,p}(\Omega)$ (resp. $W^{s,p}(\Omega)$)(2).

(d) $W^{-m,p'}(\Omega)$ (resp. $W^{-s,p'}(\Omega)$) is the dual of $W_0^{m,p}(\Omega)$ (resp. $W_0^{s,p}(\Omega)$).

Finally we define the space $B^{m,p}(\Omega)$ using the real interpolation functor, by setting

$$B^{m,p}(\Omega) = (W^{m+1,p}(\Omega) \; ; \; W^{m-1,p}(\Omega))_{1/2,p} \; ;$$

---

(1) We will sometimes use the analogous spaces for vector functions in some B space X ; the corresponding spaces are denoted by $W^{m,p}(\Omega;X)$, $W^{s,p}(\Omega;X)$ and so on.

(2) As usual, we denote by $C_0^\infty(\Omega)$ the space of functions defined in Ω , having continuous derivatives of every order and compact support.

From this definition it is clear that $D^\alpha u \in L^p(\Omega_R)$ for all finite R and $|\alpha| \le m$ and consequently we have

$$P^{s,p}(\Omega) \subseteq W^{s,p}(\Omega_R) \ , \ \forall R < +\infty \ .$$

Furthermore if $s \in]0, 1[$ and $s \neq 1/p$ we know from Lemma 2.2' that $u \in W_0^{s,p}(\Omega)$ implies $r^{-s}u \in L^p(\Omega)$ and thus

$$W_0^{s,p}(\Omega) \subseteq P^{s,p}(\Omega) \ .$$

This proves that we will obtain the extension of Theorem 1 (or Theorem 3) to small non integer values of the order of differentiation simply by checking that for a given $f \in P^{s,p}(\Omega)$ with $0 < s < 1/p$, problem (3.5), (3.6) has a solution $u \in P^{s+2,p}(\Omega)$. As for the integer values of the order of differentiation this will be done through the Euler substitution owing to

Proposition 3.4. Let $0 < s < 1/p$ then $f \in P^{s,p}(\Omega)$ iff $e^{-(s-n/p)t} g \in W^{s,p}(\mathbb{R}\times G)$ where $g(t,\sigma) = f(e^t\sigma)$,

and its Corollary:

Corollary 3.5. Let $s = m+\sigma$ with m a non negative integer and $0 < \sigma < 1/p$; then $u \in P^{s+2,p}(\Omega)$ iff $e^{-(s+2-n/p)t} v \in W^{s+2,p}(\Omega)$ where $v(t,\sigma) = u(e^t\sigma)$.

These statements are far from being obvious and will be proved in the Appendix n° 1.

3.2. Now we will solve equation (3.9) in the Sobolev spaces of small fractional order as a first step. On G we assume the following:

(H) (i) L is an isomorphism from $W^{m+2,p}(G) \cap W_0^{1,p}(G)$ onto $W^{m,p}(G)$ for $0 \le m \le \alpha_{n-1}(p)$ (m integer),

(ii) L <u>is an isomorphism from</u> $W^{s+2,p}(G) \cap W_0^{1,p}(G)$ <u>onto</u> $W^{s,p}(G)$ <u>for</u> $0 < s < \inf(\frac{1}{p}\ ;\ \alpha_{n-1}(p))$, <u>where</u> α_{n-1} <u>is a real non increasing continuous function</u> (<u>the value</u> $+\infty$ <u>being admitted</u>).

When $n = 2$ it is easily checked that (H) is true with $\alpha_1(p) \equiv +\infty$ since (i) (resp. (ii)) means that the equations

$$\begin{cases} \varphi'' = \psi \text{ in }]S_1,S_2[\\ \varphi(S_j) = 0\ ,\ j = 1,2 \end{cases}$$

have a unique solution $\varphi \in W^{m+2,p}(]S_1,S_2[)$ (resp. $W^{s+2,p}(]S_1,S_2[)$) for every given $\psi \in W^{m,p}(]S_1,S_2[)$ (resp. $W^{s,p}(]S_1,S_2[)$.

When $n = 3$ it will be possible to check (H) only after Theorem 1 will have been proved; then the result will be

$$\alpha_2(p) = \operatorname*{Inf}_{j=1}^{J} \left(\frac{\pi}{\omega_j} - 2 + \frac{2}{p}\right)$$

where ω_j is the angle between S_j and S_{j+1} at the point σ_j where they meet. Using (H) we will now solve (3.9) by applying the method of [18], [12]: in $X = L^p(\mathbb{R}\times G)$ we define two bounded operators, setting

$$D_A = \{w \mid w, \frac{\partial w}{\partial t}, \frac{\partial^2 w}{\partial t^2} \in X\}$$

$$Aw = \frac{\partial^2 w}{\partial t^2} + a\frac{\partial w}{\partial t} + bw$$

$$D_B = \{w \in L^p(\mathbb{R};\ W^{2,p}(G)) \mid w_{|\mathbb{R}\times S_j} = 0,\ 1 \leq j \leq J\}$$

$$Bw = Lw$$

The problem that we want to solve is precisely

$$\begin{cases} w \in D_A \cap D_B \\ Aw + Bw = k \end{cases}$$

where k is given in X. From the results proved in [18] we know that $A + B$ is an isomorphism from

$$W = \{w \in D_A \cap D_B;\ Aw,\ Bw \in D_A(\theta,p) \cap D_B(\theta;p)\}$$

onto $D_A(\theta;p) \cap D_B(\theta,p)$ for every $\theta \in]0,1[$ when A is the infinitesimal generator of an analytic semi-group, B is the infinitesimal generator of a contraction semi-group, such that the spectra of A and $-B$ are disjoint (we have set $D_A(\theta,p) = (D_A;X)_{1-\theta,p}$ the interpolation space defined in section 2.1). We must now check the hypothesis for A and B. Let us first study the operator A: it is well known that $\frac{d}{dt}$ is the infinitesimal generator of a group of isometries in $L^p(\mathbb{R}\times G)$ and consequently its spectrum is $i\mathbb{R}$. Then the spectrum of A is $\sigma_A = \{(iy)^2+a(iy)+b \mid y\in\mathbb{R}\} = \{b-y^2+iay \mid y\in\mathbb{R}\}$; this is a parabola which cuts the real axis at the point b. The resolvent $(A-\lambda I)^{-1}$ is given by

$$\left(\frac{d}{dt} + \frac{1}{2}\{a+\sqrt{a^2+4(\lambda-b)}\}\right)^{-1} \left(\frac{d}{dt} + \frac{1}{2}\{a-\sqrt{a^2+4(\lambda-b)}\}\right)^{-1}$$

and is bounded by

$$\left(\mathrm{Re}\ \frac{1}{2}\{a+\sqrt{a^2+4(\lambda-b)}\}\ \mathrm{Re}\ \frac{1}{2}\{a-\sqrt{a^2+4(\lambda-b)}\}\right)^{-1}$$

which behaves asymptotically as $\dfrac{1}{|\lambda|(\cos\frac{1}{2}\arg\lambda)^2}$; this implies that A is the infinitesimal generator of an analytical semi-group.

Let us now study the operator B , or equivalently L with domain $D_L = W^{2,p}(G) \cap W_0^{1,p}(G)$ in $L^p(G)$. If we suppose that $\alpha_{n-1}(p) > 0$, then owing to (H) , L is invertible and therefore its resolvent operator is compact. Thus we know from [25] that its spectrum is a sequence of eigenvalues. Now let us consider an eigenfunction φ related to one of the eigenvalues of L :

$$L\varphi = \lambda\varphi$$

with $\varphi \in D_L \subset W^{2,p}(G)$. Since G has at most two dimensions and $p > 1$, it follows from Sobolev's theorem that $\varphi \in W^{1,2}(G)$; consequently λ is among the eigenvalues of the variational problem calculated by the minimax principle. Conversely let λ be such an eigenvalue and φ one of the related eigenfunctions, then $\varphi \in W^{1,2}(G)$ and since $p < +\infty$, it follows once again from Sobolev's theorem that $\varphi \in L^p(G)$. But from (H) it follows that $\varphi \in W^{2,p}(G)$ and consequently $\varphi \in D_L$. This argument proves that provided $a_{n-1}(p) > 0$, the spectrum of B is exactly the sequence of the eigenvalues of L obtained by applying the minimax principle; in particular the spectrum of B does not depend on p : it is

$$\sigma_B = \{-\lambda_\ell;\ \ell=1,2\cdots\}$$

where λ_ℓ, $\ell = 1,2\cdots$ is a non decreasing sequence of positive numbers. We also must estimate the resolvent of B ; it will be enough to estimate $(B-\lambda)^{-1}$ for $\lambda > 0$. Let us consider a real φ in D_L , setting

$$\varphi^* = |\varphi|^{p-2}\varphi$$

we will prove that

$$\int_G L\varphi\ \varphi^*\ d\sigma \leq 0\ . \tag{3.10}$$

Let us first consider the case of dimension 2; identity (3.10) means

$$\int_{S_1}^{S_2} \varphi'' \, \varphi^* \, d\sigma \leq 0$$

and this is immediate since by integrating by parts we get

$$\int_{S_1}^{S_2} \varphi'' \, \varphi^* \, dx = -(p-1) \int_{S_1}^{S_2} |\varphi'|^2 \, |\varphi|^{p-2} \, dx$$

$$+ (\varphi' \, \varphi \, |\varphi|^{p-2}) \Big|_{S_1}^{S_2}$$

and the boundary condition is $\varphi(S_j) = 0$, $j = 1,2$.

Now in the case of dimension 3 we use the spherical coordinates in $\mathbb{R}^3$ defined as usual by

$$\begin{cases} x_1 = r \cos \theta \sin \omega \\ x_2 = r \sin \theta \sin \omega \\ x_3 = r \cos \omega \end{cases}$$

with $r > 0$, $0 \leq \theta \leq 2\pi$, $0 \leq \omega \leq \pi$, then L is known to be the operator

$$L = \frac{1}{\sin \omega} D_\omega (\sin \omega \, D_\omega) + \frac{1}{\sin^2 \varphi} D_\theta^2$$

and the Lebesgue measure on G is given by

$$d\sigma = \sin \omega \, d\omega \, d\theta \ .$$

Consequently we get

$$\int_G L\varphi\, \varphi^*\, d\sigma$$

$$= \int_G D_\omega(\sin\omega\, D_\omega)\varphi\, \varphi^*\, d\omega\, d\theta + \int_G D_\theta\left(\frac{1}{\sin\omega} D_\theta\right)\varphi\, \varphi^*\, d\omega\, d\theta$$

$$= -(p-1)\left[\int_G \sin\omega\, |D_\omega \varphi|^2\, |\varphi|^{p-2}\, d\omega\, d\theta + \int_G \frac{1}{\sin\omega}\, |D_\theta \varphi|^2\, |\varphi|^{p-2}\, d\omega\, d\theta\right]$$

$$- \sum_{j=1}^{J} \int_{S_j} \frac{\partial\varphi}{\partial\nu_j}\, \varphi\, |\varphi|^{p-2}\, ds_j$$

$$\leq - \sum_{j=1}^{J} \int_{S_j} \frac{\partial\varphi}{\partial\nu_j}\, \varphi\, |\varphi|^{p-2}\, ds_j$$

where ds_j is the length measure along S_j . The last integral vanishes since on each S_j , the boundary condition is $\varphi = 0$. This proves (3.10). To conclude we observe that inequality (3.10) means that L is a dissipative operator in $L^p(G)$; it is maximal because it follows from (H) that L is invertible; consequently L is the infinitesimal generator of a contraction semi-group in $L^p(G)$ and B is the infinitesimal generator of a contraction semi-group in X owing to [56].

Summing up we have proved that all the conditions for applying the abstract theorem of [18] are fulfilled provided

$$b \neq \lambda_\ell \;, \quad \ell = 1,2\cdots$$

In order to make explicit the result just obtained, we need a description of the spaces $D_A(\theta,p)$ and $D_B(\theta,p)$. Owing to thedefinitions of D_A and D_B we have

$$D_A(\theta;p) = W^{2\theta,p}(\mathbb{R};L^p(G))$$

(see [36] for instance) and applying Cor. 2.4 we have

$$D_B(\theta;p) = L^p(\mathbb{R};W^{2\theta,p}(G))$$

provided $2\theta < 1/p$. This implies that

$$D_A(\theta,p) \cap D_B(\theta,p) = W^{2\theta,p}(\mathbb{R}\times G) \quad \text{if} \quad 2\theta < \frac{1}{p} ,$$

and using (H) that

$$W = \{w \in W^{2,p}(\mathbb{R}\times G) \mid w_{|\mathbb{R}\times S_j} = 0 , \quad 1 \leq j \leq J ;$$

$$\frac{\partial^2 w}{\partial t^2} \in W^{2\theta,p}(\mathbb{R}\times G) , \quad Lw \in W^{2\theta,p}(\mathbb{R}\times G)\}$$

$$= W^{2+2\theta,p}(\mathbb{R}\times G) \cap W_0^{1,p}(\mathbb{R}\times G)$$

provided $2\theta < \inf\{\frac{1}{p} ; \alpha_{n-1}(p)\}$. Thus we have proved

THEOREM 3.6. *Assuming* (H) , $0 < s < \inf\{\frac{1}{p} ; \alpha_{n-1}(p)\}$ *and* $b \neq \lambda_\ell$ *for every* ℓ , *then equations* (3.9) *has a unique solution* $w \in W^{s+2,p}(\mathbb{R}\times G) \cap W_0^{1,p}(\mathbb{R}\times G)$ *for every given* $k \in W^{s,p}(R\times G)$. Consequently this means that under the same assumptions, problem (3.5) (3.6) has a unique solution $u \in P^{s+2,p}(\Omega)$ for every given $f \in P^{s,p}(\Omega)$.

Remark 3.7. From the results in [18] we also conclude that the solution $w \in W^{2,p}(\mathbb{R}\times G) \cap W_0^{1,p}(\mathbb{R}\times G)$ of equation (3.9) for a given $k \in L^p(\mathbb{R}\times G)$, is unique; however we are not yet able to prove its existence which is precisely the aim of the next paragraph.

Remark 3.8. Inequality (3.10) remains true under Neuman boundary conditions or under mixed boundary conditions and this is why our method is applicable to prove Theorem 3 in the general case (and not only under Dirichlet boundary conditions).

4. Regularity in Weighted Sobolev Spaces of Integer Order

4.1. We will extend here the result of Theorem 3.6 to the case where k is given in the Besov space of order zero $B^{o,p}(\mathbb{R}\times G)$.

For doing this we start with the same result in $W^{-1,2}(\mathbb{R}\times G)$, and apply interpolation.

Lemma 4.1. We assume that $b \neq \lambda_\ell$ for every $\ell = 1,2,\cdots$, then equation (3.9) has a unique solution $w \in W_0^{1,2}(\mathbb{R}\times G)$ for every given $k \in W^{-1,2}(\mathbb{R}\times G)$.

Proof. We use the Fourier transform with respect to t and expansion in series of eigenfunctions of L with respect to x : let $\varphi_\ell \in W_0^{1,2}(G)$, $\ell = 1,2,\cdots$ be a complete orthonormal system of functions in $L^2(G)$ such that

$$L\,\varphi_\ell = -\,\lambda_\ell\,\varphi_\ell \qquad \ell = 1,2,\cdots .$$

Every function w may be written

$$w = \frac{1}{\sqrt{2\pi}}\int_{-\infty}^{+\infty} e^{it\tau} \sum_{\ell\geq 1} \hat{w}_\ell(\tau)\,\varphi_\ell\,d\tau$$

where

$$\hat{w}_\ell(\tau) = \frac{1}{\sqrt{2\pi}}\int_{\mathbb{R}\times G} e^{-it\tau}\,\varphi_\ell(x)\,w(t,x)\,dt\,dx$$

provided w is tempered in the t variable, with values in $W^{-1,2}(G)$; furthermore

$$\int_{\mathbb{R}\times G} |w|^2\,dt\,dx = \sum_{\ell\geq 1}\int_{-\infty}^{+\infty}|\hat{w}_\ell(\tau)|^2\,d\tau$$

provided $w \in L^2(\mathbb{R}\times G)$. Now let

$$k \in W^{-1,2}(\mathbb{R}\times G) = W^{-1,2}(\mathbb{R};L^2(G)) + L^2(\mathbb{R};W^{-1,2}(G)) \;;$$

since $-L$ is a positive self adjoint operator in $L^2(G)$ such that $D_{\sqrt{-L}} = W_0^{1,2}(G)$ (cf [33]), we have

$$\sum_{\ell\geq 1}\int_{-\infty}^{+\infty}(\lambda_\ell+\tau^2)^{-1}\,|\hat{k}_\ell(\tau)|^2\,d\tau < +\infty \,.$$

On the other hand equation (3.9) is equivalent to the sequence of equations:

$$(-\tau^2 + ia\tau - \lambda_\ell + b)\hat{w}_\ell(\tau) = \hat{k}_\ell(\tau) , \qquad \ell = 1,2,\cdots$$

where we seek $\hat{w}_\ell(\tau) \in W_0^{1,2}(G)$. From the assumption $b \neq \lambda_\ell$ for all ℓ, we deduce that each of these equations has the unique solution

$$\hat{w}_\ell(\tau) = \frac{\hat{k}_\ell(\tau)}{-\tau^2+ia\tau-\lambda_\ell+b} , \qquad \forall\tau , \forall\ell ;$$

Furthermore the following estimate is true

$$\sum_{\ell\geq 1} \int_{-\infty}^{+\infty} (\lambda_\ell+\tau^2) \; |\hat{w}_\ell(\tau)|^2 \; d\tau < +\infty ,$$

because the function

$$\frac{\lambda_\ell+\tau^2}{-\tau^2+ia\tau-\lambda_\ell+b}$$

is uniformly bounded for $\tau \in \mathbb{R}$ and $\ell = 1,2,\cdots$. This proves that $w \in W_0^{1,2}(\mathbb{R}\times G)$. Q.E.D.

THEOREM 4.2. We assume (H), $\alpha_{n-1}(p) > 0$ and $b \neq \lambda_\ell$ for all ℓ; then equation (3.9) has a unique solution $w \in B^{2,p}(\mathbb{R}\times G) \cap W_0^{1,p}(\mathbb{R}\times G)$ for every given $k \in B^{0,p}(\mathbb{R}\times G)$.

Proof. Uniqueness follows from the abstract theorem in [18] (and is not essential in what follows). Then in order to prove existence we consider the Green operator G of equation (3.9) with Dirichlet boundary conditions. From Theorem 3.6 we know that G is continuous from

$$W^{s,q}(\mathbb{R}\times G) \quad \text{into} \quad W^{s+2,q}(\mathbb{R}\times G)$$

provided $0 < s < \text{Inf}\,\{\frac{1}{q}\,;\,\alpha_{n-1}(q)\}$ and from Theorem 4.2 we know that G is continuous from

$$W^{-1,2}(\mathbb{R}\times G) \quad \text{into} \quad W^{1,2}(\mathbb{R}\times G)$$

Applying Lemmas 2.1 and 2.1' together with interpolation, choosing s and q such that

$$\frac{1}{q} = \frac{1-\theta}{2} + \frac{\theta}{q}\;,\quad (\theta-1) + sq = 0$$

we conclude that G is continuous from

$$B^{o,p}(\mathbb{R}\times G) \quad \text{into} \quad B^{2.p}(\mathbb{R}\times G)\ .$$

The proof is complete.

4.2. In the next step we will prove the analogue of Theorem 4.2 with $B^{o,p}(\mathbb{R}\times G)$ replaced by $L^p(\mathbb{R}\times G)$ when $1 < p \le 2$. We take advantage of the identity between the spaces spanned by the traces of $B^{2,p}(\mathbb{R}\times G)$ and $W^{2,p}(\mathbb{R}\times G)$, (see Theorem 2.7), by considering the non homogeneous Dirichlet boundary value problem for equation (3.9) with $k = 0$. This implies that we are first able to solve equation (3.9), at least in an approximate fashion, following the method of [9], [49]. We will use for this purpose:

Lemma 4.3. Let $e(x;\tau,\xi)$ be a symbol defined in $\mathbb{R}_t \times \mathbb{R}_x^{n-1} \times \mathbb{R}_\tau \times \mathbb{R}_\xi^{n-1}$ and independent of t. Assume that for every integer $\ell \ge 0$ and every pair of multi-indices α,β there exists a constant $C = C_{|\alpha|,\ell,|\beta|}$ such that

$$|D_x^\alpha\, D_\tau^\ell\, D_\xi^\beta\, e| \le C(1+\iota^2+|\xi|^2)^{-m-\frac{\ell+|\beta|}{2}}\ .$$

Then the operator E defined by

$$u \to \frac{1}{\sqrt{2\pi}^{n}} \int_{R^n} e^{i(t\tau + x.\xi)}\, e(x;\tau,\xi)\, \hat{u}(\tau,\xi)\, d\tau\, d\xi$$

<u>is linear continuous from</u> $L^p(R^n)$ <u>into the space of all</u> u <u>such that</u> $\varphi u \in W^{m,p}(\mathbb{R}^n)$ <u>for every</u> $\varphi \in \mathcal{D}(\mathbb{R}_x^{n-1})$ [1].

In the case $p = 2$ this is a very particular case of [10]; otherwise it is proved in Appendix n° 2. From Lemma 4.3 we deduce

<u>Proposition 4.4</u>. <u>There exists a linear operator</u> E <u>continuous from</u> $L^p(\mathbb{R}\times G)$ <u>into</u> $W^{2,p}(\mathbb{R}\ G)$ <u>such that</u> $ME - 1$ <u>is continuous from</u> $L^p(\mathbb{R}\times G)$ <u>into</u> $W^{1,p}(\mathbb{R}\times G)$, <u>where</u> $M = D_t^2 + a\, D_t + L + bI$. In other words, E is a parametrix for M.

<u>Proof</u>. The result does not depend on b and therefore we may suppose that b is negative. As indicated in section 3.1, we may consider G as a bounded open subset of $\mathbb{R}^{n-1}$ after performing a stereographic transformation. Consequently L is a strongly elliptic operator of the second order in a neighborhood of $\bar{G}$; we assume that L is extended outside $\bar{G}$ in such a way that it is uniformly strongly elliptic in $\mathbb{R}^n$.

Now let $\ell(x;\xi) = \sum_{i,k=1}^{n-1} \ell_{i,k}(x)\, \xi_i\, \xi_k$ be the principal symbol of $-L$; according to what we said below there exists $\alpha > 0$ such that

$$\ell(x;\xi) \geq \alpha|\xi|^2 \ , \ \forall x \in \mathbb{R}^{n-1} \ , \ \xi \in \mathbb{R}^{n-1} \ .$$

We define E setting

$$Ef(t,x) = \frac{1}{\sqrt{2\pi}^{n}} \int_{\mathbb{R}^n} e^{i(t\tau + x.\xi)}\, \frac{\hat{f}(\tau;\xi)}{-\tau^2 + i a\tau - \ell(x;\xi) + b}\, d\tau\, d\xi \ .$$

---

1) i.e. φ does not depend on t.

It follows from Lemma 4.3 that E is continuous from $L^p(\mathbb{R}\times G)$ into $W^{2,p}(\mathbb{R}\times G)$. Then we have

$$MEf - f = Rf + Sf$$

modulo a linear continuous operator from $L^p(\mathbb{R}\times G)$ into $W^{1,p}(\mathbb{R}\times G)$, where R is the operator whose symbol is

$$\sum_{i,k=1}^{n-1} \ell_{i,k}(x)\, D_i\, D_k \frac{1}{-\tau^2+ia\tau-\ell(x;\xi)+b}$$

and S the operator whose symbol is

$$\sum_{i,k=1}^{n-1} \ell_{i,k}(x)\, i\xi_i\, D_k \frac{1}{-\tau^2+ia\tau-\ell(x;\xi)+b} \quad .$$

It is readily seen that R fulfills the hypothesis of Lemma 4.3 with $m = 2$ and that S fulfills the hypothesis of Lemma 4.3 with $m = 1$. consequently $ME - 1$ is continuous from $L^p(\mathbb{R}\times G)$ into $W^{1,p}(\mathbb{R}\times G)$. Q.E.D.

Now we are able to prove

THEOREM 4.5. *Assuming* (H) , $\alpha_{n-1}(p) > 0$, $1 < p \leq 2$ *and* $b \neq \lambda_\ell$, $\forall \ell$, *then equation* (3.9) *has a unique solution* $w \in W^{2,p}(\mathbb{R}\times G) \cap W_0^{1,p}(\mathbb{R}\times G)$ *for every given* $k \in L^p(\mathbb{R}\times G)$.

Proof. We seek w as a sum $w_1 + w_2$, where $w_1 = Ek \in W^{2,p}(\mathbb{R}\times G)$ owing to Prop. 4.4 and w_2 is the solution of

$$\begin{cases} Mw_2 = k - Mw_1 & \text{in } R \times G \\ w_2 = -w_1 & \text{on } R \times \partial G \end{cases} \quad .$$

From Theorem 2.7 we know that there exists $w_3 \in B^{2,p}(\mathbb{R}\times G)$ such that $w_3 = -w_1$ on $\mathbb{R} \times \partial G$; thus we will obtain w_2 as a sum $w_3 + w_4$ where

$$\begin{cases} Mw_4 = Mw_2 - Mw_3 = k - MEk - Mw_3 & \text{in } \mathbb{R} \times G \\ w_4 = 0 \quad \text{on } \mathbb{R} \times \partial G & . \end{cases}$$

This last problem is solved by applying Theorem 4.2 since we have

$$k - MEk - Mw_3 \in W^{1,p}(\mathbb{R}\times G) + B^{o,p}(\mathbb{R}\times G) = B^{o,p}(\mathbb{R}\times G) .$$

We conclude by observing that

$$w = w_1 + w_3 + w_4 \in W^{2,p}(\mathbb{R}\times G) + B^{2,p}(\mathbb{R}\times G) = W^{2,p}(\mathbb{R}\times G)$$

since $1 < p \le 2$ (using Lemma 2.1), and consequently w is the desired solution. Q.E.D.

4.3. In this last step we will extend the result of Theorem 4.5 to the case where $2 < p < +\infty$. This will be achieved by a duality procedure: so far we know that M is an isomorphism from $W^{2,p'}(\mathbb{R}\times G) \cap W_o^{1,p'}(\mathbb{R}\times G)$ onto $L^{p'}(\mathbb{R}\times G)$, provided $\alpha_{n-1}(p') > 0$. Let φ be the continuous linear form on $W^{2,p'}(\mathbb{R}\times G) \cap W_o^{1,p'}(\mathbb{R}\times G)$ defined by

$$\varphi(w) = \sum_{j=1}^{J} \int_{\mathbb{R}\times S_j} \frac{\partial w}{\partial \nu_j} \varphi_j \, dt \, ds_j$$

where ds_j is the Lebesgue measure on S_j , with $\varphi_j \in W^{-1/p,p}(\mathbb{R}\times S_j)$, $j = 1, 2\cdots J$, given; this linear form is well defined since

$$w \to \frac{\partial w}{\partial \nu_j}\Big|_{\mathbb{R}\times S_j}$$

is a continuous linear mapping from $W^{2,p'}(\mathbb{R}\times G)$ into $W^{1/p,p'}(\mathbb{R}\times S_j)$ whose dual space is $W^{-1/p,p}(\mathbb{R}\times S_j)$ (recall that $1/p < 1/p'$) .

Now given such a φ there exists a unique $v \in L^p(\mathbb{R}\times G)$ such that

$$(4.1) \qquad \sum_{j=1}^{J} \int_{\mathbb{R}\times S_j} \frac{\partial w}{\partial \nu_j} \varphi_j \, dt \, ds_j = \int_{\mathbb{R}\times G} Mw \, v \, dt \, d\sigma$$

for every $w \in W^{2,p'}(\mathbb{R}\times G) \cap W_0^{1,p'}(\mathbb{R}\times G)$. If we use (4.1) with $w \in \mathcal{D}(\mathbb{R}\times G)$ we get

$$< Mw \, ; \, v > = 0$$

and therefore $M^* v = 0$; then if we use (4.1) with $w \in \mathcal{D}(\overline{\mathbb{R}\times G})$ (the space of the restrictions to $\mathbb{R} \times G$ of the elements in $\mathcal{D}(\mathbb{R}^n)$) such that $w_{|\mathbb{R}\times S_j} = 0$, $1 \leq j \leq J$, we obtain readily that

$$\sum_{j=1}^{J} \int_{\mathbb{R}\times S_j} \frac{\partial w}{\partial \nu_j} (\varphi_j - v) \, dt \, ds_j = 0 \, ,$$

where the trace of v on $\mathbb{R} \times S_j$ is defined in the weak sense (since v is the solution of an elliptic second order equation $M^* v = 0$) according to [35], and consequently $w_{|\mathbb{R}\times S_j} = \varphi_j$. This proves

Lemma 4.6. Assuming (H) , $2 < p < +\infty$, $\alpha_{n-1}(p') > 0$ and $b \neq \lambda_\ell$, $\ell = 1, 2, \cdots$ then for every

$$\varphi_j \in W^{-1/p,p}(\mathbb{R}\times S_j) \, , \quad 1 \leq j \leq J$$

there exists a unique $v \in L^p(\mathbb{R}\times G)$ solution of

$$\begin{cases} M^* v = 0 & \text{in } \mathbb{R} \times G \\ v = \varphi_j & \text{on } \mathbb{R} \times S_j \, , \quad 1 \leq j \leq J \, . \end{cases}$$

Furthermore there exists a constant C such that

$$\|v\|_{L^p(\mathbb{R}\times G)} \leq C \sum_{j=1}^{J} \|\varphi_j\|_{W^{-1/p,p}(\mathbb{R}\times S_j)} .$$

However we have $M^* = D_t^2 - aD_t + L + bI$ since $a,b \in \mathbb{R}$ and $L = L^*$ and therefore M^* is M with a replaced by $-a$. For M we conclude that assuming (H), $2 < p < +\infty$, $\alpha_{n-1}(p') > 0$ and $b \neq \lambda_\ell$, $\ell = 1, 2 \cdots$ then for every $\varphi_j \in W^{-1/p,p}(\mathbb{R}\times S_j)$, $1 \leq j \leq J$ there exists a unique $w \in L^p(\mathbb{R}\times G)$ which is the solution of

$$\begin{cases} Mw = 0 & \text{in } \mathbb{R} \times G \\ w = \varphi_j & \text{on } \mathbb{R} \times S_j , \quad 1 \leq j \leq J . \end{cases}$$

This implies

Proposition 4.7. Assuming (H), $2 < p < +\infty$, $\alpha_{n-1}(p) > 0$ and $b \neq \lambda_\ell$, $\ell = 1, 2 \cdots$ then for every $u \in W^{2,p}(\mathbb{R}\times G)$ there exists a unique $w \in W^{2,p}(\mathbb{R}\times G)$ which is the solution of

$$\begin{cases} Mv = 0 & \text{in } \mathbb{R} \times G \\ w = u & \text{on } \mathbb{R} \times S_j , \quad 1 \leq j \leq J . \end{cases}$$

Proof. We have $\alpha_{n-1}(p') > 0$ since α_{n-1} is non increasing and $p' < p$; consequently we may use the previous result $\varphi_j = u_{|\mathbb{R}\times S_j} \in W^{2-1/p,p}(\mathbb{R}\times S_j)$, $1 \leq j \leq J$. Let w be the solution obtained earlier in $L^p(\mathbb{R}\times G)$, we will prove that $w \in W^{2,p}(\mathbb{R}\times S_j)$ by a method of differential quotients. Let τ_h be the translation operator along the t-axis defined by

$$(\tau_h w)(t;\sigma) = w(t+h;\sigma) , \quad h > 0 ;$$

we have clearly

$$\begin{cases} (\tau_h-1)^2 w \in L^p(\mathbb{R}\times G) \\ M(\tau_h-1)^2 w = 0 \quad \text{in } \mathbb{R}\times G \\ (\tau_h-1)^2 w = (\tau_h-1)^2\varphi_j \quad \text{in } \mathbb{R}\times S_j \, , \quad 1 \le j \le J \, . \end{cases}$$

Applying to $(\tau_h-1)^2 w$ the inequality in Lemma 4.6 we get

$$\|(\tau_h-1)^2 w\|_{L^p(\mathbb{R}\times G)} \le C \sum_{j=1}^{J} \|(\tau_h-1)^2\varphi_j\|_{W^{-1/p,p}(\mathbb{R}\times S_j)} = O(h^2)$$

since $\varphi_j \in W^{2-1/p,p}(\mathbb{R}\times S_j)$; letting $h \to 0$ we obtain $\frac{\partial^2 w}{\partial t^2} \in L^p(\mathbb{R}\times G)$. Thus we know that w , $\frac{\partial w}{\partial t}$, $\frac{\partial^2 w}{\partial t^2} \in L^p(\mathbb{R}\times G)$ and and consequently $Lw \in L^p(\mathbb{R}\times G)$ too. To conclude we use the assumption that $u \in W^{2,p}(\mathbb{R}\times G) \subset L^p(\mathbb{R};W^{2,p}(G))$, thus $L(u-w) \in L^p(\mathbb{R};L^p(G))$; from (H) we know that L is an isomorphism from $W^{2,p}(G) \cap W_0^{1,p}(G)$ onto $L^p(G)$, consequently owing to the uniqueness we have $u - w \in L^p(\mathbb{R};W^{2,p}(G))$ and finally $w \in L^p(\mathbb{R};W^{2,p}(G))$. Summing up we proved that $w \in W^{2,p}(\mathbb{R}\times G)$. Q.E.D.

Now we are able to extend Theorem 4.5 to the case where $p > 2$.

THEOREM 4.8. *Assuming* (H) , $\alpha_{n-1}(p) > 0$ *and* $\lambda_\ell \neq b$, $\ell = 1, 2 \cdots$ *then equation* (3.9) *has a unique solution* $w \in W^{2,p}(\mathbb{R}\times G) \cap W_0^{1,p}(\mathbb{R}\times G)$ *for every given* $k \in L^p(\mathbb{R}\times G)$.

Proof. We have only to consider the case $2 < p < +\infty$: Proposition 4.4 provides us with a $w_1 \in W^{2,p}(\mathbb{R}\times G)$ such that $Mw_1 - k \in W^{1,p}(\mathbb{R}\times G) \subset W^{s,p}(\mathbb{R}\times G)$ for $0 < s < \text{Inf}\,\{\frac{1}{p};\alpha_{n-1}(p)\}$.

Then Theorem 3.6 implies the existence of $w_2 \in W^{2,p}(\mathbb{R}\times G) \cap W_0^{1,p}(\mathbb{R}\times G)$ such that $Mw_2 = k - Mw_1$. Finally from Proposition 4.7 we know that there exists $w_3 \in W^{2,p}(\mathbb{R}\times G)$ such that $Mw_3 = 0$ in $\mathbb{R} \times G$ and $w_3 = - w_1$ on $\mathbb{R} \times S_j$, $j = 1, 2 \cdots J$. We obtain the desired solution by setting $w = w_1 + w_2 + w_3$. Q.E.D.

The following Theorem gives us a further regularity result.

THEOREM 4.9. Assuming (H), $\alpha_{n-1}(p) > m$ and $\lambda_\ell \neq b$, $\ell = 1, 2 \cdots$ then equation (3.9) has a unique solution $w \in W^{m+2,p}(\mathbb{R}\times G) \cap W_0^{1,p}(\mathbb{R}\times G)$ for every given $k \in W^{m,p}(\mathbb{R}\times G)$.

Proof. We use the method of differential quotients to get regularity in the t-variable and then use (H) to get regularity in the other variables.

5. Regularity in the Usual Sobolev Spaces

5.1. Returning to our original problem (3.5) (3.6), Theorem 4.9 means through Proposition 3.3, under the assumptions of Theorem 4.9, that problem (3.5) (3.6) has a unique solution $u \in P^{m+2,p}(\Omega)$ for every given $f \in P^{m,p}(\Omega)$. However this solution is obtained indirectly by first solving equation (3.9) under homogeneous Dirichlet condition. Furthermore the coefficients in equation (3.9) clearly depend on the Sobolev exponent and consequently the solution which we built, may also depend on it. We will now make precise this dependence by writing an explicit formula. As we already showed, we have

$$u(e^t\sigma) = e^{(s+2)t}\, w(t,\sigma)$$

$$f(e^t\sigma) = e^{st}\, k(t,\sigma)$$

where we have set $s = m-n/p$. Besides this, we know from [18] that

$$w = \frac{1}{2i\pi} \int_\gamma (A-zI)^{-1} (B+zI)^{-1} k\, dz$$

where γ is a curve separating the spectra of A and of $-B$ as defined in section 3.2. Consequently this means that:

$$u = \frac{1}{2i\pi} \int_\gamma (D_t^2+(n-2)D_t-zI)^{-1} (L+zI)^{-1} e^{2t} f\, dz$$

where the resolvent operators are calculated in the space

$$Y = \{u; e^{-(s+2)t} u \in L^p(R \times G)\} .$$

It is clear that

$$(D_t^2+(n-2)D_t-zI) = (D_t+\alpha_+(z)I)(D_t-\alpha_-(z)I)$$

where

$$\alpha\pm(z) = \frac{n-2}{2} \pm \sqrt{(\frac{n-2}{2})^2+z} .$$

On the other hand, let $\{\varphi_\ell\}_{\ell=1,2..}$ be an orthonormal basis in $L^2(G)$ such that every φ_ℓ is an eigenfunction for $-L$ with Dirichlet condition, related to the eigenvalue λ_ℓ :

$$-L\ \varphi_\ell = \lambda_\ell\ \varphi_\ell \in W_0^{1,2}(G) ;$$

then

$$(L+zI)^{-1}\psi = \sum_{\ell\geq 1} (z-\lambda_\ell)^{-1} (\int_G \psi\ \varphi_\ell d\sigma)\varphi_\ell .$$

Consequently we obtain the identity

$$u = \sum_{\ell\geq 1} (D_t+\alpha_+(\lambda_\ell)I)^{-1} (D_t+\alpha_-(\lambda_\ell)I)^{-1} (e^{2t} \int_G f\ \varphi_\ell dx)\varphi_\ell .$$

We now need to make explicit the operator $(D_t+\alpha I)^{-1}$; it is the convolution operator in the t variable by the elementary solution E of the operator $D_t + \alpha I$ such that

$$e^{-(s+2)t}E \in L^1(\mathbb{R}) \;;$$

therefore we have

$$(D_t+\alpha I)^{-1} = \begin{cases} e_+^{-\alpha t} * & \text{if} \quad \mathrm{Re}\alpha > -(s+2) \\ -e_-^{-\alpha t} * & \text{if} \quad \mathrm{Re}\alpha < -(s+2) \end{cases}$$

where we denote by $e_\pm^{-\alpha t}$ the function which coincides with $e^{-\alpha t}$ for $t \geq 0$ (resp. $t \leq 0$) and vanishes otherwise. To conclude we must know the sign of $\alpha_\pm(\lambda_\ell) + s + 2$.

It follows from the fact that λ_ℓ is positive, that

$$\alpha_+(\lambda_\ell) > n - 2$$

for all ℓ ; consequently

$$\alpha_+(\lambda_\ell) + s + 2 > s + n = m + \frac{n}{p}, > 0$$

since $m \geq 0$, $1 < p < +\infty$. On the contrary it is easily seen that $\alpha_-(\lambda_\ell) + s + 2$ may change sign when ℓ varies, however $\alpha_-(\lambda_\ell) + s + 2$ never vanishes under the assumptions of Theorem 4.9: indeed the hypothesis $\lambda_\ell \neq b = (s+2)(s+n)$ is equivalent to $\alpha_-(\lambda_\ell) + s + 2 \neq 0$.

Summing up, we obtain the following formula:

(5.1)
$$u = \sum_{\lambda_\ell<(s+2)(s+n)} e_+^{-\alpha_+(\lambda_\ell)t} * e_+^{-\alpha_-(\lambda_\ell)t} * (e^{2t} \int_G f\varphi_\ell \, d\sigma)\varphi_\ell$$
$$- \sum_{\lambda_\ell>(s+2)(s+n)} e_+^{-\alpha_+(\lambda_\ell)t} * e_-^{-\alpha_-(\lambda_\ell)t} * (e^{2t} \int_G f\varphi_\ell \, d\sigma)\varphi_\ell$$

It is clear from this formula that the solution u depends only on $s = m - n/p$ and furthermore it is the same for two Sobolev exponenets s' and s'' such that none of the λ_ℓ's is in the interval

$$[(s'+2)(s'+n)\ ;\ (s''+2)(s''+n)]\ .$$

Now on the contrary, if we consider two Sobolev exponents s' and s" such that $\lambda_\ell \neq (s'+2)(s'+n)$ and $\ell \neq (s''+2)(s''+n)$ for all ℓ, although there may be some λ_ℓ's in the interval

$$](s'+2)(s'+n)\ ;\ (s''+2)(s''+n)[\ ,$$

then the difference between the two solution is

$$\sum_{\substack{\lambda_\ell>(s'+2)(s'+n)\\ \lambda_\ell<(s''+2)(s''+n)}} e_+^{-\alpha_+(\lambda_\ell)t} * e^{-\alpha_-(\lambda_\ell)t} * (e^{2t} \textstyle\int_G f\varphi_\ell\, d\omega)\varphi_\ell .$$

The terms in this expression may be rewritten as follows:

$$\int_{-\infty}^{t} e^{-\alpha_+(\lambda_\ell)(t-\tau)} \int_{-\infty}^{+\infty} e^{-\alpha_-(\lambda_\ell)(\tau-\xi)}\, e^{2\xi}$$

$$\times \int_G f(e^{\xi}\omega)\varphi_\ell\,(\omega)\ d\omega\ d\xi\ d\tau\ \varphi_\ell$$

$$= e^{-\alpha_+(\lambda_\ell)t} \int_{-\infty}^{t} e^{\{\alpha_+(\lambda_\ell)-\alpha_-(\lambda_\ell)\}\tau}\ d\tau\ \varphi_\ell$$

$$\times \int_{-\infty}^{+\infty} e^{\{\alpha_-(\lambda_\ell)+2\}\xi} \int_G f(e^{\xi}\omega)\varphi_\ell\,(\omega)\ d\omega\ d\xi\ ,$$

and if we set $e^t = r, e^{\xi} = \rho$, $x = \rho\omega$ we obtain

$$\frac{1}{\alpha_+(\lambda_\ell)-\alpha_-(\lambda_\ell)} r^{-\alpha_-(\lambda_\ell)} \varphi_\ell(\sigma) \int_\Omega \rho^{\alpha_-(\lambda_\ell)+2-n} \varphi_\ell(\omega)\, f\, dx .$$

This shows that the difference between the solutions corresponding to s' and s'' is given by

$$(5.2) \qquad \sum_{(s'+2)(s'+n)<\lambda_\ell<(s''+2)(s''+n)} \left(\int_\Omega f\, w^*_\ell\, dx\right) w$$

where

$$\begin{cases} w_\ell(r\sigma) = - r^{-\alpha_-(\lambda_\ell)} \varphi_\ell(\sigma) \\ w^*(r\sigma) = r^{-\alpha_+(\lambda_\ell)} \varphi_\ell(\sigma) . \end{cases}$$

It is easily seen that w_ℓ and w^*_ℓ are harmonic functions in Ω, vanishing on Γ; furthermore $w^*_\ell \in W^{-m,p'}(\Omega_R)$ for every finite R.

In what follows we will use all these preliminary results to study the regularity of the variational solution $u \in W^{1,2}_0(\Omega)$; this solution is related to the Sobolev exponent $s = -1 - \frac{n}{2}$ which minimizes the function $s \to (s+2)(s+n)$. Consequently when we compare the variational solution with the solution corresponding to $s = m - n/p$, the difference is the sum (5.2) over the ℓ's such that

$$(5.3) \qquad \lambda_\ell < \left(m+2 - \frac{n}{p}\right)\left(m+n - \frac{n}{p}\right) .$$

5.2. Now we are able to conclude in the case $n = 2$: we have already observed that (H) is fulfilled with $\alpha_1(p) = +\infty$, when Ω is an infinite sector with interior angle ω. It is easily seen that $\lambda_\ell = \ell^2 \frac{\pi^2}{\omega^2}$, $\ell = 1,2,\cdots$ and therefore inequality (5.3) reduces to

$$\ell \frac{\pi}{\omega} < m+2 - \frac{2}{p}\ , \ \ell = 1,2,\cdots$$

Thus from lemma 2.2, proposition 3.3 and Theorem 4.9 we conclude that for a given $f \in W_0^{m,p}(\Omega)$, problem (3.5)(3.6) has its variational solution u in $W^{m+2,p}(\Omega_R)$ for every $R < +\infty$ provided $m+2 - \frac{2}{p} < \frac{\pi}{\omega}$. From Theorem 3.6 we deduce that $u \in W^{s+2,p}(\Omega_{R_2})$ for $R < +\infty$, when f is given in $W^{s,p}(\Omega)$ and $s + 2 - \frac{2}{p} < \frac{\pi}{\omega}$, $0 < s < 1/p$.

Furthermore, we are also able to calculate w_ℓ and w_ℓ^*: let us suppose that in the usual polar coordinates where $x_1 = r \cos \theta$, $x_2 = r \sin \theta$, Ω is defined by $0 < \theta < \omega$; then $w_\ell(x) = \sqrt{2} r^{\ell\pi/\omega} \sin \ell\pi \frac{\theta}{\omega}$, $w_\ell^* = \sqrt{2} r^{-\ell\pi/\omega} \sin \ell\pi \frac{\theta}{\omega}$. The function w_ℓ is regular when $\ell\pi/\omega$ is an integer; otherwise it is readily seen that $w_\ell \notin W^{m+2,p}(\Omega_R)$ $(R<+\infty)$ for $\ell\pi/\omega < m+2 - \frac{2}{p}$. Consequently when f is given in $W_0^{m,p}(\Omega)$ the variational solution u of problem (3.5)(3.6) is in $W^{m+2,p}(\Omega_R)$ $(R<+\infty)$ iff f is orthogonal to those w_ℓ^* with $\ell \frac{\pi}{\omega} < m+2 - \frac{2}{p}$ and $\ell \frac{\pi}{\omega}$ non integer; if this last condition is not fulfilled then u is in the space spanned by $W^{m+2,p}(\Omega_R)$ and the functions w_ℓ.

By localization we obtain the corresponding results where Ω is a bounded connected open set in R^2 with a polygonal boundary (here we use the same notation as in section 1.1).

THEOREM 5.1. *Let $u \in W_0^{1,2}(\Omega)$ be the solution of $\Delta u = f \in W_0^{m,p}(\Omega)$; then $u \in W^{m+2,p}(\Omega)$ if we assume $m - \frac{2}{p} + 2 < \Pi/\omega_j$, $1 \le j \le J$.*

This is the desired result (Theorem 1 for the Dirichlet problem) when $m=0$ since $W_0^{0,p}(\Omega) = L^p(\Omega)$. For a general $m > 0$, the boundary conditions on f in $W_0^{m,p}(\Omega)$ are superflous; indeed we will cancel them by using

Lemma 5.2. Let $f \in W^{m,p}(\Omega)$, then there exists a $v \in W^{m+2,p}(\Omega) \cap W_0^{1,p}(\Omega)$ such that $\Delta v - f \in W_0^{m,p}(\Omega)$ provided $\omega_j \neq \frac{\ell\pi}{k}$ where k, ℓ are integers ≥ 1 with $k < m+2-\frac{2}{p}$, $j = 1,2,\cdots J$.

It must be observed here that the exceptional angles are those for which some of the w_ℓ are regular. As a consequence of Theorem 5.1 and Lemma 5.2 we get

Corollary 5.3. Let $u \in W_0^{1,2}(\Omega)$ be the solution of $\Delta u = f \in W^{m,p}(\Omega)$; then $u \in W^{m+2,p}(\Omega)$ when $m - \frac{2}{p} + 2 < \frac{\pi}{\omega_j}$, $1 \leq j \leq J$.

Lemma 5.2 is a consequence of Theorem 2.5 through an elementary but cumbersome calculation which is done in [19] when $p = 2$; the proof when $p \neq 2$ is analogous. For the non integer orders of differentiation we have:

THEOREM 5.4. Let $u \in W_0^{1,2}(\Omega)$ be the solution of $\Delta u = f \in W^{s,p}(\Omega)$, $0 < s < 1/p$; then $u \in W^{s+2,p}(\Omega)$ when $s - \frac{2}{p} + 2 < \frac{\pi}{\omega_j}$, $1 \leq j \leq J$.

The extension of Theorem 5.1 to large m is

THEOREM 5.5. Assuming that $\frac{\ell\pi}{\omega_j} \neq m+2-\frac{2}{p}$, $\ell \geq 1$, $1 \leq j \leq N$, the range of Δ as an operator from

$$\{u \in W^{m+2,p}(\Omega) \cap W_0^{1,2}(\Omega) / \Delta u \in W_0^{m,p}(\Omega)\}$$

in $W_0^{m,p}(\Omega)$ is closed and of finite codimension equal to $\sum_{j=1}^{J} \chi_j$ where

$$\chi_j = \text{Card}\{\ell \text{ integer}; \ell \geq 1, \frac{\ell\pi}{\omega_j} \text{ non integer}, \frac{\ell\pi}{\omega_j} < m+2-\frac{2}{p}\}$$

Of course the image is the orthogonal complement of the space

$$\{v \in W^{-m,p'}(\Omega) \; ; \; \Delta v = 0 \text{ in } \Omega \; , \; v|_{\Gamma_j} = 0 \; , \; 1 \leq j \leq J\}$$

which is of dimension $\sum_{j=1}^{J} \chi_j$. Finally if f is not orthogonal to this space, the solution u is in the space spanned by $W^{m+2,p}(\Omega) \cap W_0^{1,p}(\Omega)$ and a finite number of functions $w_{j,\ell}, 1 \leq \ell < \frac{\omega_j}{\pi}(m+2-\frac{2}{p})$, $1 \leq j \leq J$; these $w_{j,\ell}$ are harmonic functions in Ω , vanishing on the boundary and $w_{j,\ell}$ is regular everywhere outside σ_j where the singularity is of the same kind as the singularity of $r^{-\ell\pi/\omega_j}\sin\ell\pi\;\theta/\omega_j$ in the infinite sector. This proves Theorem 2 for the Dirichlet boundary conditions.

5.3. Let us now turn to the case where $n = 3$, which is our essential goal. Actually, Theorem 5.4 and Corollary 5.3 are exactly what we need to compute α_2 and therefore verify (H) . By the usual Korn procedure we see that

$$\alpha_2(p) = \sum_{j=1}^{J} \left(\frac{\pi}{\omega_j} - 2 + \frac{2}{p}\right)$$

where the ω_j , $1 \leq j \leq J$ are the measures of the interior angles in G (notation as in section 3.1). Thus Theorem 4.9 asserts that when $\frac{\omega_j}{\pi}(m+2-\frac{2}{p}) < 1$, $1 \leq j \leq J$, and $\lambda_\ell \neq b$, $\ell = 1,2,\cdots$ then equation (3.9) has a unique solution $w \in W^{m+2,p}(R\times G) \cap W_0^{1,p}(R\;G)$ for every given $k \in W^{m,p}(R\times G)$. This last result also proves Theorem 3.1, since by the Korn procedure, it is easily seen to be a Corollary of Theorem 4.9 with $b = 0$.

From now on let Ω be a bounded connected open set in R^3 with a polyhedral boundary (here we use the same notation as in part 1.2).

THEOREM 5.6. Let $u \in W_0^{1,2}(\Omega)$ be the solution of $\Delta u = f$ with $f \in W_0^{m,p}(\Omega)$; then $u \in W^{m+2,p}(\Omega)$ if we assume

$$\text{(i)} \quad \frac{\omega_{j,k}}{\pi}\left(m+2-\frac{2}{p}\right) < 1 \quad \text{for all } j,k$$

$$\text{(ii)} \quad \left(m+2-\frac{3}{p}\right)\left(m+3-\frac{3}{p}\right) < \lambda_i^{(1)} \quad \text{for all } i .$$

Proof: The regularity of u inside Ω and up to the faces Γ_j is well known (see [3] for instance). Then condition (i) ensures that u is of class $W^{m+2,p}$ near the edges owing to Theorem 3.1. Finally condition (ii) is exactly the condition for the sum (5.2) to vanish near the vertex σ_i whose solid angle is G_i , consequantly $u \in W^{m+2,p}(\Omega)$. Q.E.D. Of course, using a three dimensional analogue for Lemma 5.2 (which is also a consequence of Theorem 2.5) we get:

Corollary 5.7. The result of Theorem 5.6 remains true for f given in $W^{m,p}(\Omega)$ under the same hypothesis.

For studying the same problem without condition (ii) we observe that the functions w_ℓ and w_ℓ^* defined in 5.1 are for $n = 3$

$$w_\ell(r\sigma) = r^{-\frac{1}{2}+\sqrt{\frac{1}{4}+\lambda_\ell}}\,\varphi_\ell(\sigma)$$

$$w_\ell^*(r\sigma) = r^{-\frac{1}{2}-\sqrt{\frac{1}{4}+\lambda_\ell}}\,\varphi_\ell(\sigma) .$$

The difficulty here is that it is hard to know the regularity of w_ℓ near the vertex of the cone. Clearly w_ℓ is not regular when the power of r is not an integer; however this is a further requirement on the eigenvalues λ_ℓ . The condition for u to be of class $W^{m+2,p}$ near the vertex is therefore that f be orthogonal to w_ℓ^* for those indices ℓ for which w_ℓ is not in $W^{m+2,p}$ near the vertex. By localization we get:

THEOREM 5.8. Assuming that $\frac{\omega_{j,k}}{\pi}(m+2-\frac{2}{p}) < 1$ for all j,k and $(m+2-\frac{3}{p})(m+3-\frac{3}{p}) \neq \lambda_i^{(\ell)}$ for all i,ℓ the range of Δ as an operator from

$$\{u \in W^{m+2,p}(\Omega) \cap W_0^{1,2}(\Omega) \mid \Delta u \in W^{m,p}(\Omega)\}$$

in $W_0^{m,p}(\Omega)$ is closed and of finite codimension less than or equal to $\sum_{i=1}^{I} \chi_i$ where $\chi_i = \text{Card}\{\ell \mid \ell \text{ integer} \geq 1,$ $(m+2-\frac{3}{p})(m+3-\frac{3}{p}) > \lambda_i^{(\ell)}\}$.

Thus the proof of Theorem 3 in the case of Dirichlet boundary conditions is complete. When we consider a general $f \in W_0^{m,p}(\Omega)$, the solution lies in the span of $W^{m+2,p}(\Omega) \cap W_0^{1,p}(\Omega)$ and the exceptional functions $w_i^{(\ell)}$, defined for $\lambda_i^{(\ell)} < (m+2-\frac{3}{p})(m+3-\frac{3}{p}), i = 1,\cdots,I$ as follows; $w_i^{(\ell)}$ is harmonic in Ω, vanishes on the boundary, is regular outside σ_i and behaves like

$$r^{-\frac{1}{2}+\sqrt{\frac{1}{4}+\lambda_i^{(\ell)}}} \varphi_i^{(\ell)}(\sigma)$$

where $\varphi^{(\ell)}$ is an eigenfunction of L on G_i with Dirichlet boundary conditions, related to $-\lambda_i^{(\ell)}$ ($\{r,\sigma\}$ being the spherical coordinates with origin in σ_i).

We will end this section by proving that in several cases (of practical interest), condition (i) in Theorem 5.6 implies condition (ii) (and therefore $\chi_i = 0$ in Theorem 5.8 and no exceptional functions $w_i^{(\ell)}$ are needed to describe the solutions). We assume that the infinite cone Ω is such that G is contained in one of the dihedra formed by two successive faces of Ω (see fig. 2). Let us denote by ω the measure of the corresponding angle; then after rotating the coordinates axes we have $G \subset G_\omega$ where

$$G_\omega = \{(\cos\theta\sin\varphi;\sin\theta\sin\varphi\ ;\cos\varphi);\ 0 < \theta < \omega, 0 < \varphi < \pi\}$$

and consequently we have $\lambda_1 > \mu_1$ where $-\mu_1$ is the least eigenvalue of L under Dirichlet boundary conditions in G_ω (see [11]). Then it is shown in [20] that $\mu_1 > \pi^2/\omega^2$.

Now if we assume that each of the cones corresponding to the vertices σ_i of Ω, fulfills the geometric condition introduced above, then $\lambda_i^{(1)} > \pi^2/\omega_{j,k}^2$ for some j,k. Consequently condition (ii) is true when

$$\frac{\pi^2}{\omega_{j,k}^2} \geq (m+2-\frac{3}{p})(m+3-\frac{3}{p}) .$$

Then condition (i) implies

$$\frac{\pi^2}{\omega_{j,k}^2} > (m+2-\frac{2}{p})^2$$

and this shows that (i) implies (ii) at least when

$$(m+2-\frac{2}{p})^2 \geq (m+2-\frac{3}{p})(m+3-\frac{3}{p}) ;$$

this is true when $p \leq 2$. Summing up we have proved:

<u>Corollary 5.9</u>. <u>We assume that for each vertex</u> σ_i <u>of</u> Ω, <u>the infinite cone of solid angle</u> G_i <u>is contained in one of the infinite dihedra constituted by two successive faces of</u> Ω <u>passing by</u> σ_i. <u>We also assume that</u>

(i) $\frac{\omega_{j,k}}{\pi}(m+2-\frac{2}{p}) < 1$ <u>for all</u> j,k <u>then</u> $u \in W_0^{1,2}(\Omega)$ <u>the solution of</u> $\Delta u = f \in W_0^{m,p}(\Omega)$, <u>is in</u> $W^{m+2,p}(\Omega)$, <u>provided</u> $1 < p \leq 2$.

We observe finally that the geometric condition is fulfilled when Ω is convex.

Appendix No. 1.

We sketch here the calculations which prove Proposition 3.4 when $n = 2$, which is the only case where we used it. We identify $\mathbb{R}^2$ with the complex plane, setting $x = re^{i\theta}$; Ω is an infinite sector, but an extension of the function f in the variable θ reduces the problem to the particular case where Ω is replaced by $\mathbb{R}^2$. From Definition 3.4 we know that an f such that $r^{-s}f \in L^p(\mathbb{R}^2)$, is in $P^{s,p}(\mathbb{R}^2)$ iff the integral

$$I = \iint_{R^2\times R^2} |f(x) - f(y)|^p \frac{dx\ dy}{|x-y|^{2+sp}}$$

converges. If we set $x = e^{\xi+i\theta}$ and $y = e^{(\xi+\eta)+i(\theta+k)}$ we obtain

$$I = 2\iint_{\substack{\eta\le 0\\ |k|\le\pi}} e^{(2-sp)\xi}|g(\xi,\theta) - g(\xi+h;\theta+k)|^p \frac{e^{2\eta}\ d\xi\ d\theta\ d\eta\ dk}{\{(e^{\eta}-1)+|k|\}^{2+sp}}.$$

Let us denote by I_0 the same integral with the integration in η performed only on the interval $-1 \le \eta \le 0$; the difference between I and I_0 is finite iff the following integrals converge

$$I' = \iint_{\substack{\eta\le -1\\ |k|\le\pi}} e^{(2-sp)\xi}|g(\xi,\theta)|^p \frac{e^{2\eta}\ d\xi\ d\theta\ d\eta\ dk}{\{(e^{\eta}-1)+k\}^{2+sp}}$$

$$I'' = \iint_{\substack{\eta\le -1\\ |k|\le\pi}} e^{(2-sp)\xi}|g(\xi+\eta;\theta+k)|^p \frac{e^{2\eta}\ d\xi\ d\theta\ d\eta\ dk}{\{(e^{\eta}-1)+k\}^{2+sp}}.$$

We will show that since $r^{-s}f \in L^p(\mathbb{R}^2)$ then I' and I'' are finite. Indeed performing the integration in k and η, we see that

$$I' = C' \iint e^{(2-sp)\xi} |g(\xi;\theta)|^p \, d\xi \, d\theta < +\infty$$

since

$$\iint_{\eta\leq -1} \frac{e^{2\eta} \, d\eta \, dk}{\{(e^{\eta}-1)+k\}^{2+sp}} = C' < +\infty \quad .$$

Then in I" we replace θ by $\theta + k$ and set $\varepsilon = \xi + \eta$, thus we get

$$I'' = C'' \iint_{\eta\leq -1} e^{(2-sp)\varepsilon} \, |g(\varepsilon,\theta)|^p \, d\varepsilon \, d\theta < +\infty$$

since

$$\iint_{\eta\leq -1} \frac{e^{sp\eta} \, d\eta \, dk}{\{(e^{\eta}-1)+k\}^{2+sp}} = C'' < +\infty \quad .$$

Let us now study the meaning of the finiteness of I_0 ; we assert that it means the convergence of

$$J = 2 \iint_{\substack{-1\leq\eta\leq 0 \\ |k|\leq\eta}} |e^{(2/p-s)\xi} \, g(\xi;\theta) - e^{(2/p-s)(\xi+\eta)} \, g(\xi+\eta;\theta+k)|^p$$

$$\times \frac{e^{2\eta} \, d\xi \, d\theta \, d\eta \, dk}{\{(e^{\eta}-1)+|k|\}^{2+sp}}$$

because the difference between I_0 and J is

$$2 \iint_{\substack{-1\leq\eta\leq 0 \\ |k|\leq\pi}} |v(\xi+\eta;\theta+k)|^p \, e^{(2-sp)(\xi+\eta)} \, (1-e^{-(2-sp)\eta})$$

$$\times \frac{e^{2\eta} \, d\xi \, d\theta \, d\eta \, dk}{\{(e^{\eta}-1)+|k|\}^{2+sp}} \, .$$

This last integral is convergent since by replacing θ by $\theta + k$, setting $\varepsilon = \xi + \eta$ and integrating in k we see that it is proportional to

$$\iint_{\substack{-1\le\eta\le 0\\ |k|\le\pi}} |v(\varepsilon;\theta)|^p\, e^{(2-sp)\varepsilon}\,(1-e^{-(2-sp)\eta})\,\frac{e^{2\eta}\,d\varepsilon\,d\theta\,d\eta}{(e^{\eta}-1)^{1+sp}}$$

$$= C'' \iint |v(\varepsilon;\theta)|^p\, e^{(2-sp)\varepsilon}\,d\varepsilon\,d\theta < +\infty$$

where

$$C' = \int_{-1}^{0} (1-e^{-(2-sp)\eta})\,\frac{e^{2\eta}}{(e^{\eta}-1)^{1+sp}}\,d\eta < +\infty$$

(we recall that $sp < 1$ by assumption).

Summing up we know that I is finite iff J is finite and the convergence of J is equivalent to the convergence of

$$\iint_{\substack{-1\le\eta\le 0\\ |k|\le\pi}} |e^{(2/p-s)\xi}\, g(\xi;\theta) - e^{(2/p-s)(\xi+\eta)}\, g(\xi+\eta;\theta+k)|^p$$

$$\times \frac{d\xi\,d\theta\,d\eta\,dk}{\{|\eta|+|k|\}^{2+sp}}$$

which means that $e^{(2/p-s)\xi}\, g \in W^{s,p}(\mathbb{R}\times]-\pi;+\pi[)$. Q.E.D.

Appendix No. 2.

Let us indicate briefly here how Lemma 4.3 can be proved for $p \neq 2$: we only need to show that for every $\varphi \in D(\mathbb{R}^{n-1})$ the pseudo-differential operator with symbol $\varphi(x)\, e(x;\tau,\xi)$ is continuous from $L^p(\mathbb{R}^n)$ into $W^{m,p}(\mathbb{R}^n)$. This will be achieved using an expansion of φe of the form

$$\varphi(x)\, e(x;\tau,\xi) = \sum_{k\ge 1} e_k(\tau,\xi)\, \psi_k(x)$$

where the e_k are such that

$$(1+\tau^2+|\xi|^2)^{+m/2}\, e_k(\tau,\xi)$$

is an L_p multiplier in $\mathbb{R}^n$ and the ψ_k are smooth functions. For instance we may choose the ψ_k to be an orthonormal sequence of eigenfunctions of the Laplace operator with Dirichlet boundary condition in some bounded open subset with a regular boundary, U in $\mathbb{R}^{n-1}$, such that the support of φ is contained in U. Therefore e_k is given by the following formula

$$e_k(x;\xi) = \int_U \varphi(x)\; e(x,\tau,\xi)\; \psi_k(x)\; dx$$

and we will estimate e_k. Let us denote by $-\mu_k$ the eigenvalue of $+\Delta$ related to ψ_k:

$$\Delta\psi_k = -\mu_k\,\psi_k\,, \quad k = 1, 2 \cdots$$

therefore we have

$$\psi_k = \frac{1}{\mu_k^q}\,(-\Delta)^q\,\psi_k$$

for every integer q ; and integrating by parts (we recall that φ has compact support in U) we get

$$e_k(x;\xi) = \frac{1}{\mu_k^q}\int_U (-\Delta)^q\,\{\varphi(x)\; e(x,\tau,\xi)\}\;\psi_k(x)\; dx\,.$$

Consequently using the assumptions in Lemma 4.3 we see that

$$|D_\tau^\ell\, D_\xi^\beta\, e_k(\tau,\xi)| \le \frac{C}{\mu_k^q}\,(1+|\xi|^2+\tau^2)^{-m-\frac{\ell+|\beta|}{2}}$$

for every ℓ and β. These inequaltities are exactly what we need for applying Mikhlin's theorem (see [40] for instance) to $(1+|\xi|^2+\tau^2)^{m/2}\, e_k$; thus the operator with symbol e_k is continuous from $L^p(\mathbb{R}^n)$ into $W^{m,p}(\mathbb{R}^n)$ with a norm which is $O(1/\mu_k^q)$ when $k \to +\infty$, and then we have

$$\|\varphi E f\|_{W^{m,p}(\mathbb{R}^n)} \leq C\Big(\sum_{k\geq 1} \|\psi_k\|_{W^{m,\infty}(U)} \frac{1}{\mu_k^q}\Big) \|f\|_{L^p(\mathbb{R}^n)}$$

$$\leq C'\Big(\sum_{k\geq 1} \|\psi_k\|_{W^{\sigma,2}(U)} \frac{1}{\mu_k^q}\Big) \|f\|_{L^p(\mathbb{R}^n)}$$

owing to Sobolev's theorem provided $\sigma - m > \frac{n-1}{2}$. If we assume that σ is even we get

$$\|\psi_k\|_{W^{\sigma,2}(U)} \leq C'' \|\Delta^{\sigma/2} \psi_k\|_{L^2(U)} = C'' \mu_k^{\sigma/2}$$

and consequently

$$\|\varphi E f\|_{W^{m,p}(\mathbb{R}^n)} \leq C'''\Big(\sum_{k\geq 1} \mu_k^{\sigma/2-q}\Big) \|f\|_{L^p(\mathbb{R}^n)} \;;$$

we know that the last series converges when q is chosen large enough since it is proved in [11] that

$$\liminf_{k \to +\infty} \mu_k\, k^{-\frac{2}{n-1}} > 0 \,.$$

This proves the continuity of φE from $L^p(\mathbb{R}^n)$ into $W^{m,p}(\mathbb{R}^n)$.

Bibliography

1. Adams, Aronszajn, Smith, Theory of Bessel potentials part II, Annales de l'Institut Fourier, Grenoble 17, 2, 1967, 1-135; Aronszjan, Mulla, Szeptycki, On Spaces of potentials connected with L^p classes, Annales de l'Institut Fourier, Grenoble 13, 2, 1963, 211-306.
2. Agmon, On the eigenfunctions and on the eigenvalues of general elliptic boundary value problems, Comm. on Pure and Applied Math., Vol. 15, 1962, 119-147.
3. Agmon, Douglis, Nirenberg, Estimates near the boundary for solutions of elliptic partial differential equations satisfying general boundary conditions, Comm. on Pure and Applied Math., Vol. 17, 1959, 623-727.
4. Agranovitch, Vichik, Elliptic problems with parameters and parabolic problems of general type, Uspehi Mat. Nauk, Vol. 19, 3, 1964, 53-161 (and Russian Math. Survey, Vol. 19, 1964, 53-157).
5. Avantaggiati, Troisi, Problemi al contorno di tipo ellittico in un dominio piano limitato e dotato di punti angolosi, Atti della Accademia Nazionale dei Lincei Vol. XII, Fasc. 4, 1974, 271-308 and Spazi di Sobolev con peso e problemi ellittici in un angolo I, II, III, Annali di Matem. Pura ed Applicata, Vol. 95, 1973, 361-348, Vol. 97, 1973, 207-252, Vol. 99, 1974, 1-64.
6. Ben M'Barek, Régularité L^p de la solution d'un problème de transmission dans un polygone, Doctoral dissertation (3rd cycle), University of Nice, 1975, to appear.
7. Besov, The continuation of functions in L_p^ℓ and W_p^ℓ , Trudy Mat. Instituta Steklova, 89, 1967, 5-17 (and Proceedings of the Steklov Institute of Math. 89, 1968, 1-15).
8. Birman Skvorcov, On the square summability of highest derivatives of the solution the Dirichlet problem in a domain with piecewise smooth boundary, Izv. Vyss. Ucebn Zaved Matematicka n° 5,(30),1962, 11-20.
9. Calderon, Boundary value problems for Elliptic equations, Joint Soviet American Symposium on Partial Differential Equations, 1963, Novosibirsk.
10. Calderon, Vaillancourt, On the boundedness of pseudo differential operators, J. Math. Soc. Japan, Vol. 23, n° 2, 1971, 374-378.
11. Courant, Hilbert, Method of Mathematical Physics, Interscience pub., 1962.
12. Da Prato, Grisvard, Sommes d'opérateurs linéaires et équations différentielles opérationnelles, to appear in J. de Mathématiques Pures et Appliquées, 1975.

13. Eskin, The conjugacy problem for equations of principal type with two independent variables, Trudy Mosk. Mat. Obsh. t. 21, 1970, 245-292 (and Trans. Moscow Mat. Soc. Vol. 21, 1970, 263-316).
14. Fufaev, On the Dirichlet problem in a domain with angles, Doklady Akad Nauk 131, n° 1, 1960, 37-39.
15. Gagliardo, Caratterizzazioni della tracce sulla frontiera relative ad alcune classi di funzioni in n variabili, Rendiconti del Seminario Matematico della Università di Padova, Vol. XXVII, 1957, 284-30 .
16. Gandulfo, Problèmes elliptiques dans un quadrant du plan, C.R. Acad. Sc. Paris, t. 269, 1969, I, 836-839 and II, 896-899.
17. Grisvard, Commutativité de deux foncteurs d'interpolation et applications, J. Math. Pures et Appl., t. 45, 1966, 143-290.
18. Grisvard, Equations opérationnelles abstraites dans les espaces de Banach et problèmes aux limites dans des ouverts cylindriques, Annali della Scuola Normale Superiore di Pisa, Vol. XXI, Fasc. III, 1967, 307-347.
19. Grisvard, Alternative de Fredholm relative au problème de Dirichlet dans un polygone ou un polyèdre, première partie, Bollettino U.M.I. (4) 5, 1972, 132-164.
20. Grisvard, Alternative de Fredholm relative au problème de Dirichlet dans un polygone ou un polyèdre, seconde partie, to appear in Annali della Scuola Normale Superiore di Pisa, 1975.
21. Grisvard, Théorèmes de traces relatifs à un polyèdre, C.R. Acad. Sc. Paris, t. 278, 1581-1583 and related detailed paper to appear.
22. Hanna, Smith, Some remarks on the Dirichlet problem in piecewise smooth domains, Comm. Pure and Applied Math, Vol. XX, 1967, 577-593.
23. Ibuki, Dirichlet problem for elliptic equations of the second order in a singular domain of $\mathbb{R}^2$, J. Math. Kyoto Univ. 14-1, 1974, 55-71.
24. Kadlec, On the regularity of the solution of the Poisson problem on a domain with boundary locally similar to the boundary of a convex open set, Czechoslovak Math. J., t. 14 (89), 1964, 386-393.
25. Kato, Perturbation Theory for Linear Operators, Springer Verlag, Band 132.
26. Kellogg, Singularities in interface problems, Numerical Solution of Partial Differential Equations II, Academic Press, 1971, 351-400.
27. Komec, Elliptic boundary problems for pseudodifferential operators on manifolds with conical points, Mat. Sbornik, tom 86 (128), 1971, 268-298 (and Math. U.S.S.R. Sbornik,

Vol. 15, 1971, nº 2, 261-297).

28. Komec, Elliptic boundary value problems on manifolds with a piecewise smooth boundary, Mat. Sbornik, tom 92 (134), 1973, 89-134 (and Math. U.S.S.R. Sbornik, Vol. 21, 1973, nº 1, 91-135).
29. Kondratiev, Boundary value problems for elliptic equations in domains with conical or angular points, Trudy Moskovkogo Mat. Obschetsva, t. 16, 1967, 209-292 (and Transactions of the Moscow Mat. Soc., 1967, 227-313).
30. Kondratiev, The smoothness of a solution of Dirichlet's problem for 2nd order elliptic equations in a region with a piecewise smooth boundary, Differential Equations, Vol. 6, nº 10, 1970, 1831-1843 (and Differential Equations, Vol. 6, 1970, 1392-1401).
31. Ladyzenskaja Uralceva, Equations aux Dérivées Partielles de Type Elliptique, Dunod, Paris, 1968.
32. Lemrabet, Sur la régularité de la solution d'un problème de transmission, C.R. Acad. Sc. Paris, t. 279, 265-268 and related detailed paper to appear in J. Math. Pures et Appl. 1976.
33. Lions, Equations différentielles opérationneles et problèmes aux limites, Springer Verlag, Band 111.
34. Lions, Magenes, Problèmes aux limites non homogènes, Dunod, Paris, 1968.
35. Lions, Magenes, Problemi ai limiti non omogenei, Annali della Scuola Normale Superiore di Pisa, (III), Vol. XV, 1961, 39-101 and (V), Vol XV1, 1962, 1-44.
36. Lions, Peetre, Sur une classe d'espaces d'interpolation, Publications Mathématiques de l'I.H.E.S., Paris, Vol. 19, 1963, 5-68.
37. Lorenzi, A mixed problem for the Laplace equation in a right angle with an oblique derivative given on a side of the angle, Università degli Studi di Firenze, 1973.
38. Merigot, Régularité des dérivées de la solution du problème de Dirichlet dans un secteur plan, Le Mathematiche, Vol. XXVII, Fasc. 2, 1972, 1-36 and Etude du problème $\Delta u = f$ dans un polygone plan. Inégalités a priori, Bollettino U.M.I. (4), 10, 1974, 577-597.
39. Merigot, Solutions en normes L^p des problèmes elliptiques dans des polygones plans, Doctoral dissertation, University of Nice, 1974, to appear.
40. Mikhlin, Singular integrals in several variables and integral equations (Russian), Moscow, 1962.
41. Moussaoui, Régularité de la solution d'un problème à dérivée oblique, C.R. Acad. Sc. Paris, t. 279, 869-872.

42. Mazya, Plamenevski, The oblique derivative problem in a domain with a piecewise smooth boundary, Funckcional Anal. i Priložen 5, 1971, n° 3, 102-103 (and Functional Anal. Appl. 5, 1971, 256-258).
43. Necas, Les Méthodes Directes en Théorie des Équations Elliptiques, Masson, 1967.
44. Nikolski, Properties on the boundary of functions defined in domains with angular points (Russian), Mat. Sbornik, t. 43, 1957, 127-144.
45. Peetre, Sur le nombre de paramètres dans la définition de certains espaces d'interpolation, Ricerche di Matematica, Vol. XII, 1963, 248-261.
46. Peetre, Mixed problem for higher order elliptic equations in two variables, Annali della Scuola Normale Superiore di Pisa (I), Vol. XV, 1961, 337-353, (II), Vol. XVII, 1963, 1-12.
47. Rademacher, Uber partielle und totale differenzierbarkeit von funcktionen mehrerer variablen, und über die transformation der doppelintegrale, Math. Annalen 79, 1918, 340-359.
48. Sadallah, Régularité de la solution de l'équation de la chaleur dans un domaine plan non rectangulaire, C.R. Acad. Sc. Paris, 1975 and related detailed paper to appear in Supplemento de Bollettino dell' U.M.I., 1976.
49. Seeley, Singular integrals and boundary value problems, Amer. J. of Math., 80, 1966, 781-809.
50. Shamir, Mixed boundary value problem for elliptic equation in the plane; the L_p theory, Annali della Scuola Normale Superiore di Pisa, Vol. XVII, 1963, 117-139.
51. Shamir, Regularization of mixed second order elliptic problems, Israel J. of Maths., Vol. 6, 1968, 150-168.
52. Strang, Fix, An Analysis of the Finite Element Method, Prentice Hall, 1973.
53. Taibleson, On the theory of Lipschitz spaces of distributions on Euclidian n space, J. of Math. and Mech., Vol. 13, n° 3, 1964, 293-306.
54. Volkov, On the differential properties of solutions of boundary value problems for the Laplace and Poisson equations on a rectangle, Trudy Mat. Instituta Steklova 77, 1965, 89-112 (and Proc. of the Steklov Institute of Maths, 1967, 101-126).
55. Volkov, On the differential properties of solutions of boundary value problems for the Laplace equations on a polygon, Trudy Mat. Instituta Steklova 77, 1965, 113-142 (and Proc. of the Steklov Institute of Maths, 1967, 127-159).
56. Yosida, Functional Analysis, Springer Verlag, Band 123.
57. Yeiga (da), On the $W^{2,p}$ regularity for solutions of mixed problems, to appear in the J. de Maths pures et appl. 1975.

NUMERICAL SOLUTION OF NONLINEAR PARTIAL DIFFERENTIAL EQUATIONS OF MIXED TYPE*

Antony Jameson**

1. Introduction

The purpose of this paper is to review some recently developed numerical methods for the solution of nonlinear equations of mixed type. These methods have been used to calculate transonic flows with shock waves, and the discussion will be restricted to this topic, although some of the ideas could presumably be useful in other applications. Some typical transonic flow patterns are sketched in Figure 1. The type changes from elliptic in the region of subsonic flow to hyperbolic in the region of supersonic flow. If the flow is subsonic at infinity, the supersonic flow is confined to one or more bubbles standing above the profile. If the flow is supersonic at infinity, there is a subsonic pocket behind a detached bow wave, and oblique shock waves appear at the trailing edge, sometimes forming a fishtail pattern. Any proposed method should therefore be capable of handling a variety of complex flow patterns.

Three main elements can be recognized in the treatment of such a problem.

(1) The formulation of a suitable mathematical model, such as a differential equation or variational principle.

---

*Work supported by NASA under Grants NGR 33-016-167 and NGR 33-016-201 and by ERDA under Contract AT(11-1)-3077.

**Courant Institute of Mathematical Sciences

(2) The construction of a discrete approximation to the continuous problem.

(3) The solution of the resulting set of nonlinear equations for the undetermined parameters (typically nodal values of the discrete model).

The first question will not be discussed at length in this paper. The emphasis will be on the numerical methods for solving the two equations which have chiefly been used in transonic flow calculations, the transonic potential flow equation and the transonic small disturbance equation.

The potential flow equation can be derived from the Euler equations for inviscid compressible flow by introducing the assumption that the flow is irrotational, so that we can define a potential ϕ . Then we find that in smooth regions of two dimensional flow ϕ satisfies the quasilinear equation

$$(1.1) \qquad (a^2-u^2)\phi_{xx} - 2uv\phi_{xy} + (a^2-v^2)\phi_{yy} = 0$$

where u and v are the velocity components

$$(1.2) \qquad u = \phi_x \ , \ v = \phi_y$$

and a is the local speed of sound. Given the ratio of specific heats γ and the stagnation speed of sound a_0 , a can be determined from the energy relation

$$(1.3) \qquad a^2 = a_0^2 - \frac{\gamma-1}{2} q^2$$

where q is the speed $\sqrt{u^2+v^2}$. Equation (1.1) is hyperbolic when the local Mach number $M = q/a > 1$. On the profile the solution should satisfy the Neumann boundary condition

$$(1.4) \qquad \frac{\partial\phi}{\partial n} = 0$$

where n is the normal direction. At infinity the flow approaches a uniform stream with a Mach number M_∞. The density ρ and pressure p can be determined by the relations

$$\rho^{\gamma-1} = M_\infty^2 a^2 \tag{1.5}$$

and

$$p = \rho^\gamma / \gamma M_\infty^2 . \tag{1.6}$$

Multiplied by ρ/a^2, equation (1.1) is equivalent to the equation for conservation of mass

$$\frac{\partial}{\partial x}(\rho u) + \frac{\partial}{\partial y}(\rho v) = 0 . \tag{1.7}$$

Multiplied by $\rho u/a^2$, on the other hand, it is equivalent to the equation for conservation of the x component of momentum

$$\frac{\partial}{\partial x}(p+\rho u^2) + \frac{\partial}{\partial y}(\rho uv) = 0 . \tag{1.8}$$

For a flow in a finite region Ω the conservation law (1.7) can be derived from the Bateman variational principle that

$$I = - \iint_\Omega p \, dx \, dy$$

is stationary [1].

Smooth transonic flows are known to exist only in special cases [2]. In general shock waves appear. Thus we must admit weak solutions with suitable discontinuities. Since an irrotational flow is isentropic it is consistent to replace shock waves by jumps across which entropy is conserved. In this case it is not possible to conserve both mass and momentum. A fairly good approximation to shock waves of moderate strength is obtained by requiring the mass to be conserved. The corresponding momentum deficiency then yields an approximation to the wave drag [3]. Thus we seek a solution in which ϕ is continuous, and ϕ_x and ϕ_y are piecewise continuous, satisfying the conservation law (1.7) and the jump condition

(1.9) $$[\rho\phi_y] - \frac{dy}{dx}[\rho\phi_x] = 0$$

where [ ] denotes the jump, and dy/dx is the slope of the discontinuity.

If we construct a difference approximation in conservation form to the conservation law (1.7), then we can expect the jump condition (1.9) to be satisfied in the limit as the mesh width approaches zero [4]. Since the quasilinear form (1.1) does not distinguish between the conservation laws (1.7) and (1.8), difference approximations to (1.1) will not necessarily converge to a solution which satisfies the jump condition (1.9) unless a shock fitting procedure is used.

A useful simplification is provided by the small disturbance theory. Suppose that the profile is given in the form $y = \tau f(x)$ and τ is small. If we expand the solution in powers of τ under the assumption that $1-M_\infty^2 \sim \tau^{2/3}$, and retain only the leading term, we obtain the transonic small disturbance equation [5]. Let K be the similarity constant $(1-M_\infty^2)/\tau^{2/3}$. Then a typical form is

(1.10) $$A\phi_{xx} + \phi_{yy} = 0$$

where

(1.11) $$A = K - (\gamma+1)\phi_x$$

In this equation the y coordinate has been scaled by the factor $\tau^{1/3}$, and ϕ is the disturbance potential, scaled by the factor $\tau^{-2/3}$. The Neumann boundary condition is now transferred to the axis

$$\phi_y = \frac{df}{dx} \quad \text{at} \quad y = 0 . \tag{1.12}$$

If the small disturbance equation is written in the conservation form

$$\frac{\partial}{\partial x}\left(K\phi_x - \frac{\gamma+1}{2}\phi_x^2\right) + \phi_{yy} = 0 \tag{1.13}$$

then the corresponding jump condition

$$[\phi_y] - \frac{dy}{dx}\left[\phi_x - \frac{\gamma+1}{2}\phi_x^2\right] = 0 \tag{1.14}$$

yields a consistent approximation to shock waves [5].

2. Formulation of Finite Difference Methods

The methods proposed in this paper use finite difference approximations to the differential equation. Their formulation is based on an idea introduced by Murman and Cole [6]. That is to use central difference formulas in the subsonic zone, where the governing equation is elliptic, and upwind difference formulas in the supersonic zone, where it is hyperbolic. Thus the numerical scheme has a directional bias. This corresponds to the upwind region of dependence of the flow in the supersonic zone, and also serves the purpose of enforcing the entropy condition that discontinuous expansions must be excluded. If we consider the transonic flow past a profile with fore and aft symmetry such as an ellipse, the desired solution of the potential equation is not symmetric. Instead it exhibits a smooth acceleration over the front half of the profile followed by a discontinuous recompression through a shock wave. In the absence of a directional bias in the numerical scheme the fore and aft symmetry would be preserved in any solution which could be obtained, resulting in the appearance of improper discontinuities. It is not so easy to introduce the desired bias in a finite element formulation when there is no particular coordinate direction that can be treated separately as the time-like direction. Thus the construction of a unified finite element method for the subsonic and supersonic zones appears difficult.

The dominant term in the discretization error introduced by the upwind differencing acts like an artificial viscosity. We can turn this idea around. Instead of using a switch in the difference scheme to introduce a viscosity, we can explicitly add a viscosity which produces an upwind bias in the difference scheme at supersonic points. This simplifies the construction of difference schemes in conservation form. Suppose that we have a central difference approximation in conservation form. Then the conservation form will be preserved as long as the added viscosity has a divergence form. The effect of the viscosity is simply to alter the conserved quantities by terms proportional to the mesh width Δx, which vanish in the limit as Δx approaches zero. By including a switching function in the viscosity to make it vanish in the subsonic zone we continue to obtain the sharp representation of shock waves which results from switching the difference scheme.

The finite difference approximation produces a set of nonlinear difference equations. There remains the problem of finding a convergent iterative scheme for solving these equations. Suppose that in the $(n+1)^{st}$ cycle the residual R_{ij} at the point $i\,\Delta x$, $j\,\Delta y$ is evaluated by inserting the result $\phi_{ij}^{(n)}$ of the nth cycle in the difference approximation.

Then the correction $C_{ij} = \phi_{ij}^{(n+1)} - \phi_{ij}^{(n)}$ is to be calculated by solving an equation of the form

$$NC + \sigma R = 0 \tag{2.1}$$

where N is a discrete linear operator, and σ is a scaling function. In a relaxation method N is restricted to a lower triangular or block triangular form so that the elements of C can be determined sequentially. In the analysis of such a scheme it is helpful to introduce a time dependent analogy. The vector R is an approximation to $L\phi$, where L is the operator appearing in the differential equation. If we consider C as representing $\Delta t\, \phi_t$, where t is an artificial time coordinate, and N is an approximation to a differential operator $(1/\Delta x)F$, then equation (2.1) is an approximation to

$$F\phi_t + \sigma \frac{\Delta x}{\Delta t} L\phi = 0 \tag{2.2}$$

Thus we should choose N so that this is a convergent time dependent process.

With this approach the formulation of a relaxation method for solving an equation of mixed type is reduced to three main steps:

(1) Construct a central difference approximation to the differential equation.

(2) Add a numerical viscosity to produce the desired directional bias in the hyperbolic region.

(3) Add time dependent terms to embed the steady state equation in a convergent time dependent process.

Methods constructed along these lines have proved extremely reliable. Their main shortcoming is a rather slow rate of convergence.

In order to speed up the convergence we can extend the permissible class of operators N. In particular, if N is taken as the Laplacian, we can solve the resulting discrete Poisson equation by a fast direct method at each iteration. This method converges rapidly in subsonic flow but diverges for transonic flows. However, if we use several relaxation steps after each Poisson step, the two methods in combination give fast convergence.

3. The Relaxation Method for the Small Disturbance Equation

The treatment of the small disturbance equation is simplified by the fact that the characteristics are locally symmetric about the x direction. Thus the desired directional bias can be introduced simply by switching to upwind differencing in the x direction at all supersonic points. To preserve the conservation form some care must be exercised in the method of switching. Let p_{ij} be a central difference approximation to the x derivatives at the point $i\,\Delta x$, $j\,\Delta y$:

$$(3.1)\quad p_{ij} = K\{(\phi_{i+1,j}-\phi_{ij}) - (\phi_{ij}-\phi_{i-1,j})\}/\Delta x^2$$

$$- (\gamma+1)\{(\phi_{i+1,j}-\phi_{ij})^2 - (\phi_{ij}-\phi_{i-1,j})^2\}/2\Delta x^3$$

$$= A_{ij}(\phi_{i+1,j}-2\phi_{ij}+\phi_{i-1,j})/\Delta x^2$$

where

$$(3.2)\qquad A_{ij} = K - (\gamma+1)(\phi_{i+1,j}-\phi_{i-1,j})/2\Delta x$$

Also let q_{ij} be a central difference approximation to ϕ_{yy}

(3.3) $$q_{ij} = (\phi_{i,j+1} - 2\phi_{ij} + \phi_{i,j-1})/\Delta y^2$$

Define a switching function μ with the value unity at supersonic points and zero at subsonic points

(3.4) $$\mu_{ij} = \begin{cases} 0 & \text{if } A_{ij} > 0 \\ 1 & \text{if } A_{ij} < 0 \end{cases} .$$

Then we approximate equation (1.13) by

(3.5) $$p_{ij} + q_{ij} - \mu_{ij}p_{ij} + \mu_{i-1,j}p_{i-1,j} = 0 .$$

This is equivalent to Murman's conservative scheme [7]. In the supersonic zone p_{ij} is replaced by the upwind formula $p_{i-1,j}$. At points where the flow enters and leaves the supersonic zone μ_{ij} and $\mu_{i-1,j}$ have different values, giving special parabolic and shock point operators.

The added terms $-\mu_{ij}p_{ij} + \mu_{i-1,j}p_{i-1,j}$ are an approximation to $\partial P/\partial x$ where

$$P = \mu \, \Delta x \frac{\partial}{\partial x} (K\phi_x - \frac{\gamma+1}{2} \phi_x^2) .$$

This may be regarded as an artifical viscosity of order Δx. The use of a divergence form for the viscosity preserves the conservation form of the difference scheme, and it can be shown that the difference approximation converges to a solution satisfying the correct jump condition [7].

The nonlinear difference equations (3.1)-(3.5) may be solved by a generalization of the line relaxation method for elliptic equations. At each point we calculate the coefficient A_{ij} and the residual R_{ij} by substituting the result $\phi_{ij}^{(n)}$ of the previous cycle in the difference equations. Then we set $\phi_{ij}^{(n+1)} = \phi_{ij}^{(n)} + C_{ij}$ where the correction C_{ij} is determined by solving the linear equations

(3.6)

$$(C_{i,j+1}-2C_{ij}+C_{i,j-1})/\Delta y^2 + (1-\mu_{ij})A_{ij}(-\frac{2}{\omega}C_{ij}+C_{i-1,j})/\Delta x^2$$

$$+ \mu_{i-1,j}A_{i-1,j}(C_{ij}-2C_{i-1,j}+C_{i-2,j})/\Delta x^2 + R_{ij} = 0$$

on each successive vertical line. In these equations ω is the over-relaxation factor for subsonic points with a value in the range $1 \leq \omega \leq 2$. In a typical line relaxation scheme for an elliptic equation, provisional values $\tilde{\phi}_{ij}$ are determined on the line $x = i\,\Delta x$ by solving the difference equations with the latest available values $\phi_{i-1,j}^{(n+1)}$ and $\phi_{i+1,j}^{(n)}$ inserted at points on the adjacent lines. Then new values $\phi_{ij}^{(n+1)}$ are determined by the formula $\phi_{ij}^{(n+1)} = \phi_{ij}^{(n)} + \omega(\tilde{\phi}_{ij} - \phi_{ij}^{(n)})$. Equation (3.6) is derived by modifying this process. New values $\phi_{ij}^{(n+1)}$ are used instead of provisional values $\tilde{\phi}_{ij}$ to evaluate ϕ_{yy} at both supersonic and subsonic points. At supersonic points ϕ_{xx} is also evaluated using new values. At subsonic points ϕ_{xx} is evaluated from $\phi_{i-1,j}^{(n+1)}$, $\phi_{i+1,j}^{(n)}$ and a linear combination of $\phi_{ij}^{(n+1)}$ and $\phi_{ij}^{(n)}$ equivalent to $\tilde{\phi}_{ij}$. In the subsonic zone the scheme acts like a line relaxation scheme, with a comparable rate of convergence. In the supersonic zone it is equivalent to a marching scheme, once the coefficients A_{ij} have been evaluated. Since the supersonic difference scheme is implicit, no limit is imposed on the step length Δx as A_{ij} approaches zero near the sonic line. The transition at the sonic line is effected smoothly because ϕ_{yy} is treated in the same manner throughout the flow. If provisional values $\tilde{\phi}_{ij}$ were used to evaluate ϕ_{yy} at subsonic points, there would be a discontinuity at the sonic line in the treatment of ϕ_{yy} .

To illustrate the application of the Murman difference formulas consider uniform flow in a parallel channel. Then $\phi_{yy} = 0$, and with a suitable normalization $K = 0$, so that the equation reduces to

$$- \frac{\partial}{\partial x} (\phi_x^2/2) = 0$$

with ϕ and ϕ_x given at $x = 0$, and ϕ given at $\phi = L$. Since ϕ_x^2 is constant, ϕ_x simply reverses sign at a jump. Provided we enforce the entropy condition that ϕ_x decreases through a jump, there is a unique solution with a single jump whenever $\phi_x(0) > 0$ and $\phi(0) + L\phi_x(0) \geq \phi(L) \geq \phi(0) - L\phi_x(0)$. Let $u_{i+1/2} = (\phi_{i+1} - \phi_i)/\Delta x$ and $u_i = (u_{i+1/2} + u_{i-1/2})/2$. Then the difference equations can be written as

$$u_{i+1/2}^2 = u_{i-1/2}^2 \quad \text{when} \quad u_i \leq 0 \ , \ u_{i-1} \leq 0 \qquad \text{(elliptic)}$$

$$u_{i-1/2}^2 = u_{i-3/2}^2 \quad \text{when} \quad u_i > 0 \ , \ u_{i-1} > 0 \qquad \text{(hyperbolic)}$$

$$u_{i+1/2}^2 = u_{i-3/2}^2 \quad \text{when} \quad u_i \leq 0 \ , \ u_{i-1} > 0 \qquad \text{(shockpoint)}$$

$$0 = 0 \quad \text{when} \quad u_i > 0 \ , \ u_{i-1} \leq 0 \qquad \text{(parabolic)}$$

These admit the correct solution, illustrated in Figure 3a, with a constant slope on the two sides of the shock. The shock point operator allows a single link with an intermediate slope, corresponding to the shock lying in the middle of a mesh cell.

The difference equations also admit, however, various improper solutions. Figure 3b illustrates a sawtooth solution with u^2 constant everywhere except in one cell ahead of a shock point. Figure 3c illustrates another improper solution in which the shock is too far forward. At the last interior point there is then an expansion shock which is admitted by the parabolic operator. Since the difference equations have more than one root we must depend on the iterative scheme to find the desired root. The scheme should ideally be designed so that the correct solution is stable under a small perturbation, and improper solutions are unstable. Using a scheme similar to (3.6), the instability of the sawtooth solution has been confirmed in numerical experiments. The solutions with an expansion shock at the downstream boundary are stable, on the other hand, if the compression shock is more than the width of a mesh cell too far forward. Thus there is a continuous range of stable improper solutions, while the correct solution is an isolated stable equilibrium point.

4. Difference Schemes for the Potential Flow Equation in Quasilinear Form

It is less easy to construct difference approximations to the potential flow equation with a correct directional bias because the upwind direction is not known in advance. If, however, the supersonic flow is confined to a bubble above the profile, it may be possible to use a coordinate system in which the x coordinate is more or less aligned with the flow in the supersonic zone. For this purpose we can use a conformal mapping to make the profile coincide with an x coordinate line [8,9]. A simple difference approximation to the quasilinear form (1.1) can then be constructed in the following manner. The velocity components u and v are evaluated throughout the flow field by central difference formulas, and the speed of sound is determined by equation (1.3). Then at subsonic points we use central difference formulas for ϕ_{xx}, ϕ_{xy} and ϕ_{yy}, while at supersonic points we switch to upwind difference formulas for ϕ_{xx} and ϕ_{xy}.

The upwind difference formulas can be regarded as approximations to $\phi_{xx} - \Delta x\, \phi_{xxx}$ and $\phi_{xy} - (\Delta x/2)\phi_{xxy}$. Thus they introduce an effective artificial viscosity

$$\Delta x\left\{(u^2-a^2)\phi_{xxx} + uv\phi_{xxy}\right\} = \Delta x\left\{(u^2-a^2)u_{xx} + uvv_{xx}\right\}.$$

When the flow is not perfectly aligned with the x coordinate there exist supersonic points at which $u^2 < a^2 < u^2 + v^2$. One characteristic lies ahead of the y coordinate line at such a point, so that the difference scheme does not have the correct region of dependence. Also the artificial viscosity $\Delta x(u^2-a^2)\phi_{xxx}$ introduced by the upwind difference formula for ϕ_{xx} is then negative. Despite this fact, schemes of this type have proved quite satisfactory in practice for flows with supersonic zones of moderate size.

To treat more general flows it is necessary to derive a method of rotating the upwind differencing to conform with the flow direction [10]. For this purpose suppose that s and n are streamwise and normal Cartesian coordinates in a reference frame locally aligned with the flow. Then equation (1.1) is equivalent to

$$(4.1) \qquad (a^2-q^2)\phi_{ss} + a^2\phi_{nn} = 0 .$$

Since u/q and v/q are the local direction cosines

$$(4.2) \qquad \phi_{ss} = \frac{1}{q^2}\,(u^2\phi_{xx} + 2uv\phi_{xy} + v^2\phi_{yy})$$

and

$$(4.3) \qquad \phi_{nn} = \frac{1}{q^2}\,(v^2\phi_{xx} - 2uv\phi_{xy} + u^2\phi_{yy}) .$$

Now we use central differencing at subsonic points as before, but at supersonic points we switch to upwind differencing for ϕ_{ss} . Thus if $u > 0$, $v > 0$, ϕ_{ss} is approximated at a point $i \Delta x$, $j \Delta y$ in the supersonic zone by using the formulas $(\phi_{ij}-2\phi_{i-1,j}+\phi_{i-2,j})/\Delta x^2$ and $(\phi_{ij}-\phi_{i-1,j}-\phi_{i,j-1}+\phi_{i-1,j-1})/\Delta x \Delta y$ to represent ϕ_{xx} and ϕ_{xy} , and a similar formula to represent ϕ_{yy} . This reduces exactly to the Murman scheme when either $u = 0$ or $v = 0$. Also the upwind differencing introduces an effective artificial viscosity

$$(1 - \frac{a^2}{q^2}) \left\{ \Delta x \, (u^2 u_{xx} + uvv_{xx}) + \Delta y \, (uvu_{yy} + v^2 v_{yy}) \right\}$$

which is symmetric in x and y .

5. Difference Schemes for the Potential Flow Equation in Conservation Form

In the construction of a discrete approximation to the conservation form (1.7) of the potential flow equation, it is convenient to accomplish the switch to upwind differencing by the explicit addition of an artificial viscosity in the manner proposed in Section 2. Let S_{ij} be a central difference approximation to the left-hand side of equation (1.7)

$$(5.1) \qquad S_{ij} = \frac{(\rho u)_{i+1/2,j} - (\rho u)_{i-1/2,j}}{\Delta x} + \frac{(\rho v)_{i,j+1/2} - (\rho v)_{i,j-1/2}}{\Delta y} .$$

In forming S_{ij} we have to evaluate the velocities at the midpoints of the mesh intervals, using formulas such as $u_{i+1/2,j} = (\phi_{i+1,j}-\phi_{ij})/\Delta x$ and $u_{i,j+1/2} = (\phi_{i+1,j}+\phi_{i+1,j+1} - \phi_{i-1,j}-\phi_{i-1,j+1})/4\Delta x$. The density is evaluated from equation (1.5). Then we shall solve an equation of the form

$$S_{ij} + T_{ij} = 0 \tag{5.2}$$

where T_{ij} is the artificial viscosity. The viscosity will be constructed in a divergence form

$$T = \frac{\partial P}{\partial x} + \frac{\partial Q}{\partial y}$$

to preserve the conservation form of equation (1.7).

Both simple and rotated schemes can be devised [11]. Tne term $(\partial/\partial x)(\rho u)$ can be expanded in a smooth region as $\rho(1-u^2/a^2)\phi_{xx} - \rho(uv/a^2)\phi_{xy}$. In the simple scheme $Q = 0$, and $\partial P/\partial x$ is constructed as an upwind approximation to $-\Delta x(\partial/\partial x)\mu\phi_{xx}$, where $\mu = \min\{0, \rho(1 - u^2/a^2)\}$. Thus at supersonic points the term $\rho(1 - u^2/a^2)\phi_{xx}$ is canceled and replaced by its value at the adjacent upwind point.

The rotated scheme is designed to introduce viscosity terms similar to those introduced by the rotated scheme for the quasilinear form. Let the switching function μ be defined as

$$\mu = \max\{0, (1 - a^2/q^2)\} \quad . \tag{5.3}$$

Then P and Q are constructed as approximations to

$$-\mu\{(1-\varepsilon)|u|\Delta x\ \rho_x + \varepsilon\ \Delta x^2 u\rho_{xx}\}$$

and

$$-\mu\{(1-\varepsilon)|v|\Delta y\ \rho_y + \varepsilon\ \Delta y^2 v\rho_{yy}\}$$

where ε is a parameter controlling the accuracy. If $\varepsilon = 1 - \lambda\,\Delta x$ and λ is a constant the scheme is second order accurate. If $\varepsilon = 0$ it is first order accurate, and at supersonic points where $u > 0$, $v > 0$, P then approximates

$$\Delta x \frac{\rho}{a^2} (1 - \frac{a^2}{q^2}) (u^2 u_x + uvv_x) .$$

In these expressions the derivatives of ρ are represented by upwind difference formulas. Thus the formula for the viscosity becomes

$$(5.4) \quad T_{ij} = - \frac{P_{i+1/2,j} - P_{i-1/2,j}}{\Delta x} - \frac{Q_{i,j+1/2} - Q_{i,j-1/2}}{\Delta y}$$

where if $u_{i+1/2,j} > 0$

$$P_{i+1/2,j} = u_{i+1/2,j} \mu_{ij} \{\rho_{i+1/2,j} - \rho_{i-1/2,j} - \varepsilon(\rho_{i-1/2,j} - \rho_{i-3/2,j})\}$$

and if $u_{i+1/2,j} < 0$

$$(5.5) \quad P_{i+1/2,j} = u_{i+1/2,j} \mu_{i+1,j} \{\rho_{i+1/2,j} - \rho_{i+3/2,j} - \varepsilon(\rho_{i+3/2,j} - \rho_{i+5/2,j})\}$$

while $Q_{i,j+1/2}$ is defined by a similar formula.

6. Analysis of Relaxation Schemes by the Time Dependent Analogy

As in the treatment of the small disturbance equation, relaxation methods can be used to solve the nonlinear difference equations generated by the various approximations to the potential flow equation. In the simplest case, when the equation is expressed in quasilinear form and the upwind differencing is restricted to the x coordinate, this presents no particular difficulty. In each cycle the coefficients a^2-u^2 , $2uv$ and a^2-v^2 are first calculated using the result $\phi_{ij}^{(n)}$ of the previous cycle. Then new values $\phi_{ij}^{(n+1)}$ are determined by solving a set of linear equations on each successive vertical line $x = i\ \Delta x$, with the latest available values $\phi_{i-2,j}^{(n+1)}$, $\phi_{i-1,j}^{(n+1)}$ and $\phi_{i+1,j}^{(n)}$ on the adjacent upstream and downstream lines substituted in the difference formulas for ϕ_{xx}, ϕ_{xy} and ϕ_{yy}. This gives a block triangular form to the matrix N in equation (2.1). If the coefficients of the second derivatives were frozen the method would reduce to a marching scheme in the supersonic zone, since the values on each vertical line would then depend only on the updated values just determined on the upstream lines.

When the rotated difference scheme is used, the difference equation at a supersonic point includes contributions to ϕ_{nn} from adjacent downwind points. The time dependent analogy suggested in Section 2 then provides a useful insight into the nature of the relaxation process, which no longer resembles a marching scheme in the supersonic zone [10]. The typical form of a central difference approximation to ϕ_{xx} at a point on the line $x = i\ \Delta x$ is

$$\left\{\phi_{i-1,j}^{(n+1)} - (1+r\ \Delta x)\phi_{ij}^{(n+1)} - (1-r\ \Delta x)\phi_{ij}^{(n)} + \phi_{i+1,j}^{(n)}\right\}/\Delta x^2$$

where an updated value is used at $x = (i-1)\Delta x$, an old value is used at $x = (i+1)\Delta x$ because the new value is not yet available, and a linear combination depending on a parameter r is used at $x = i\ \Delta x$. Introducing the correction $C_{ij} = \phi_{ij}^{(n+1)} - \phi_{ij}^{(n)}$, this is equivalent to

$$(\phi_{i-1,j}^{(n)} - 2\phi_{ij}^{(n)} + \phi_{i+1,j}^{(n)})/\Delta x^2 - (C_{ij} - C_{i-1,j})/\Delta x^2 - rC_{ij}/\Delta x .$$

The structure of the operator N in equation (2.1) is thus determined by the particular combination of new and old values used in the various difference formulas contributing to ϕ_{ss} and ϕ_{nn}.

If we write the equivalent time dependent equation (2.2) in a locally aligned s-n coordinate system, we now find that its principal part can be expressed as

$$(M^2-1)\phi_{ss} - \phi_{nn} + 2\alpha\phi_{st} + 2\beta\phi_{nt} = 0 \tag{6.1}$$

where M is the local Mach number, and the coefficients α and β depend on the split between new and old values of ϕ in the difference formulas. The substitution $T = t - \alpha s/(M^2-1) + \beta n$ reduces this equation to the diagonal form

$$(M^2-1)\phi_{ss} - \phi_{nn} - \left(\frac{\alpha^2}{M^2-1} - \beta^2\right)\phi_{TT} = 0 .$$

If $M > 1$ either s or n is timelike, depending on the sign of the coefficient of ϕ_{TT}, while T is spacelike. Since s is the timelike direction of the steady state equation, it ought also to be the timelike direction when this equation is embedded in a time dependent process. Thus when $M > 1$ the coefficients α and β should satisfy the compatibility condition

$$\alpha > \beta\sqrt{M^2-1} \tag{6.2}$$

The characteristic cone of equation (6.1) touches the s-n plane. As long as condition (6.2) holds it slants upstream in the reverse time direction, as illustrated in Figure 2. The iterative scheme will then have a proper region of dependence as long as we sweep the flow field in a direction such that the updated region always includes the upwind line of tangency between the characteristic cone and the s-n plane. It can be seen from Figure 2 that the region of dependence of a subsonic point contains the t axis, with the result that it is important to include a damping term $\gamma\,\phi_t$

to attenuate the influence of the initial guess. The coefficient γ is controlled by the choice of an overrelaxation factor [12]. The situation is different at a supersonic point. If the coefficients of equation (6.1) were constant with $M > 1$, the region of dependence would cease to intersect the initial data after a sufficient time interval. Instead it would intersect a surface containing the Cauchy data of the steady state problem. Thus no damping due to ϕ_t is required in the supersonic zone for the process to reach a steady state.

These considerations lead to the following method for deriving the operator N. We substitute new values $\phi_{ij}^{(n+1)}$ whenever they are available in the central difference formulas at subsonic points, and also in the central difference formulas contributing to ϕ_{nn} at supersonic points. In order to satisfy the compatibility condition (6.2) at supersonic points, however, we do not use new values in the upwind difference formulas contributing to ϕ_{ss}. Instead, if $u > 0$, ϕ_{xx} is represented by

$$(2\phi_{ij}^{(n+1)} - \phi_{ij}^{(n)} - 2\phi_{i-1,j}^{(n+1)} + \phi_{i-2,j}^{(n)})/\Delta x^2 .$$

This can be regarded as an approximation to $\phi_{xx}-2(\Delta t/\Delta x)\phi_{xt}$. Similar formulas are used for ϕ_{xy} and ϕ_{yy} with the result that the approximation to ϕ_{ss} introduces a term $2(M^2-1)((u/q)(\Delta t/\Delta x) + (v/q)(\Delta t/\Delta y))\phi_{st}$ in the equivalent time dependent equation. Finally to make sure that (6.2) is satisfied when M is close to unity we add a term to augment further the coefficient of ϕ_{st}. If $u > 0$ and $v > 0$ this term is

$$\frac{\omega S}{\Delta x}\left\{\frac{u}{\Delta x}(C_{ij}-C_{i-1,j}) + \frac{v}{\Delta y}(C_{ij}-C_{i,j-1})\right\}$$

where ω_S is a relaxation factor with a value ≥ 0. The best rate of convergence is obtained by using the smallest possible value of ω_S. Often it is sufficient to take $\omega_S = 0$.

Similar schemes are easily constructed for the difference approximations to the conservation form (1.7). Since the conservation form is equivalent to the quasilinear form multiplied by ρ/a^2, we have only to multiply the operator N by ρ/a^2 to produce a time dependent process for the conservation form which converges at about the same rate as the process for the quasilinear form [11]. An advantage of this procedure is that the iterative scheme does not have to be modified to reflect every variation in the difference equations. We can use the same operator N, for example, for all values of the viscosity parameter ε.

7. Accelerated Iterative Method

If we consider iterative schemes of the class defined by equation (2.1), we can expect to improve the rate of convergence by choosing N as the closest possible approximation to the operator used to evaluate the residual R. In this we are constrained by the need to limit the number of operations required for each cycle. In recent years fast direct methods have been developed for solving finite difference approximations to Poisson's equation on a rectangle [13,14]. On an $N \times N$ square these require a number of operations proportional to $N^2 \log N$. The coefficient A in the small disturbance equation (1.10) is an approximation to $1-M^2$, where M is the local Mach number, so if M is small the Laplacian is a fair approximation to the operator on the left-hand side of the equation. This suggests the use of the discrete Laplacian for N in equation (2.1), with the scaling function σ replaced by a fixed relaxation factor ω. An iteration of this kind was proposed by Martin and Lomax [15].

In order to estimate the rate of convergence which might be expected, consider the Prandtl Glauert equation, which is obtained by replacing A by $1-M_\infty^2$ in equation (1.10). Let H and V be positive definite operators representing $-\partial^2/\partial x^2$ and $-\partial^2/\partial y^2$ with the appropriate boundary conditions. Also let ϕ be the solution, and let $e^{(n)} = \phi^{(n)}-\phi$ be the error after n cycles. With $\omega = 1$ the iteration then gives

$$(H+V)e^{(n+1)} = M_\infty^2 He^{(n)} \; .$$

Thus

$$\|H^{1/2} e^{(n+1)}\| \leq M_\infty^2 \|K\| \; \|H^{1/2} e^{(n)}\|$$

where K is the symmetric operator $H^{1/2}(H+V)^{-1}H^{1/2}$. If we use a Euclidean norm then

$$\|K\| = \max \frac{(x,Kx)}{(x,x)} = \max \frac{(y,Hy)}{(y,Hy) + (y,Vy)}$$

where $H^{1/2}y = K^{1/2}x$. Thus $\|K\| < 1$. This estimate serves to indicate that for subsonic flows the scheme should converge at a rate independent of the mesh size.

If we consider the case of linearized supersonic flow, with A replaced by $1-M_\infty^2$ and $M_\infty > 1$, on the other hand, it can be shown that the iteration would diverge. In this case, if E is a shifting operator which replaces ϕ_{ij} by $\phi_{i-1,j}$, the iteration gives

$$(H+V)e^{(n+1)} = \{I+(M_\infty^2-1)E\}He^{(n)}$$

and $H+V$ does not dominate $\{I+(M_\infty^2-1)E\}H$.

Thus we cannot expect the Poisson iteration to converge when it is used to calculate a transonic flow with a supersonic zone of appreciable size. If, however, it could be supplemented with another method which gives fast convergence in the supersonic zone, the two in combination might produce an effective iterative scheme. In fact the usual line relaxation scheme for the small disturbance equation is just such a method, since it acts like a marching scheme in the supersonic zone. Thus it would give the solution in the supersonic zone in a single sweep, if it were not for the change from cycle to cycle in the nonlinear coefficients A_{ij} and the data at the sonic line. This leads to the idea of using a two stage iteration [16], in which the first stage is a Poisson step, and the second stage consists of p relaxation steps to sweep the errors from the supersonic zone. The best value of p is most easily determined by numerical experiments. These have confirmed that in calculations using the small disturbance equation, a single relaxation step after each Poisson step is sufficient to give fast convergence. An alternative approach is to use a desymmetrized operator N of a form which still permits the use of a fast direct method. Good results have been reported by Martin, using a scheme of this type formulated for an equivalent system of first order equations [17].

The two stage iteration has the advantage that it is easily extended to treat the potential flow equation. This can be scaled so that the Laplacian represents it linear part by dividing the quasilinear form (1.1) by a^2 or the conservation form (1.7) by ρ. Thus we define a Poisson iteration by letting N be the Laplacian in equation (2.1), and setting $\sigma = \omega/a^2$ for the quasilinear form, or ω/ρ for the conservation form, where ω is a relaxation factor with a value in the range $1 \leq \omega < 2$. This gives rapid convergence for subsonic flows. For transonic flows we again use p

relaxation steps after each Poisson step. The method has proved particularly effective when it is used with the simple difference scheme in quasilinear form, with upwind differencing restricted to one coordinate. The best rate of convergence is then usually obtained with one or two relaxation steps after each Poisson step. The rotated difference schemes require a larger number of relaxation steps because the relaxation method no longer acts like a marching scheme in the supersonic zone. Typically the best rate of convergence is then obtained with $p \sim 5\text{-}8$.

8. Three Dimensional Calculations

A similar approach can be used for three dimensional transonic flow calculations [18,19]. The three dimensional small disturbance equation

$$(8.1) \qquad \frac{\partial}{\partial x}\left(K\phi_x - \frac{\gamma+1}{2}\phi_x^2\right) + \phi_{yy} + \phi_{zz} = 0$$

can be approximated by the obvious generalization of the scheme proposed in Section 3. Let p_{ijk} be a central difference approximation to the first term and q_{ijk} a central difference approximation to $\phi_{yy} + \phi_{zz}$. Then we evaluate the residual as

$$(8.2) \qquad R_{ijk} = p_{ijk} + q_{ijk} - \mu_{ijk}p_{ijk} + \mu_{i-1,j,k}p_{i-1,j,k}$$

where the switching function μ_{ijk} is unity at supersonic points and zero at subsonic points.

The time dependent analogy is again helpful in devising a relaxation method for solving the difference equations. If we solve on vertical lines, for example, the use of mixed new and old values in the approximation to ϕ_{zz} introduces a term containing ϕ_{zt} in the equivalent time dependent equation. In order to make sure that we obtain a wave equation in which

x is timelike at supersonic points, as it is in the steady state equation, a compensating term $\alpha\, \phi_{xt}$ should then be added, with $\alpha > 0$.

An alternative method is to relax the equations on successive longitudinal lines with the following scheme. Let the residual R_{ijk} and the coefficient

(8.3) $$A_{ijk} = K - (\gamma+1)(\phi_{i+1,j,k}-\phi_{i-1,j,k})2\Delta x$$

be evaluated using the result $\phi_{ijk}^{(n)}$ of the nth cycle. Then the correction $C_{ijk} = \phi_{ijk}^{(n+1)} - \phi_{ijk}^{(n)}$ is determined by solving the equation

(8.4) $$\begin{aligned}&(1-\mu_{ijk})A_{ijk}(C_{i+1,j,k}-2C_{ijk}+C_{i-1,j,k})/\Delta x^2\\ &- (\alpha-2\mu_{i-1,j,k}A_{i-1,j,k})(C_{ijk}-C_{i-1,j,k})/\Delta x^2\\ &- (C_{ijk}-C_{i,j-1,k})/\Delta y^2 - (C_{ijk}-C_{i,j,k-1})/\Delta z^2\\ &- (1-\mu_{ijk})(\frac{2}{\omega} - 1)(1/\Delta y^2+1/\Delta z^2)C_{ijk} + R_{ijk} = 0\end{aligned}$$

where ω is the subsonic over-relaxation factor, and α is a parameter controlling the coefficient of ϕ_{xt} in the equivalent time dependent equation.

A rotated difference scheme for the three dimensional transonic potential flow equation can be devised by writing it locally as

$$(a^2-q^2)\phi_{ss} + a^2(\Delta\phi-\phi_{ss}) = 0 .$$

In this equation Δ is the Laplacian and

$$\phi_{ss} = \frac{1}{q^2}\{u^2\phi_{xx}+v^2\phi_{yy}+w^2\phi_{zz}+2uv\phi_{xy}+2vw\phi_{yz}+2uw\phi_{xz}\}$$

where u, v and w are the velocity components ϕ_x, ϕ_y and ϕ_z and q is the speed $\sqrt{u^2+v^2+w^2}$. Now upwind differencing is used for ϕ_{ss} in the supersonic zone as before. Horizontal or vertical line relaxation schemes can be devised in the same manner as in the two dimensional case.

9. Typical results

Methods constructed along these lines have been quite widely used in the last few years. They have proved particularly effective for calculating two dimensional and axially symmetric flows [5-10, 20-21]. Some typical results are presented here.

As a check on the accuracy attainable with these methods some results are first given for the Tricomi equation

$$y\,\phi_{xx} + \phi_{yy} = 0$$

for which exact polynomial solutions can easily be constructed. The basic scheme described in Section 3 was applied to this equation in the rectangle $-1 \le x \le 1$, $-1 \le y \le 1$, with Dirichlet boundary conditions on the sides $y = \pm 1$ and $x = \pm 1$ with $y \ge 0$, Cauchy data on the side $x = -1$ with $y < 0$, and no data on the side $x = 1$ with $y < 0$. Central difference formulas were used for ϕ_{yy} everywhere and for ϕ_{xx} when $y > 0$. When $y < 0$, ϕ_{xx} was approximated by the upwind formula

$$\frac{1}{\Delta x^2}\left\{\phi_{ij} - 2\phi_{i-1,j} + \phi_{i-2,j} + \varepsilon(\phi_{ij}-3\phi_{i-1,j}+3\phi_{i-2,j}-\phi_{i-3,j})\right\}$$

which is first order accurate if $\varepsilon = 0$ and second order accurate if $\varepsilon = 1$. An equal mesh spacing $\Delta x = \Delta y = h$ was used in each coordinate direction, and exact values of ϕ were provided on the boundaries carrying Dirichlet data, and also at $x = (-1 + h)$, $x = -(1 + 2h)$ on the boundary carrying Cauchy data. The solution $\bar{\phi}(h)$ of the difference equations was obtained on three grids with $h = 1/16$, $1/32$ and $1/64$ by the iterative method described in Section 3. On each grid the iterations were continued until the residual, normalized by multiplying by h^2 , was $< 10^{-12}$ at every interior point. This generally required about 300 cycles on the fine mesh. The results were then compared with the exact solution. A typical polynomial solution is

$$\phi = x^4 y - x^2 y^4 + y^7/21 .$$

Table 1 shows the errors $\|\phi - \bar{\phi}(h)\|$ and $\|\phi_x - \bar{\phi}_x(h)\|$ for this case, where $\bar{\phi}_x(h)$ was estimated by a central difference formula. The norm was defined as $\|a\| = \left(\frac{1}{N} \sum_i \sum_j a_{ij}^2\right)^{1/2}$, where the sum is over the interior mesh points, and N is the number of these points. For the error in ϕ_x points on the line $x = 1-h$ were excluded from the sum to avoid estimating ϕ_x by differencing between interior points and points on the boundary $x = 1$.

Table 1

Errors in solution of the Tricomi equation

Mesh Width	Errors for $\varepsilon = 0$		Errors for $\varepsilon = 1$	
	$\|\phi-\bar{\phi}(h)\|$	$\|\phi_x-\bar{\phi}_x(h)\|$	$\|\phi-\bar{\phi}(h)\|$	$\|\phi_x-\bar{\phi}_x(h)\|$
$\frac{1}{16}$	$.428\times10^{-1}$	$.111$	$.429\times10^{-2}$	$.937\times10^{-2}$
$\frac{1}{32}$	$.242\times10^{-1}$	$.631\times10^{-1}$	$.107\times10^{-2}$	$.259\times10^{-2}$
$\frac{1}{64}$	$.130\times10^{-1}$	$.350\times10^{-1}$	$.218\times10^{-3}$	$.675\times10^{-3}$

The errors can be seen to decrease in the expected manner as the mesh width is decreased. For either $\varepsilon = 0$ or 1 they are roughly consistent with the estimate

$$\phi = \bar{\phi}(h) + Ah^{\varepsilon+1} + O(h^{\varepsilon+2}) \tag{9.1}$$

where the function $A(x,y)$ determines the distribution of the dominant error. Assuming this to be the case Richardson extrapolation was used to give the estimate $\phi = \tilde{\phi}(h) + O(h^{\varepsilon+2})$ where

$$\tilde{\phi}(h) = \bar{\phi}(y) + \frac{1}{2^{\varepsilon+1}-1}\left(\bar{\phi}(h) - \bar{\phi}(2h)\right) .$$

The derivative was also extrapolated by a similar formula. The results are shown in Table 2.

Table 2

Result of Richardson Extrapolation

Mesh Width of Finer Mesh	Errors for $\varepsilon = 0$ $\lVert\phi-\tilde{\phi}(h)\rVert$	$\lVert\phi_x-\tilde{\phi}_x(h)\rVert$	Errors for $\varepsilon = 1$ $\lVert\phi-\tilde{\phi}(h)\rVert$	$\lVert\phi_x-\tilde{\phi}_x(h)\rVert$
$\frac{1}{32}$	$.673\times10^{-2}$	$.210\times10^{-1}$	$.117\times10^{-3}$	$.544\times10^{-3}$
$\frac{1}{64}$	$.211\times10^{-2}$	$.719\times10^{-2}$	$.123\times10^{-4}$	$.804\times10^{-4}$

The success of the extrapolation provides further confirmation of the error estimate (9.1).

Figures 4-6 show typical solutions of the transonic potential flow equation for flows past airfoils. Curvilinear coordinates were used for these calculations. These were generated by mapping the exterior of the profile conformally onto the interior of a unit circle, and introducing polar coordinates r and θ in the circle [8-11]. This simplifies the representation of the Neumann boundary condition. It also provides a regular and finite mesh suitable for the application of a fast Poisson solver to accelerate the iterative scheme. To check the influence of the mesh width on the result each calculation was performed on three grids, first with 64 cells in the θ direction and 16 cells in the r direction, then with 128×32 cells and finally with 256×64 cells. On the second two grids the interpolated solution on the previous grid was used to provide the initial guess. A convenient measure of the local flow condition is the pressure coefficient $C_p = (p-p_\infty)/\frac{1}{2}\rho_\infty q_\infty^2$, where the subscript ∞ denotes free stream values. Each figure shows the calculated pressure coefficient over the surface of the profile. The critical pressure coefficient at which the flow has sonic velocity is marked by a horizontal line on the axis. Lift and drag coefficients CL and CD are also shown. These were obtained by integrating the surface pressure.

Figure 4 shows an airfoil designed by Garabedian to produce shock free flow [22, page 44]. The quasilinear form (1.1) was treated using the simple difference scheme described in Section 4, with upwind differencing restricted to the θ direction. The accelerated iterative method described in Section 7 was used, with 2 relaxation sweeps after each Poisson step. The Poisson solution was calculated using the Buneman algorithm [13] in the θ direction. The largest residual R_{ij} at any point of the field, normalized by multiplying by $\Delta\theta^2$, was used as a measure of convergence. It required 24 cycles to reduce the largest residual from $.89\times10^{-2}$ to $.91\times10^{-9}$ on the 64×16 grid, 21 cycles to reduce it from 16×10^{-2} to $.84\times10^{-9}$ on the 128×32 grid, and 25 cycles to reduce it from $.38\times10^{-3}$ to 81×10^{-9} on the 256×64 grid. The entire calculation took 262 seconds on a CDC 6600. A Poisson step takes about the same amount of time as 2 relaxation sweeps, so in this case the calculation on the 128×32 grid was equivalent to about 84 relaxation sweeps. Typically it takes about 4000 sweeps to reduce the largest residual to 10^{-9} on a 128×32 grid by relaxation alone.

It has been found that the jump at a normal shock wave is consistently underestimated by calculations which do not use conservation form [7,11]. The Mach number behind the shock wave is generally too close to unity. This can be corrected by using conservation form. The schemes described in Section 5 have proved less accurate than the nonconservation schemes of Section 4, however, in the treatment of shock free flows, particularly on coarse grids. They also have the disadvantage of requiring more computer time. Two examples of results obtained with the rotated difference scheme in conservation form, equations (5.1)-(5.4), are shown in Figures 5 and 6. The NACA 64A410, shown in Figure 5, is a typical example of an airfoil which produces a shock wave at quite a low Mach number. The accelerated iterative scheme was used in this calculation, with 8 relaxation steps after each Poisson step. 49 cycles were required to reduce the largest residual to 10^{-9} on the 256×64 grid. Figure 6 shows a symmetric airfoil (suitable for a vertical tail) designed by Hicks and Murman to give a low wave drag at Mach .80 [23]. The degeneration to a flow containing two shock waves is typical of an airfoil designed to operate efficiently at transonic speeds when either the Mach number or (in the case of a lifting airfoil) the lift coefficient is slightly reduced. The forward shock can be seen to be completely eliminated on the 64×16 grid. This calculation was performed by relaxation alone. 2000 cycles were used to reduce the largest residual to $.22 \times 10^{-6}$ on the 256×64 grid. In both calculations the viscosity parameter ε was zero, giving first order accuracy. The accuracy can generally be improved by using values of ε between 0 and 1. In the easier cases it is possible to set $\varepsilon = 1$. This may, however, lead to divergence for the more sensitive flows, such as the flow past a shock free airfoil near its design point.

Figure 7 shows an example of a three dimensional calculation for a wing with a roughly elliptic plan form. In this case a curvilinear coordinate system was generated in two stages [19]. First parabolic coordinates were introduced in planes containing the wing section by the square root transformation

$$X_1 + iY_1 = (x-x_0(z) + i(y-y_0(z))^{1/2} \ , \ Z_1 = z \ .$$

The singular line $x_0(z) + i\, y_0(z)$ was located just behind the leading edge in order to unwrap the wing to form a shallow bump $Y_1 = S(X_1,Z_1)$. Then a shearing transformation

$$X = X_1 \ , \ Y = Y_1 - S(X_1,Z_1) \ , \ Z = Z_1$$

was used to map the wing surface to a coordinate surface. The calculation was performed on a mesh with $192 \times 16 \times 32$ cells in the X, Y and Z directions. After a preliminary calculation on a mesh with $96 \times 8 \times 16$ cells, 100 relaxation cycles were used on the fine mesh to reduce the largest residual (multiplied by ΔX^2) to $\sim 10^{-5}$. The figure shows the wing configuration, and the upper and lower surface pressure distributions in separate plots. On the upper surface there is a single shock wave near the center of the wing, but two shock waves near each tip where less lift is produced.

When solutions of the transonic potential flow equation are compared with experimental data it is important to allow for viscous effects, which are dominant in the boundary layer adjacent to the surface. It has been found that remarkably good agreement can frequently be obtained simply by correcting the profile to allow for the displacement effect of the boundary layer [22]. This procedure, which is effective in the regime where the shock waves are not strong enough to cause a separated flow, often proves more successful with the quasi-linear form than with the conservation form. The shock jumps observed in practice in the presence of a boundary layer are weaker than the jumps predicted by the Rankine Hugoniot theory for normal shock waves. The attenuation of the shock jumps which results from the use of the quasilinear form provides a partial simulation of this effect even when no correction is made for the boundary layer.

10. Conclusion

Finite difference methods with an upwind bias in the hyperbolic region are now quite well established as practical tools for transonic flow calculations. Three dimensional calculations are presently restricted by limitations of computer memory capacity and time.

Much work remains to be done to improve these methods. In particular no estimates of global error bounds have been obtained. As long as the difference scheme is in conservation form, the solution of the difference equations should satisfy the proper jump conditions in the limit as the mesh width approaches zero. With the mesh widths realizable in practice, however, there is not enough resolution to provide a good representation of an oblique shock wave. If a reliable shock fitting scheme could be devised the results should be improved. If such a technique could be combined with the use of higher order accurate difference formulas it should be possible to use relatively coarse grids. This would open the way to more extensive three dimensional applications. Relaxation has proved a reliable but slow method for solving the nonlinear difference equations. Faster iterative methods are now in hand for treating two dimensional flows. Methods of comparable efficiency are needed for three dimensional calculations.

REFERENCES

1. Bateman, H., "Notes on a differential equation which occurs in the two dimensional motion of a compressible fluid and the associated variational problem," Proc. Roy. Soc. Series A, Vol. 125, 1029, 598-618.
2. Morawetz, C. S., "On the nonexistence of continuous flows past profiles," Comm. Pure Appl. Math. Vol. 9, 1956, 445-468.
3. Steger, J. L., and Baldwin, B. S., "Shock waves and drag in the numerical culculation of isentropic transonic flow," NASA TN D-6997, 1972.
4. Lax, Peter, and Wendroff, Burton, "Systems of conservation laws," Comm. Pure Appl. Math., Vol. 13, 1960, 217-237.
5. Cole, Julian D., "Twenty years of transonic flow," Boeing Scientific Research Laboratories Report D1-82-0878, July 1969.
6. Murman, E. M., and Cole, J. D., "Calculation of plane steady transonic flows," AIAA Journal, Vol. 9, 1971, 114-121.
7. Murman, Earll M., "Analysis of embedded shock waves calculated by relaxation methods," AIAA Conference on Computational Fluid Dynamics, Palm Springs, July 1973.
8. Garabedian, P. R., and Korn, D. G., "Analysis of transonic airfoils," Comm. Pure Appl. Math., Vol. 24, 1972, 841-851.
9. Jameson, Antony, "Transonic flow calculations for airfoils and bodies of revolution," Grumman Aerodynamics Report 370-71-1, December 1971.
10. Jameson, Antony, "Iterative solution of transonic flows over airfoils and wings, including flows at Mach 1," Comm. Pure Appl. Math., Vol. 27, 1974, 283-309.
11. Jameson, Antony, "Transonic potential flow calculations using conservation form," Second AIAA Conference on Computational Fluid Dynamics, Hartford, June 1975.
12. Garabedian, P. R., "Estimation of the relaxation factor for small mesh size," Math. Tables Aids Comp. Vol. 10, 1956, 183-185.
13. Buzbee, B. L., Golub, G. H., and Nielsen, C. W., "On direct methods of solving Poisson's equation," SIAM J. Numerical Analysis, Vol. 7, 1970, 627-656.

14. Fischer, D., Golub, G., Hald, O., Leiva, C., and Widlund, O., "On Fourier Toeplitz methods for separable elliptic problems," Math. Computation, Vol. 28, 1974, 349-368.
15. Martin, E. Dale, and Lomax, Harvard, "Rapid finite difference computation of subsonic and transonic aerodynamic flows," AIAA Paper 74-11, 1974.
16. Jameson, Antony, "Accelerated iteration schemes for transonic flow calculations using fast Poisson solvers," New York University ERDA Report COO-3077-82, 1975.
17. Martin, E. Dale, "A fast semi-direct method for computing transonic aerodynamic flows," Second AIAA Conference on Computational Fluid Dynamics, Hartford, June 1975.
18. Bailey, F. R., and Ballhaus, W. F., "Relaxation methods for transonic flows about wing-cylinder combinations and lifting swept wings," Third International Congress on Numerical Methods in Fluid Dynamics, Paris, July 1972.
19. Jameson, Antony, "Three dimensional flows around airfoils with shocks," IFIP Symposium on Computing Methods in Applied Sciences and Engineering, Versailles, Dec. 1973, Springer Verlag, Lecture Notes on Computer Science, Vol. 11, 185-212.
20. South, J. C., and Jameson, A., "Relaxation solutions for inviscid axisymmetric flow over blunt or pointed bodies," AIAA Conference on Computational Fluid Dynamics, Palm Springs, July 1973.
21. Arlinger, B. G., "Calculation of transonic flow around axisymmetric inlets," AIAA Paper 75-80, January 1975.
22. Bauer, F., Garabedian, P., Korn, D., and Jameson, A., Supercritical Wing Sections II, Springer Verlag, New York, 1975.
23. Hicks, R. M., Murman, E. M., and Vanderplaats, G. N., "An assessment of airfoil design by numerical optimization," NASA TMX-3092, 1974.

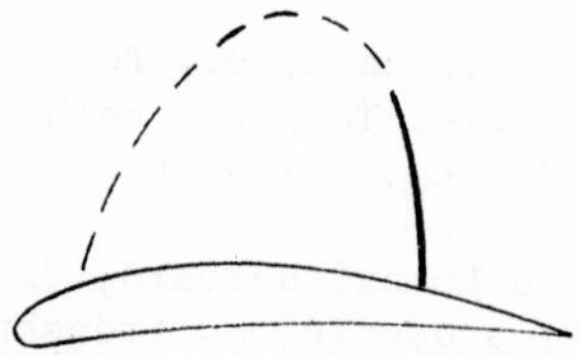

(a) SUBSONIC -- ONE SHOCK WAVE

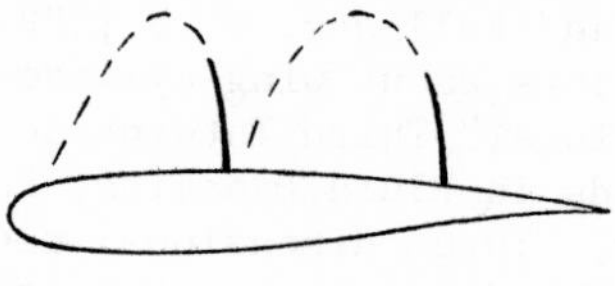

(b) SUBSONIC -- TWO SHOCK WAVES

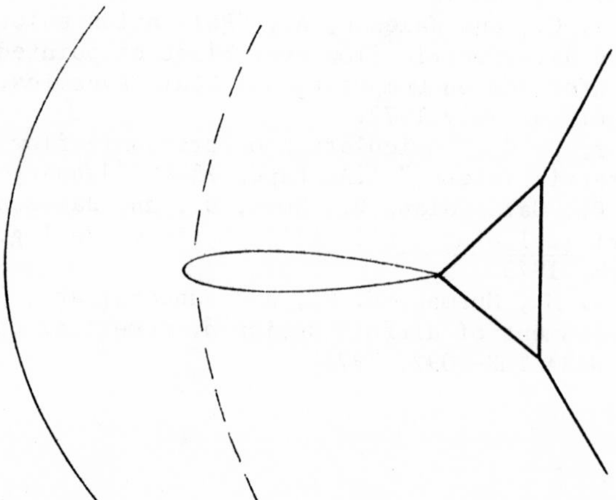

(c) SUPERSONIC -- BOW WAVE AND FISH TAIL SHOCK

FIGURE 1

TRANSONIC FLOW PATTERNS

—— SHOCK WAVE --- SONIC LINE

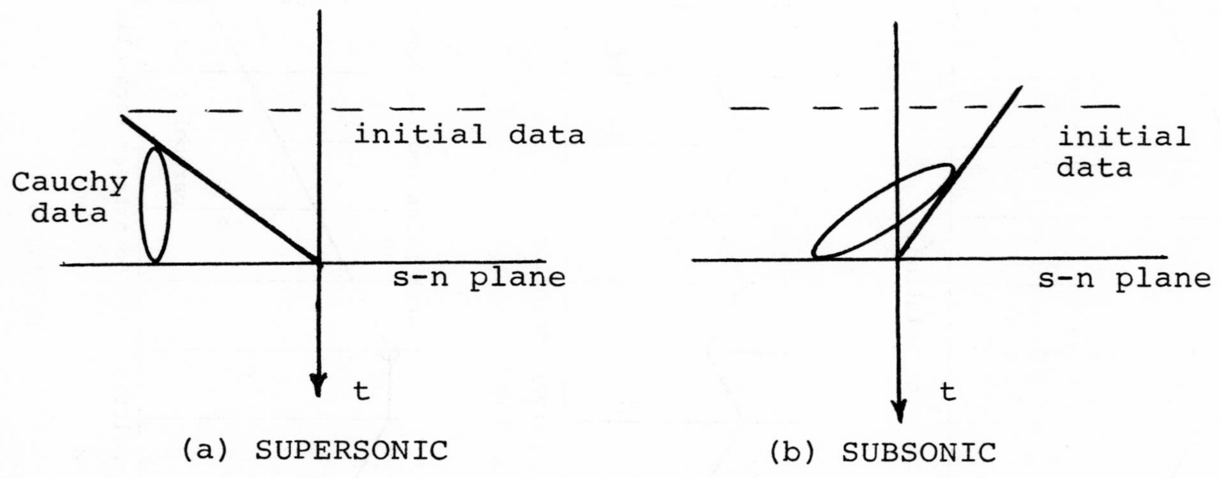

FIGURE 2

CHARACTERISTIC CONE OF EQUIVALENT TIME DEPENDENT EQUATION

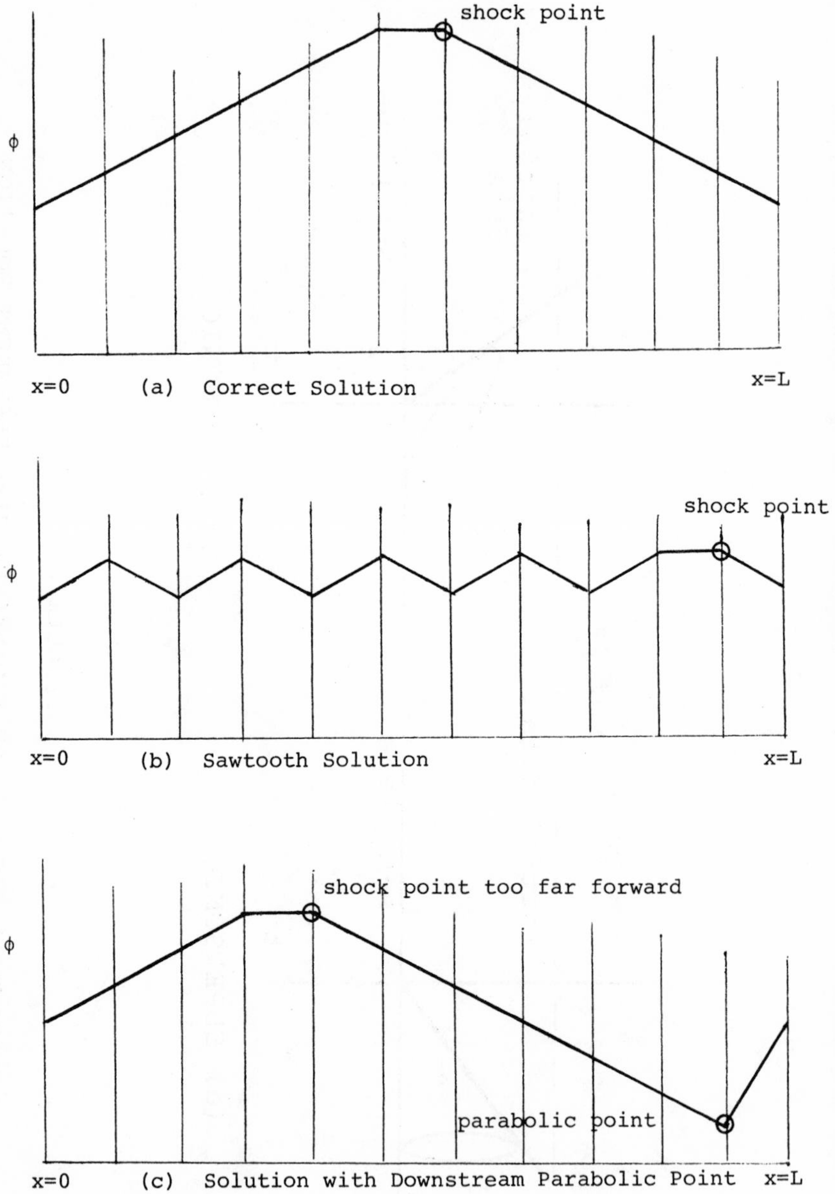

FIGURE 3

ONE DIMENSIONAL FLOW IN A CHANNEL

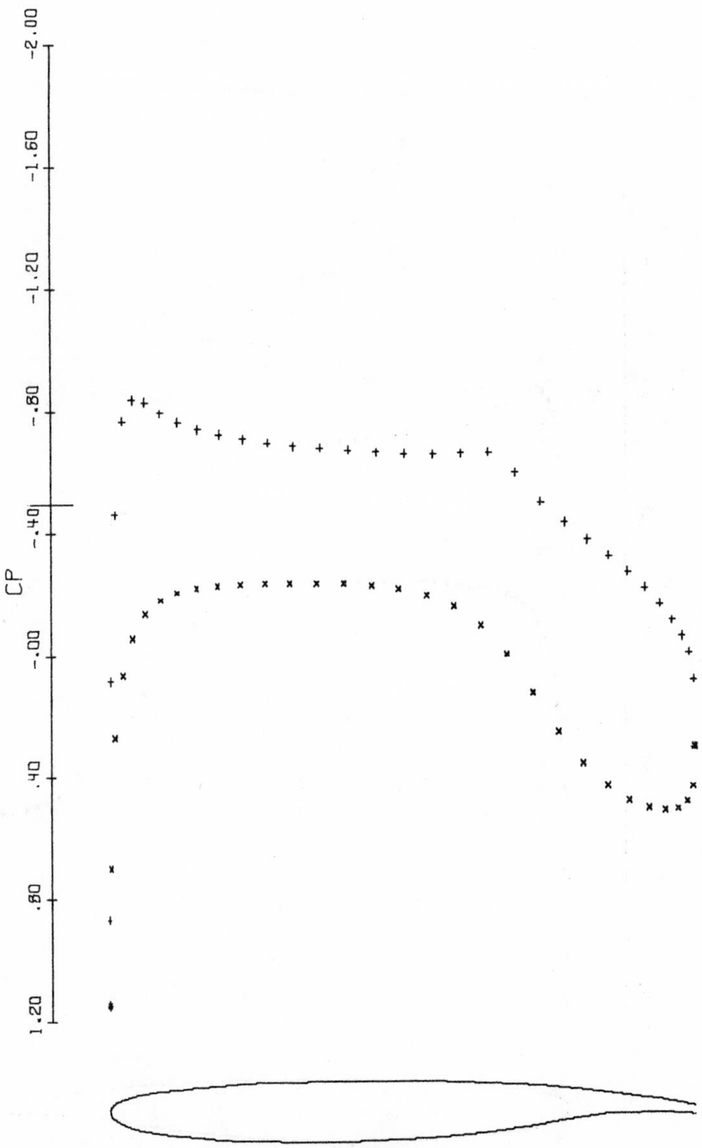

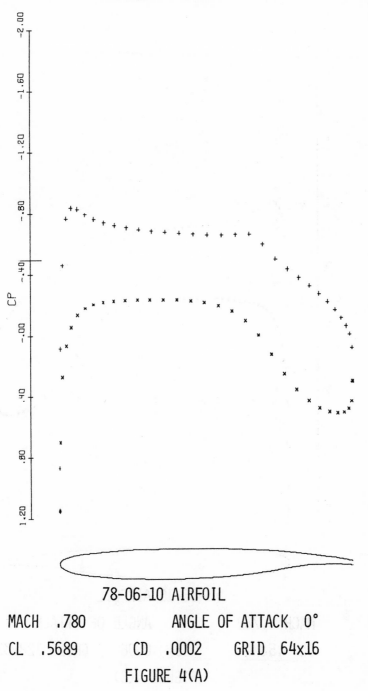

78-06-10 AIRFOIL

MACH .780 ANGLE OF ATTACK 0°

CL .5689 CD .0002 GRID 64x16

FIGURE 4(A)

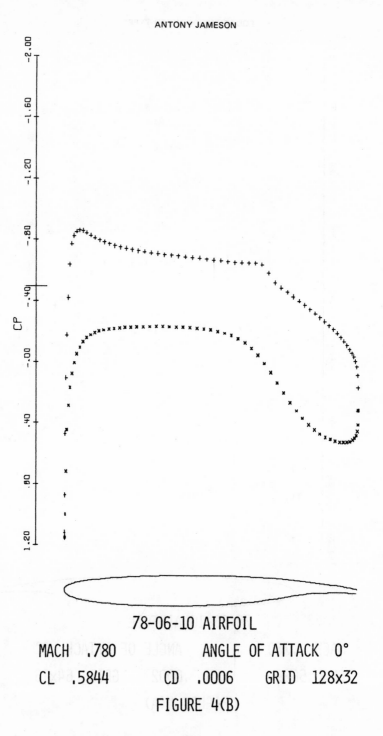

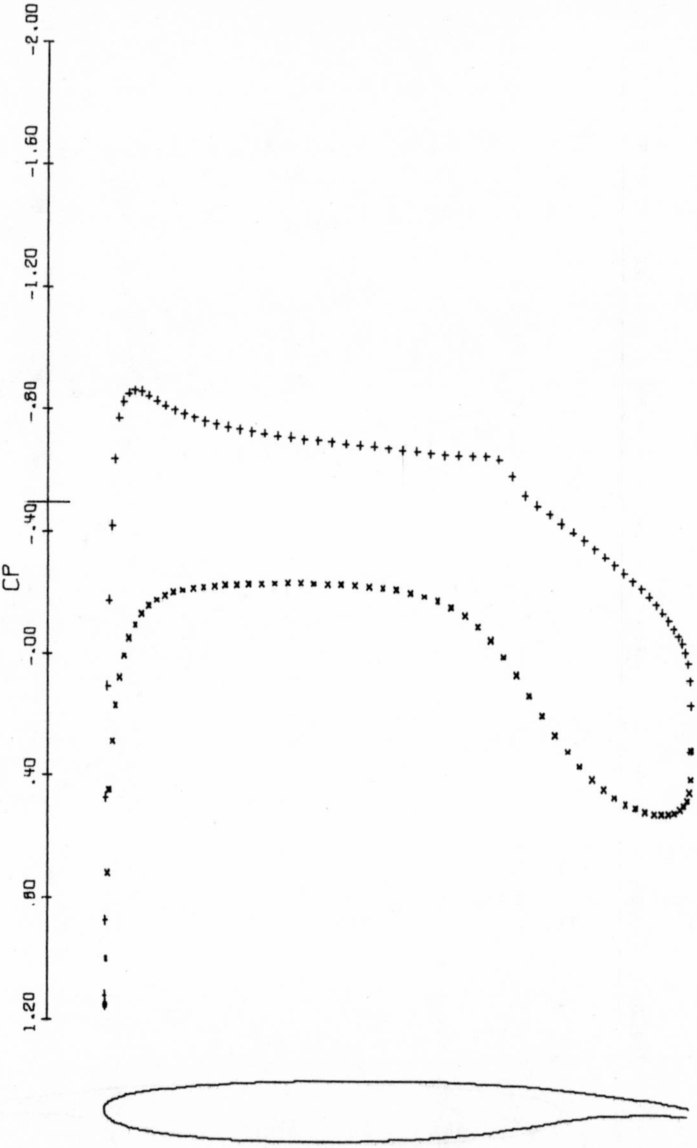

78-06-10 AIRFOIL

MACH .780 ANGLE OF ATTACK 0°

CL .5844 CD .0006 GRID 128x32

FIGURE 4(B)

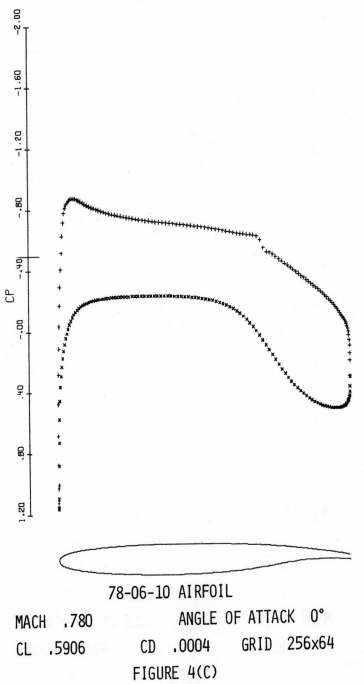

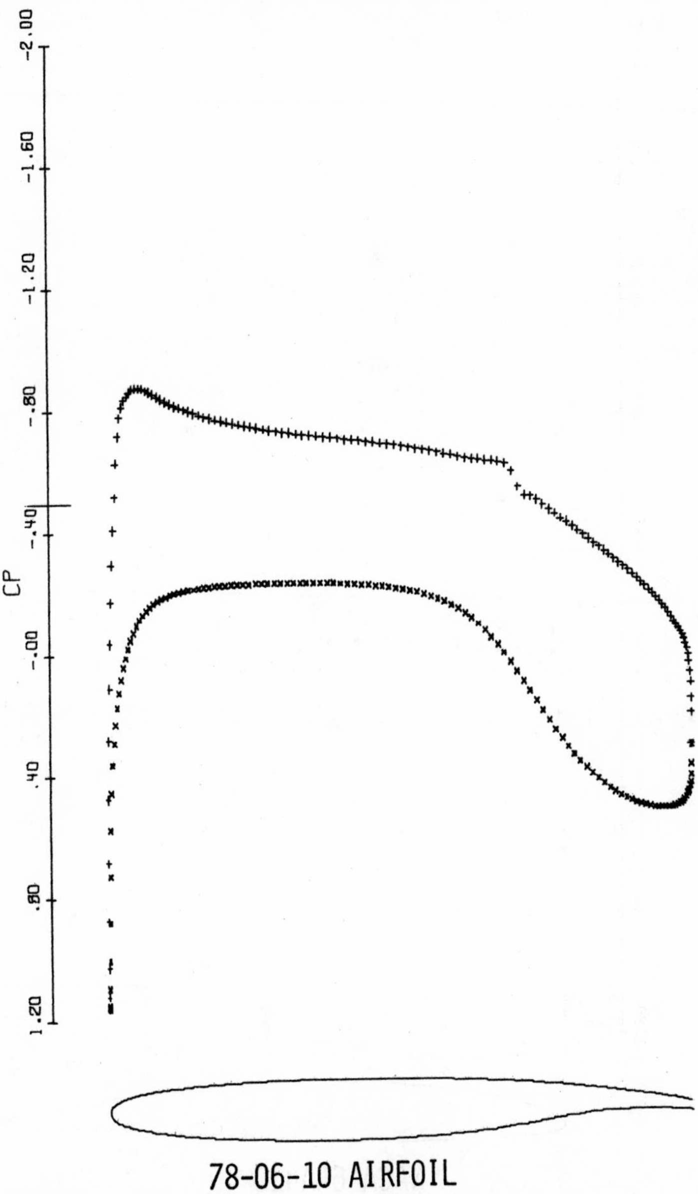

78-06-10 AIRFOIL

MACH .780 ANGLE OF ATTACK 0°

CL .5906 CD .0004 GRID 256x64

FIGURE 4(C)

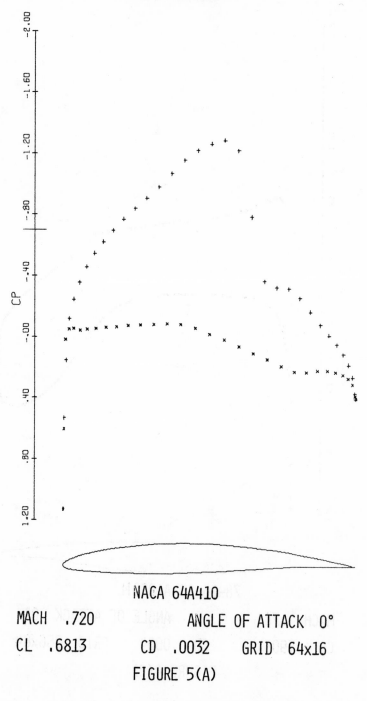

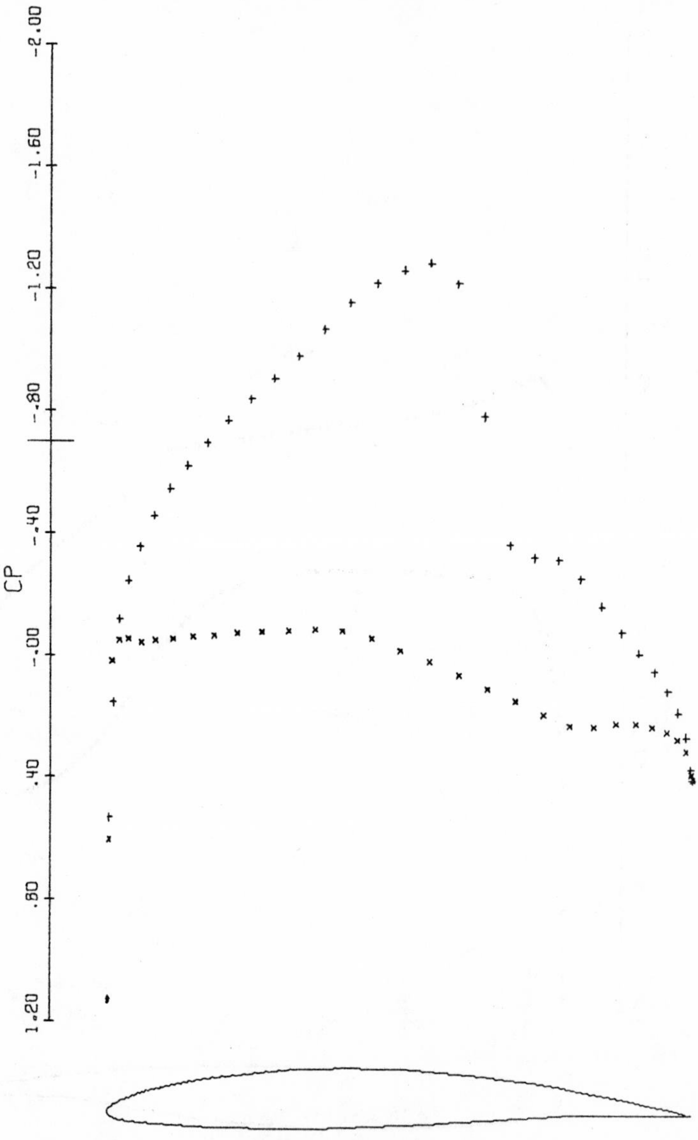

NACA 64A410

MACH .720 ANGLE OF ATTACK 0°

CL .6813 CD .0032 GRID 64x16

FIGURE 5(A)

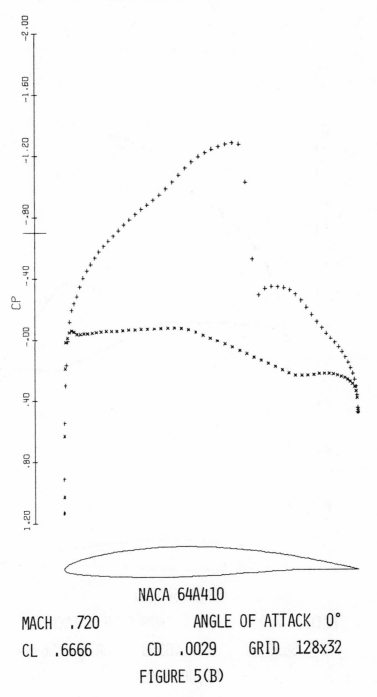

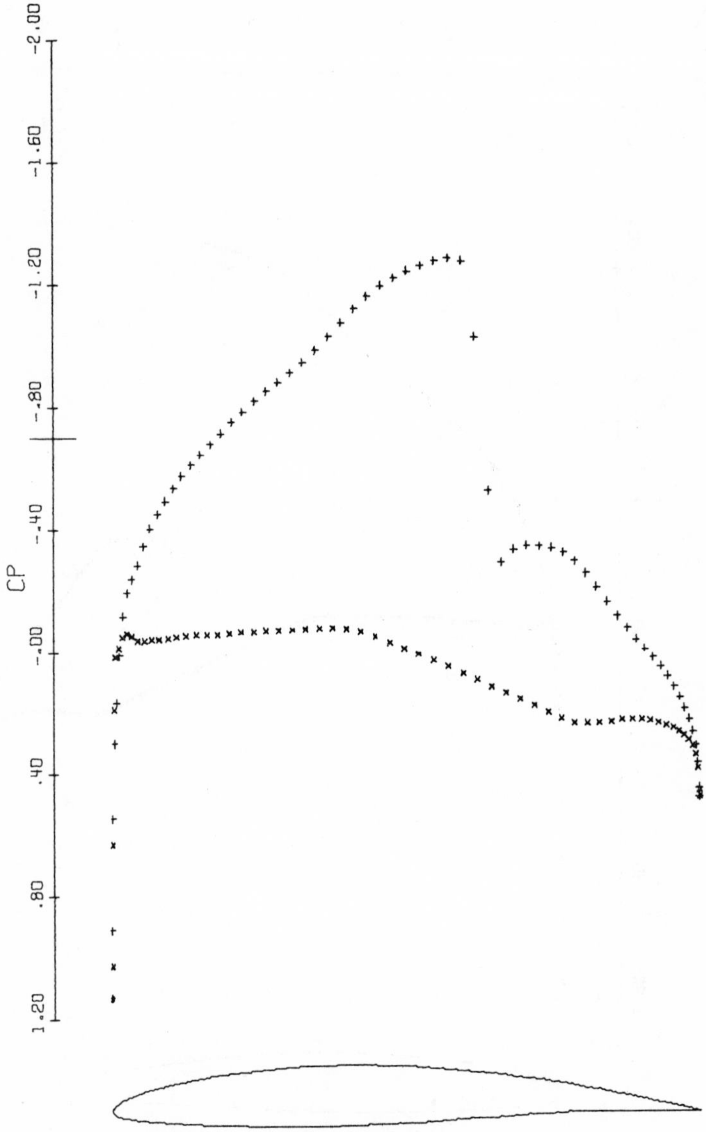

NACA 64A410

MACH .720 ANGLE OF ATTACK 0°

CL .6666 CD .0029 GRID 128x32

FIGURE 5(B)

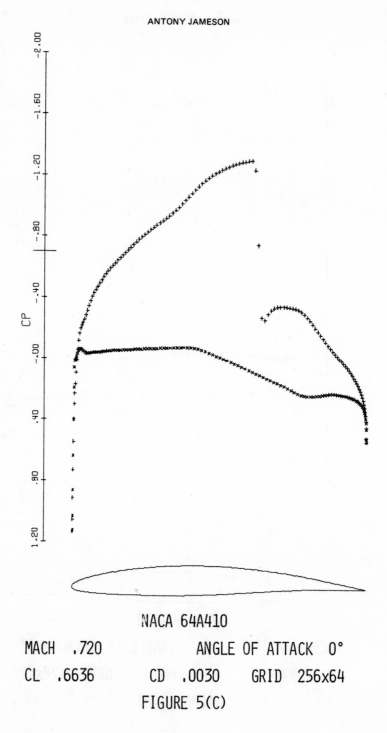

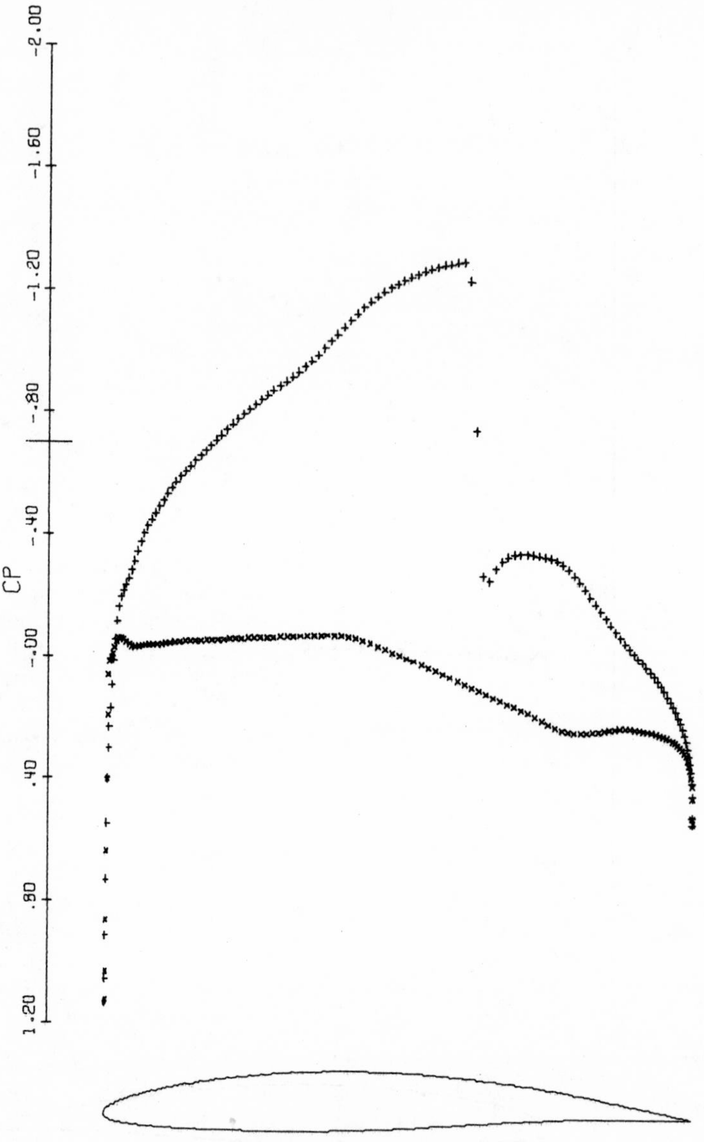

NACA 64A410

MACH .720 ANGLE OF ATTACK 0°

CL .6636 CD .0030 GRID 256x64

FIGURE 5(C)

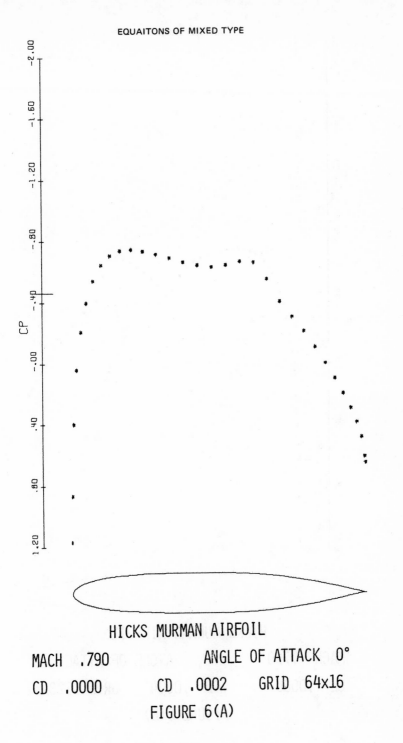

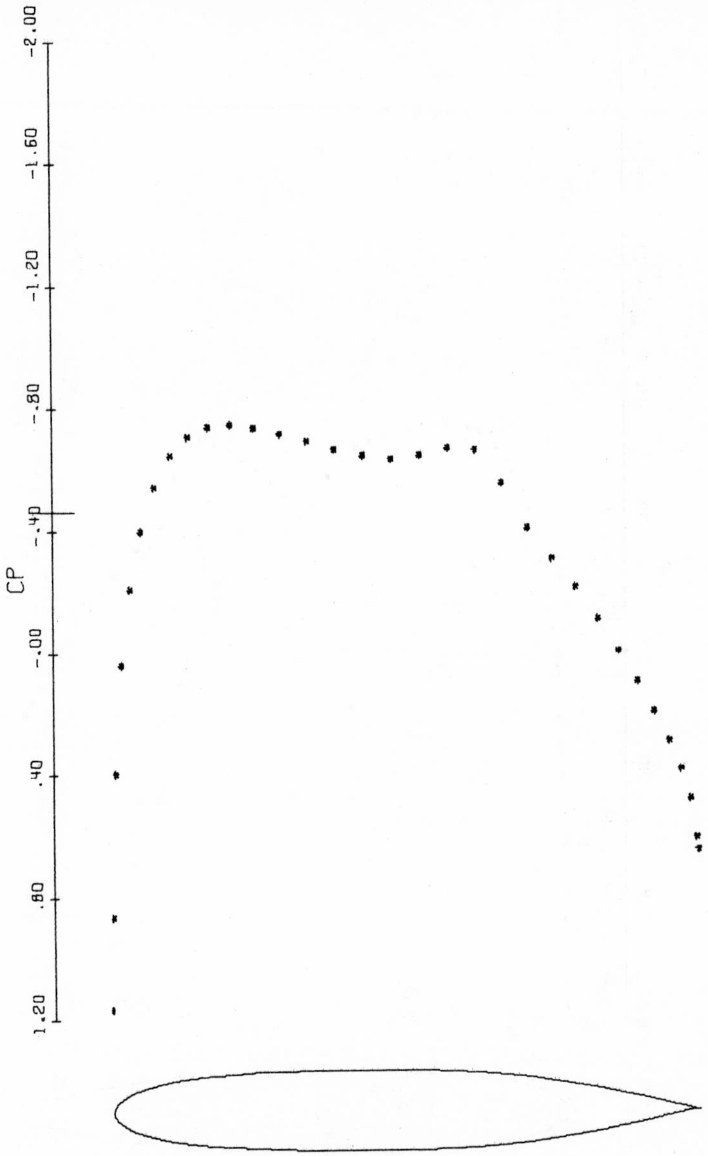

HICKS MURMAN AIRFOIL

MACH .790 ANGLE OF ATTACK 0°

CD .0000 CD .0002 GRID 64x16

FIGURE 6(A)

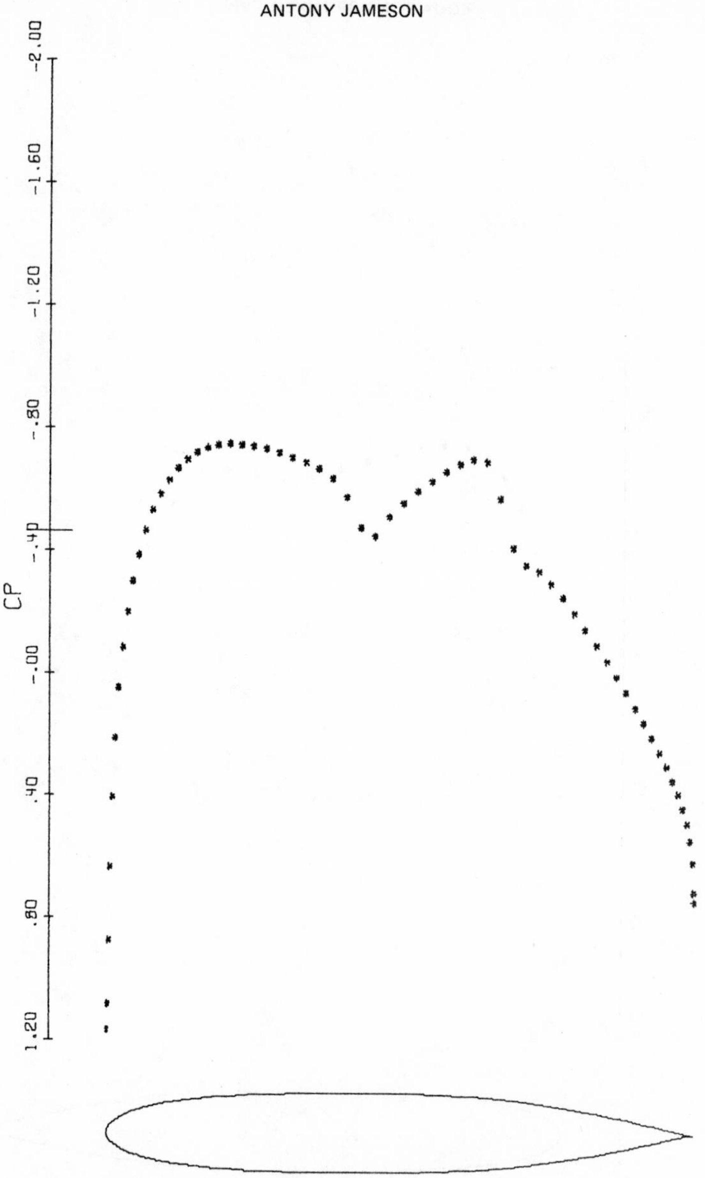

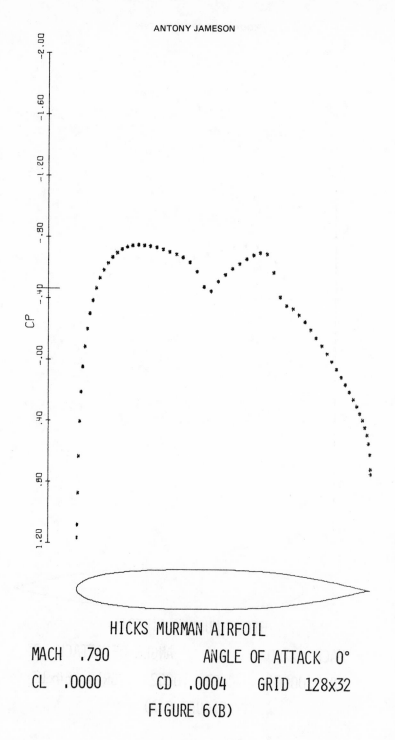

HICKS MURMAN AIRFOIL

MACH .790 ANGLE OF ATTACK 0°

CL .0000 CD .0004 GRID 128x32

FIGURE 6(B)

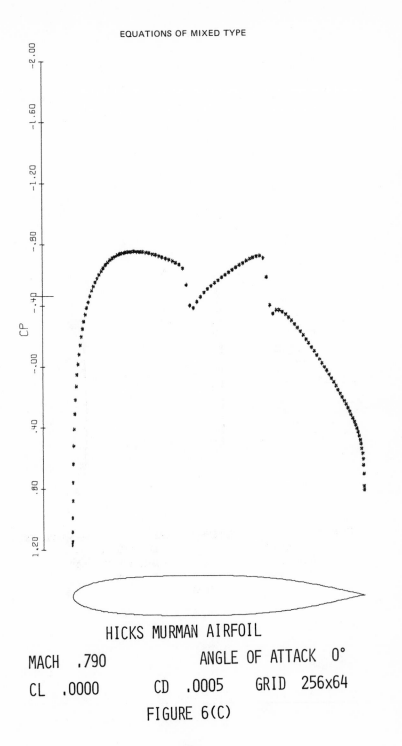

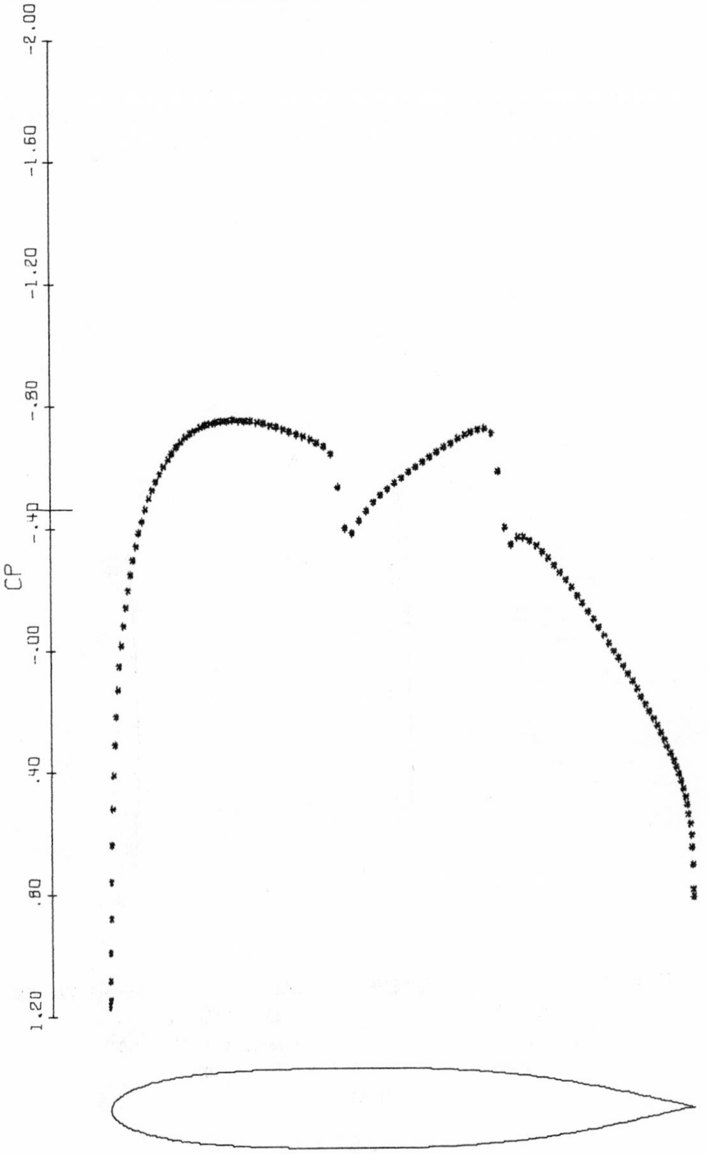

HICKS MURMAN AIRFOIL

MACH .790 ANGLE OF ATTACK 0°

CL .0000 CD .0005 GRID 256x64

FIGURE 6(C)

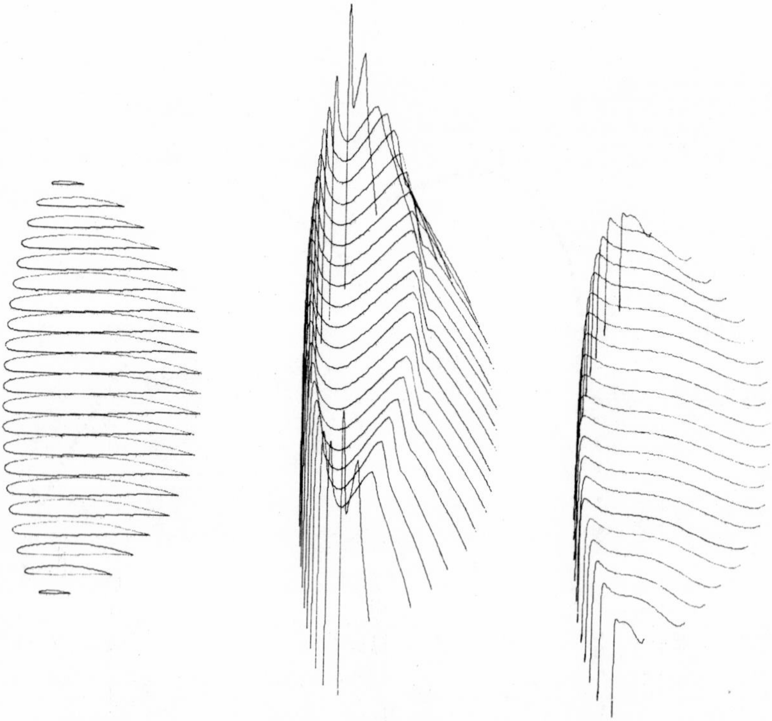

JONES SECTION
MACH .700
CL .8358

10 TO 1 ELLIPSE
ANGLE OF ATTACK 2.00°
CD .0147

FIGURE 7

NUMERICAL ANALYSIS OF THE NEUTRON TRANSPORT EQUATION*

R. B. Kellogg**

1. Introduction

Let R be a bounded domain in E^3 with points $x = (x_1,x_2,x_3)$, and let ω denote the unit sphere in E^3 . Thus, ω consists of the points $\Omega = (\Omega_1,\Omega_2,\Omega_3)$ with length $|\Omega| = (\Omega_1^2+\Omega_2^2+\Omega_3^2)^{1/2} = 1$. Let n be the outward pointing normal to the boundary ∂R of R . We consider the neutron transport equation

$$(1.1) \quad \Omega\cdot\nabla_x u(x,\Omega) + \sigma(x)u(x,\Omega) - \sigma_s(x)Su(x,\Omega) = f(x,\Omega) \quad ,$$

where

$$(1.2) \qquad Su(x,\Omega) = \int_\omega k(x,\Omega',\Omega)u(x,\Omega')d\Omega' \quad ,$$

and with, e.g., the boundary condition

$$(1.3) \qquad u(x,\Omega) = g(x,\Omega) \quad , \quad x \in \partial R \quad , \quad n\cdot\Omega < 0 \quad .$$

The equation (1.1) is of fundamental importance in the theory nuclear reactors. The unknown, $u(x,\Omega)$, represents the density of neutrons at $x \in R$ and travelling in direction Ω . The operator S represents the effect of a collision of a neutron with an atom in the reactor. The boundary condition

---

* Research supported in part by the National Science Foundation under grant GP-20555.

** IFDAM, University of Maryland

(1.3) represents an external source of neutrons into the reactor. For a derivation of this equation, and a discussion of its physical significance, and other problems associated with it, see [1].

The basic existence theorems for the problems of transport theory have been obtained in a series of papers, e.g. [2]. In this paper we shall survey some regularity results for solutions of the transport equation, and we shall discuss some numerical methods for solving the equation. Our emphasis will be on the questions of theoretical numerical analysis that arise, but in the last section we shall formulate a new iterative method for solving transport problems. Various unsolved problems and conjectures will also be stated.

2. Regularity of the solution

The solutions of the transport equation display singularities, even in the most simple situations. Intuitively, this may be thought of as being caused by the interaction of the first derivative terms of (1.1), which form a hyperbolic operator, and the scattering integral in (1.1). Since the integral in (1.2) is taken over all Ω, we see that for any x on a smooth part of ∂R, there are directions Ω such that the characteristic of the hyperbolic part of (1.1) is tangent to ∂R at x. This gives rise to a singularity in the solution along ∂R, which we term the boundary singularity of the problem.

The boundary singularity was first noticed for solutions of the transport equation in a half space (the Milne problem) by Mark [3] and Davison [4]. They found that the scalar flux behaves like $z \ell n\, z$ near $z = 0$. In [5,6] this result is generalized to the slab transport equation, where an arbitrarily high expansion of the scalar and angular flux is given.

The transport problem in slab geometry with isotropic scattering and isotropic source may be written

$$\mu u_z(z,\mu) + \sigma(z)u(z,\mu) - \sigma_s(z)Su(z) = f(z) ,$$

$$a < z < b , \ |\mu| < 1 ,$$

$$\varphi(z) = Su(z) = \frac{1}{2}\int_{-1}^{1} u(z,\mu)d\mu ,$$

with the vacuum boundary conditions

$$u(a,\mu) = 0 , \ \mu > 0$$

$$u(b,\mu) = 0 , \ \mu < 0 .$$

To state the results of [5], we assume that $\sigma(z)$ and $\sigma_s(z)$ are constant in a family of subintervals whose union comprises (a,b) , and that $f(z)$ has Holder continuous derivatives of sufficiently high order in each of these intervals. Let $z = 0$ be either a point of discontinuity of σ , σ_s , or f , or the left end point of the interval (a,b) . Then we have the asymptotic expansion.

$$\text{(2.1)} \qquad \varphi(z) \sim \sum_{q=0}^{M} \sum_{p=0}^{q} a_{pq} z^q (\ell n \ z)^p , \ z \to 0 +$$

where the upper limit, M , is determined by the requirement that f has Holder continuous derivatives of the M-th order on the subintervals of the problem. (A similar result holds, but with possibly different coefficients, when $z \to 0 -$.) . The relation (2.1) may be differentiated up to M times to give an asymptotic expression for the derivatives of φ . In particular, we see that $\varphi'(z) \sim a_{11} \ \ell n \ z$ near $z = 0$.

The coefficient a_{11} is given by

$$a_{11} = \frac{1}{2} \times \text{jump in } \sigma_s\varphi + f \text{ at } z = 0 \ .$$

Thus, in general $a_{11} \neq 0$ and the singularity is present in the problem.

An asymptotic expansion is given for the angular flux, $u(z,\mu)$, in [6].

To exhibit the boundary singularity in higher dimensions, we consider a model problem in E^3 . Let $R \subset E^3$ be a convex bounded domain with a smooth boundary ∂R , let $\sigma(x) \equiv \sigma$ and $\sigma_s(x) \equiv \sigma_s$ be constant in R , and let $k(x,\Omega',\Omega) \equiv 1/4\pi$. Thus, we are in the case of isotropic scattering. Then $Su = \varphi$ is the scalar flux, and we may, assuming $g = 0$, obtain from (1.1), (1.2), (1.3), the integral equation

$$(2.3) \qquad \varphi(x) = \sigma_s \int_R \frac{\varphi(y)e^{-\sigma|x-y|}}{|x-y|^2} dy + \int_R \frac{e^{-\sigma|x-y|}}{|x-y|^2} f\left(y, \frac{x-y}{|x-y|}\right) dy \ .$$

Upon differentiating formally with respect to x_i , and setting $\partial_i = \partial/\partial x_i$, we obtain

$$(2.4) \qquad \partial_i\varphi(x) = v_1(x) + v_2(x) \ ,$$

where

$$v_1(x) = \int_R |x-y|^{-3}\kappa_i(y,x-y)e^{-\sigma|x-y|}dy \quad ,$$

$$v_2(x) = -\sigma \int_R |x-y|^{-2}e^{-\sigma|x-y|}\,\frac{x_i-y_i}{|x-y|}\left\{\varphi(y) + f\left(y, \frac{x-y}{|x-y|}\right)\right\} dy \;,$$

$$\varkappa_i(y,z) = -2\sigma_s\varphi(y)\,\frac{z_i}{|z|} - 2f\left(y, \frac{z}{|z|}\right)\frac{z_i}{|z|}$$

$$+ \frac{|z|^2-z_i^2}{|z|^2}\, f_{,i}\left(y, \frac{z}{|z|}\right) - \sum_{j\neq i} \frac{z_iz_j}{|z|^2}\, f_{,j}\left(y, \frac{z}{|z|}\right).$$

In the formula for $\varkappa_i$, $f(y,z)$ has been defined for all $z \neq 0$ by

$$f(y,z) = f\left(y, \frac{z}{|z|}\right) \;,$$

and $f_{,i}(y,z)$ represents the derivative of $f(y,z)$ with respect to z_i . We see by inspection that $\varkappa_i(y,z)$ is homogeneous of degree 0 in z , and it may be verified that

$$\int_\omega \varkappa_i(y,z)dz = 0 \quad .$$

Thus, using the theory of singular integrals [7] it may be shown that $v_1(x)$ is well defined for $x \in R$, and $\partial_i\varphi(x)$ is indeed given by (2.4), for $x \in R$. The formula for $v_2(x)$ involves an integral with an absolutely convergent integrand, so there is no difficulty with the existence of the integral.

As $x \to \partial R$, the cancellation property that enables the singular integral in $v_1(x)$ to be defined no longer hold. In fact, one may establish the formula

$$\partial_i u(x) = [\ln|x-x^\circ|] \int_{\Omega\cdot n<0} \varkappa_i(x^\circ,\Omega)\,d\Omega + O(1) \quad , \quad x \to x^\circ \in \partial R \; .$$

In practical multidimensional problems the presence of material interfaces gives rise to further singularities in the solution of transport problems. As an example of this, we consider the transport equation in a plane polygonal region R , and with isotropic scattering and isotropic source [8]. We suppose that R is divided into a number of polygonal sub-regions R_i , and that in each subregion the coefficients of the equation are constant. In this case the transport equation may be written

$$\Omega_1\partial_1 u + \Omega_2\partial_2 u + \sigma u - \sigma_s\varphi = f \quad , \tag{2.5}$$

and the integral equation for the scalar flux is

$$\varphi(x) = \int_R g(y)\,\frac{F(\beta(x,y))}{|x-y|}\,dy \quad ,$$

where

$$\beta(x,y) = |x-y| \int_0^1 \sigma(tx+(1-t)y)\,dt \; ,$$

$$g(x) = \sigma_s(x)\varphi(x) + f(x) \; ,$$

$$F'(x) = -(1/2\pi)K_0(x) \; ,$$

$K_0(x)$ being a Bessel function of the second kind. (The function $F(x)$ is sometimes known as the transport literature as the Bickley function.) Formally differentiating the integral equation, we obtain

(2.6) $$\partial_i\varphi(x) = -\int_R g(y) \frac{F(\beta(x,y))(x_i-y_i)}{|x-y|^3} dy + \int_R g(y)F'(\beta(x,y)) \frac{\partial_{i,x}\beta(x,y)}{|x-y|} dy \quad .$$

The first integral in (2.6) must be interpreted as a singular integral, and gives rise to a logarithmic singularity as $x \to \partial R_i$, for each subregion R_i. In fact one can obtain the inequality

$$|\partial_i\varphi(x)| \leq c[|\ell n\ \delta(x)| + 1] ,$$

where $\delta(x)$ is the distance to the nearest boundary of a subregion. As in the model problem discussed above, this boundary singularity is, in general, present.

The presence of material interfaces gives rise to further singularities in the solution. We illustrate these with the two region problem shown in Figures 1 and 2. In general, $\partial_1\varphi(x)$ has a jump discontinuity at the point $x = (0,-a)$, $a > 0$, and the angular flux, $u(x,\Omega)$ has a jump discontinuity at $x = (0,-a)$, $a > 0$, if Ω points in the negative x_2 direction. Furthermore,

$$|\partial_1 u(x,\Omega)| \leq c\varepsilon^{-1} |\ell n\ \varepsilon| , \quad x = (\varepsilon,-a) , \quad \varepsilon \to 0 ,$$

and the inequality is attained if Ω is chosen suitably. These singularities are caused by the shadowing effect of the subregion on the solution at (0,-a), and are hence called the <u>shadow singularities</u> of the problem. Mathematically,

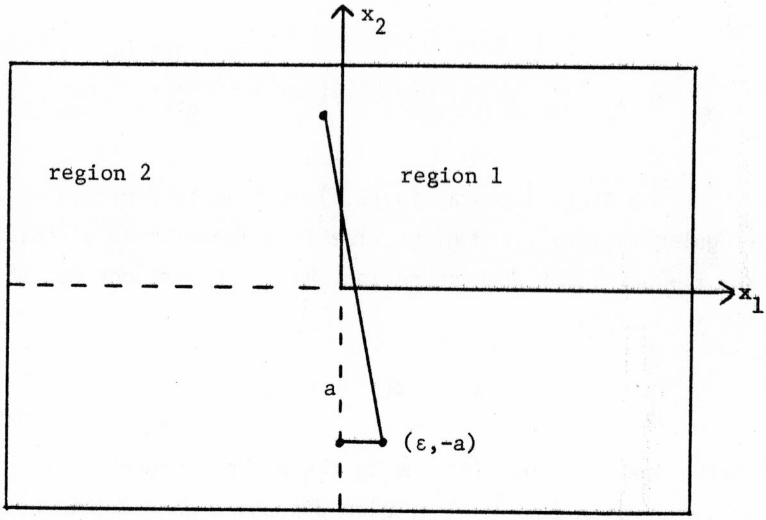

Figure 1
The shadow singularity.

the shadow singularity in $\partial_1\varphi(x)$ is caused by the singular behavior in the second term of (2.6) as $x \to (0,-a)$.

As a third type of singularity, we note [9] that $\partial_i u(x,\Omega)$ is discontinuous at any point (x,Ω) such that x lies in a subregion of the problem and Ω points away from a vertex of the problem (see Figure 2). We call this a vertex singularity of the problem.

The shadow and vertex singularities are also manifested in derivatives of $u(x,\Omega)$ with respect to the angular variables.

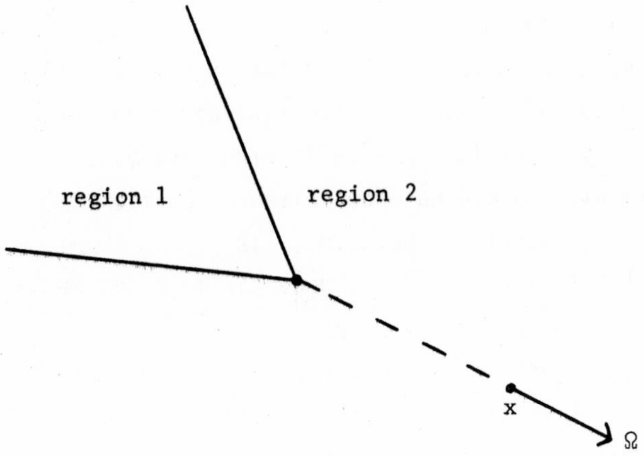

Figure 2
The vertex singularity.

From these examples, we see that there are many singularities in the derivatives of the solution of the transport equation, and that much remains to be done in understanding and classifying these singularities.

3. Discretization of the angular variable

We shall discuss briefly two methods for discretizing the angular variable in transport problems: the spherical harmonics method and the discrete ordinate method. With the spherical harmonics method (see [1] for a general discussion and [10] for a careful exposition of the method), $u(x,\Omega)$ is expanded in a finite series of spherical harmonics, and Galerkin's method is used to obtain a set of differential equations for the unknown coefficients $u_k(x)$. If the Galerkin method is applied in the context of Vladimirov's self-adjoint formulation of the transport equation, the spherical harmonics approximation emerges as the best possible approximation in a certain sense [2,10]. The boundary value problem that must be solved to obtain the $u_k(x)$ is given by an elliptic system of partial differential equations with covering boundary conditions [11]. An error analysis of the spherical harmonics method has been given by Davison for various special geometries (see, e.g., [12]), but there seems to have been little error analysis done for more complex problems.

The discrete ordinate method (or S_n method) involves selecting a set of vectors $\Omega^m \in \omega$, and a set of quadrature weights $w_m > 0$ for use in integrating over ω. The scattering operator S is then replaced by the operator

$$(3.1) \qquad (\tilde{S}u)(x,\Omega^m) = \sum_n w_n k(x,\Omega^n,\Omega^m)u(x,\Omega^n) .$$

Using $\tilde{S}$ instead of S in (1.1), one obtains a first order hyperbolic system of differential equations for the unknown functions $u(x,\Omega^m)$. In the subcritical case this hyperbolic system, together with the boundary conditions (1.3), give a positive symmetric system in the sense of Friedrichs [13]. The solution of this system is the discrete ordinate approximation to $u(x,\Omega)$.

In certain simple situations (in slab geometry), it is known that with a particular choice of quadrature formula the spherical harmonics solution can be obtained from the discrete ordinate solution. For more general geometries, the two methods are not equivalent, and it seems likely that each method is superior to the other for certain classes of problems. The discrete ordinate solutions, however, are subject to the "ray effect" [14,15], in which the discrete character of the quadrature formula causes inaccuracies in certain types of problems. To illustrate

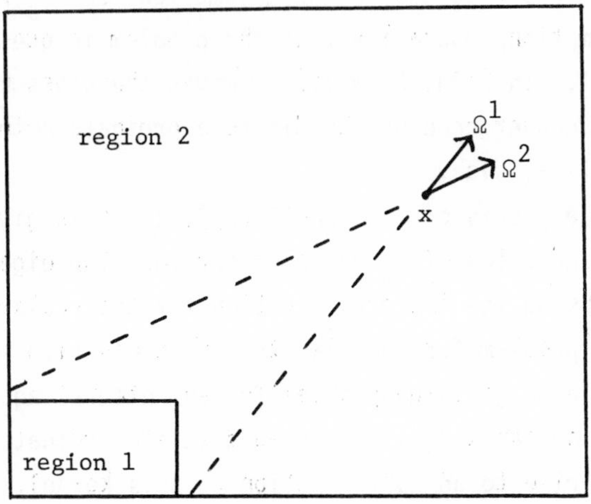

Figure 3
The ray effect.

the ray effect, we consider the two region problem of Figure 3, and we suppose there are only two ordinates in the positive octant, as shown. With these ordinates, the point x will not "see" region I, and if σ_s is small in region II, a source in region I will have little effect on the solution at x. To eliminate this inaccuracy, there have been proposed several iterative techniques in which an extra term is added to the discrete ordinate equations, so that the solution of the resulting system satisfies a lower order set of spherical harmonics equations [15,16,17].

We now survey error estimates for the discrete ordinate approximations. A number of papers have been written on this subject, e.g., [18,19,20,21]. In [18], H. B. Keller gives an inequality, in the case of slab geometry, which bounds the error in the discrete ordinate approximation by the error in the quadrature formula. In [20], N. Madsen obtains a similar result for the three dimensional problem. These results are established under certain extra assumptions on $\sigma_s(x)$. These extra assumptions guarantee that the problem in question is subcritical. In [19], P. Nelson removes these assumptions and proves convergence of the discrete ordinate method for any subcritical slab.

In the thesis of J. Bennett [21], there is given error estimates and rates of convergence for the slab eigenvalue problem. Using the integral equation for the scalar flux, the eigenvalue problem for the slab transport equation may be written as an eigenvalue problem for an integral equation with a weakly singular kernel. A given discrete ordinate approximation gives rise to an approximation to this kernel. By using the Fourier transform to analyse the difference of these kernels, Bennett is able to give an asymptotic analysis of the error in the eigenvalue. He finds, for example, that if an M point composite Newton-Cotes quadrature formula is used, the error in the eigenvalue is $O(N^{-M})$, where N is the number of discrete ordinates.

There are several unsolved mathematical problems regarding the discrete ordinate methods. Thus, it would be interesting to modify the inequalities of [18,20] to allow a weighted norm of the quadrature error on the right hand side. By chosing the weight to be small near singularities of the problem, one would have the possibility of mitigating the effect of these singularities in the error analysis. Also, it would be interesting to extend the analysis of Bennett to source problems, and to problems in other geometries. Finally, there has been considerable work on quadrature formulas on a sphere. (See, e.g., [22].) Does this work suggest any special formulas as being particularly useful for transport problems?

4. Discretization of the spatial variable

A number of finite difference methods, and some finite element methods [23] have been used for the numerical solution of transport problems. We shall not survey these, but will instead discuss some features of a particular finite difference scheme. We shall restrict our remarks to the plane transport equation with isotropic scattering; that is, to (2.5).

We write the discrete ordinate approximation to (2.5) as

$$(4.1) \qquad \Omega_1^m \partial_1 u^m + \Omega_2^m \partial_2 u + \sigma u^m - \sigma_s \varphi^M = f^m , \qquad 1 \le m \le M ,$$

where there are M discrete ordinates, $1 \le m \le M$, and where the approximate scalar flux is written

$$(4.2) \qquad \varphi^M(x) = \frac{1}{4\pi} \sum_m w_m u(x, \Omega^m) .$$

Let there be given a rectangular grid of mesh points (x_{1i}, x_{2j}), $0 \le i \le I$, $0 \le j \le J$, with uniform spacing $h = x_{1i} - x_{1i-1}$, $k = x_{2j} - x_{2j-1}$. Then we shall consider the difference approximation

$$(4.3)\quad \frac{\Omega_1^m}{2h}(v_{ij}^m+v_{i,j-1}^m-v_{i-1,j}^m-v_{i-1,j-1}^m)$$

$$+\frac{\Omega_2^m}{2k}(v_{ij}^m+v_{i-1,j}^m-v_{i,j-1}^m-v_{i-1,j-1}^m)$$

$$+\frac{1}{4}\bar{\sigma}_{ij}\,[A^m(v_{ij}^m+v_{i-1,j-1}^m)+B^m(v_{i-1,j}^m+v_{i,j-1}^m)]$$

$$=\frac{1}{4}\bar{\sigma}_{sij}\sum_n w_n[A^n(v_{ij}^n+v_{i-1,j-1}^n)+B^n(v_{i-1,j}^n+v_{i,j-1}^n)]$$

$$+\bar{f}_{ij}^m\ .$$

where $\bar{\sigma}_{ij}$, $\bar{\sigma}_{sij}$, and $\bar{f}_{ij}^m$ are appropriate averages of the corresponding functions, and A^m, B^m are numbers which satisfy $A^m + B^m = 2$. Typical choices for these parameters are

$$(4.4)\qquad A^m = 1\ ,\quad B^m = 1\ ,\quad 1 \le m \le M\ ,$$

or

$$(4.5)\qquad \begin{cases} A^m = 2\ ,\quad B^m = 0 & \text{if } \Omega_1^m\Omega_2^m > 0\ , \\ A^m = 0\ ,\quad B^m = 2 & \text{if } \Omega_1^m\Omega_2^m < 0\ , \end{cases}$$

or more generally,

$$(4.6)\qquad A^m = \begin{cases} a^m\ , & \Omega_1^m\Omega_2^m > 0\ , \\ 2-a^m\ , & \Omega_1^m\Omega_2^m < 0\ , \end{cases}$$

where $a^m \in [1,\infty)$ are a given set of numbers.

If it is assumed that the domain R is a rectangle, and that the sides of R are mesh lines of the problem, then the boundary conditions can be represented in an obvious way and a complete description of the approximate problem is obtained.

The difference equation (4.1), the various choices that can be made for the parameters A^m and B^m, and some numerical results have been discussed in the literature [9,24]. In particular, such questions as the stability and accuracy of the difference scheme, whether the scheme guarantees a positive solution, and whether the difference scheme may play a role in mitigating the ray effect, are important considerations in choosing A^m and B^m. We shall not repeat these discussions.

It seems that a fundamental question regarding the use of (4.3) or other difference approximations has not been treated in the literature: does the solution of the difference equations converge to the solution of the original transport problem as $h \to 0$, $k \to 0$, $M \to \infty$. One may conjecture that, using the results of Bardos [25] on the density of smooth solutions in the space of all solutions, and the results of Madsen [24] on the convergence of the difference approximation when the solution is smooth, a proof of this result could be obtained. Beyond such a convergence result lies the problem of obtaining error estimates for the solution of transport problems that generalize the error estimates of [24], which pertain to smooth solutions. Finally, it would be interesting to determine the optimal balance between the mesh spacing in the space variables and accuracy of the quadrature formula in the angular variables.

There has been some attempt to construct approximate schemes that take into account the singularities in the solution of transport problems. In [26], Abu-Shumays and Bareiss consider the numerical solution of the integral form of the transport equation in slab geometry using the finite element method. They enrich the subspace of trial functions with functions containing logarithmic singularities in order to deal with the boundary singularities described in §2. In [27], T. Kulikowska discusses difference schemes for the transport equation in one dimensional spherical geometry. The singular nature of the solution at interfaces is considered. In [9], J. Arkuszewski, T. Kulikowskaya and J. Mika consider the effect of the vertex singularities in the numerical solution of the plane transport equation. The difference equation (4.3) is modified to take into account the jump discontinuities in the derivatives of u at appropriate values of Ω . Numerical results are given which exhibit the increased accuracy that is obtained.

It seems clear that much more remains to be done on the related problems of constructing higher order schemes for the transport equation and the numerical treatment of the singularities in the solution.

5. Iterative methods

After selecting a quadrature formula and a finite difference approximation, the approximate solution of a neutron transport problem is reduced to the solution of a system of linear algebraic equations. Because of the large number of unknowns involved (the plane transport equation involves 4 independent variables), iterative methods are often used to solve such problems. We shall briefly outline a few iterative methods and propose a new, alternating direction procedure for solving the difference equations arising from the plane transport equation. For convenience, we shall discuss these methods only for the case of isotropic scattering. Furthermore, we shall not discuss the problem of accelerating the convergence of the methods with appropriate overrelaxation devices or parameter choices.

In discussing iterative methods, it is sometimes convenient to phrase the iterative method in terms of the original transport problem, before discretizing the independent variables. (Of course, the method will only be useful if it can be applied to the problem in discretized form.) In this spirit, we write the transport equation (1.1) in the operator form

$$(5.1) \qquad Lu - \sigma_s Su = f \ .$$

We shall consider the case of vacuum boundary conditions, so $g(x,\Omega) = 0$ in (1.3). The unbounded operator L may be defined in a domain that contains the boundary conditions, and in this sense, L^{-1} is a bounded operator on $L_2(R\times\omega)$ [2].

The simplest iterative method, based on the Neumann series expansion for $(L-\sigma_s S)^{-1}$, is defined by the relation

$$(5.2) \qquad Lu^{n+1} = \sigma_s Su^n + f \ , \quad j = 1,2,\cdots$$

In terms of the scalar flux $\varphi = Su$, we may write the method as

$$(5.3) \qquad \varphi^{n+1} = K\sigma_s\varphi^n + Kf \ , \quad K = SL^{-1} \ .$$

The convergence of the method in the subcritical case has been established, e.g., in [2].

The Neumann series method may be modified by utilizing the solution of a low order spherical harmonics method at each step of the iteration. This and related iterative methods have been studied independently in the U.S. (the synthetic method, [28,29]) and in the U.S.S.R. (the KP method), see [30], and the references therein). To describe the method, we select a low order spherical harmonics or discrete ordinates approximation to the transport equation, and we write the corresponding integral equation for the approximate scalar flux in the form

$$\varphi_L = K_L \sigma_s \varphi_L + K_L f \quad .$$

Then the synthetic method is defined by the iterative process

$$R^{n+1} = K[f+\sigma_s \varphi^n] - \varphi^n \tag{5.4a}$$

$$\varphi^{n+1} = \varphi^n + (I-K_L\sigma_s)^{-1} R^{n+1} \quad . \tag{5.4b}$$

Thus, to implement (5.4a,b), it is required to solve the low order problem.

Methods of the type (5.4a,b) suffer several disadvantages that make it worth while to seek other schemes. J. A. Davis and L. A. Hageman [31] have constructed another iterative method by decomposing the scattering operator, as follows. The unknown, $u(x,\Omega)$, is divided into two parts,

$$u_+(x,\Omega) = u(x,\Omega) \quad , \quad \Omega_2 > 0 \quad ,$$

$$u_-(x,\Omega) = u(x,\Omega) \quad , \quad \Omega_2 < 0 \quad .$$

Thus, $u_\pm$ is defined on the hemisphere $\omega_\pm$ defined by $\pm\Omega_2 > 0$. In the same way we define scattering operators $S_\pm$ by

$$S_\pm u(x,\Omega) = \frac{1}{4\pi} \int_{\omega_\pm} u(x,\Omega')d\Omega' \quad . \tag{5.5}$$

Then the Davis-Hageman method may be written

$$\text{(5.6a)} \qquad Lu_+^{n+1} = \sigma_s S_+ u_+^{n+1} + \sigma_s S_- u_-^n + f \ ,$$

$$\text{(5.6b)} \qquad Lu_-^{n+1} = \sigma_s S_- u_-^{n+1} + \sigma_s S_- u_-^n + f \ .$$

In (5.5a), the operator $L - \sigma_s S_+$ is considered as acting on functions $u_+(x,\Omega)$ which are defined on $R\times\omega_+$. Equation (5.6a) is then solved for u_+^{n+1} . In a similar manner (5.6b) may be solved for the functions u_-^{n+1} defined on $R\times\omega_-$. The method may be regarded as a block Gauss Seidel method, and the iteration operator is cyclic of index 2. The operator formulation of the method, and the convergence in the continuous case, is discussed in [32].

The Davis-Hageman method has the disadvantage that it is difficult to solve (5.6a,b) for $u_\pm^{n+1}$, and in fact an extra set of "inner" iterations is required for this purpose. As an alternative, we formulate an alternating direction method for the plane transport equation that avoids this difficulty. We shall establish the convergence of the discretized form of this method.

The alternating direction method is based on a different decomposition of the scattering operator than the one given in (5.5). To define this decomposition, we write $\Omega = (\Omega_1,\Omega_2,\Omega_3)$ in spherical coordinates:

$$\Omega_1 = \sin\psi\cos\theta, \Omega_2 = \sin\psi\sin\theta, \Omega_3 = \cos\psi \ ,$$

and we let Ω' have spherical coordinates θ' and ψ' . We let $\omega_1(\theta)$ be the portion of the unit sphere defined by

$$\omega_1(\theta) = \{\Omega' : \theta < \theta' < \pi-\theta\} \ ,$$

and we let $\omega_2(\theta) = \omega\backslash\omega_1(\theta)$ be the complementary set. In terms of the projection of Ω onto the (x_1,x_2) plane, the sets $\omega_1(\theta)$, $\omega_2(\theta)$ are shown in the figure.

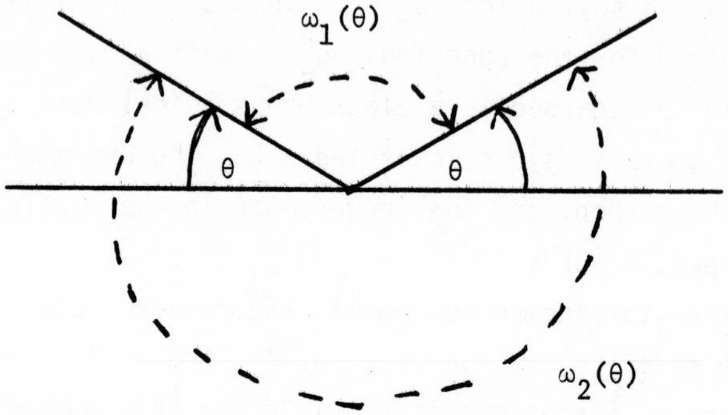

Figure 4
Decomposition of S

We define

$$S_j u(x,\Omega) = \frac{1}{4\pi} \int_{\omega_j(\theta)} u(x,\Omega')d\Omega', \; j = 1,2 \; ,$$

$$L_j u(x,\Omega) = \Omega_j \partial_j u(x,\Omega) + \frac{1}{2}\sigma u \quad .$$

Then the alternating direction method is given by the equations

(5.7a) $$(rI+L_1-\sigma_s S_1)u^{n+1/2} = (rI-L_2+\sigma_s S_2)u^n + f \; ,$$

(5.7b) $$(rI+L_2-\sigma_s S_2)u^{n+1} = (rI-L_1+\sigma_s S_1)u^{n+1/2} + f .$$

In these equations, $r > 0$ is a fixed iteration parameter.

To analyse this method, we require some facts about the operators S_j. It is trivial that

(5.8) $$S_1 + S_2 = S ,$$

and a calculation yields

(5.9) $$S_1^* = S_2 ,$$

where the adjoint operator S_1^* is taken with respect to the inner product in the real Hilbert space $L_2(\omega)$. Using these we may prove the

<u>Lemma</u>. $(S_j u,u) \le \frac{1}{2}(u,u)$.

Proof. We have, using (5.9) and (5.8),

$$(S_1 u,u) = \frac{1}{2}(Su,u)$$

$$= \frac{1}{8\pi}\int_\omega\int_\omega u(\Omega)u(\Omega')d\Omega d\Omega' \quad .$$

Applying Schwarz's inequality to the inner integral,

$$(S_1 u,u) \le \frac{1}{2}\int u(\Omega)^2 d\Omega$$

$$= \frac{1}{2}(u,u) \quad .$$

Using the above lemma, and supposing that $\sigma_s(x) < \sigma(x)$, one may establish the "accretive" inequality

$$((L_j+\sigma_s S_j)u,u) \geq 0 \ , \ j=1,2 \ . \tag{5.10}$$

This inequality is basic in establishing the convergence of alternating direction iterations, and it is reasonable to ask whether the iterations (5.7) converge in the continuous case. (An analogous result has been given for the Laplace operator [33]). There are certain analytical difficulties associated with this, and the answer is not clear.

We now consider the alternating direction method as applied to our discrete problem. We shall use the difference equation (4.3) with the coefficients satisfying (4.4). We select a set of ordinates Ω^m and quadrature weights $w_m > 0$ that are symmetrically placed in each octant. We define the approximate scattering operator $\tilde{S}$, as in (3.1), by

$$\tilde{S}u(x) = \frac{1}{4\pi} \sum_n w_n u(x,\Omega^n) \ .$$

The operator $\tilde{S}$ may be decomposed into $\tilde{S}_1$ and $\tilde{S}_2$ by the formulas

$$\tilde{S}_1 u(x,\Omega^m) = \frac{1}{4\pi} \sum_{\Omega^n \in \omega_1(\theta_m)} w_n u(x,\Omega^n) \ , \tag{5.11}$$

$$\tilde{S}_2 = S - \tilde{S}_1 \ . \tag{5.12}$$

In (5.11), we have written $\Omega^m = (\sin\psi_m \cos\theta_m \ , \ \sin\psi_m \sin\theta_m \ , \ \cos\psi_m)$. Thus, the sum in (5.11) is taken over all n such that $\theta_m < \theta_n < \pi - \theta_m$.

For each x, the operators $\tilde{S}$, $\tilde{S}_j$, may be regarded as M-square matrices. It may then be verified that $\tilde{S}_1 = \tilde{S}_2^T$, where T denotes the transpose. By following the proof of Lemma 1, we obtain the discrete version,

Lemma 2. $\frac{1}{4\pi} \sum_m w_m [S_j u(x,\Omega^m)] u(x,\Omega^m)$

$$\leq \frac{1}{8\pi} \sum_m w_m u(x,\Omega^m)^2 \quad , \quad j = 1,2 \quad .$$

We shall let the domain R be a rectangle, and we shall use the difference approximation (4.3), (4.4), together with the vacuum boundary conditions on ∂R. We define discrete operators $\tilde{L}_j$, $j = 1,2$, by

$$(\tilde{L}_1 v)^m_{ij} = \frac{\Omega^m_1}{2h} (v^m_{ij} + v^m_{i,j-1} - v^m_{i-1,j} - v^m_{i-1,j-1})$$

$$+ \frac{1}{8} \bar{\sigma}_{ij} [v^m_{ij} + v^m_{i-1,j-1} + v^m_{i,j-1} + v^m_{i-1,j}] \quad ,$$

$$(\tilde{L}_2 v)^m_{ij} = \frac{\Omega^m_2}{2k} (v^m_{ij} + v^m_{i-1,j} - v^m_{i,j-1} - v^m_{i-1,j-1})$$

$$+ \frac{1}{8} \bar{\sigma}_{ij} [v^m_{ij} + v^m_{i-1,j-1} + v^m_{i,j-1} + v^m_{i-1,j}] \quad .$$

These definitions complete the specification of the alternating direction method in the finite dimensional case. We have

Theorem. Let $r > 0$ be fixed, and let $\sigma_s(x) < \sigma(x)$. The iterates $(v^m_{ij})^n$, $n = 1,2,\cdots$, exist and converge to the solution v^m_{ij} of the difference equations.

Sketch of the proof. Let us say that a matrix is accretive if it has a positive definite symmetric part. If the matrices $\tilde{L}_j - \sigma_s \tilde{S}_j$, $j = 1,2$, are accretive, it follows from (5.7a,b) that the iterates $(v^m_{ij})^n$ are well defined. Furthermore, it is well known (see, e.g., [33]) that in the accretive case, the alternating direction iterates converage to the solution of the difference equations as $n \to \infty$. Thus it suffices to show that $\tilde{L}_j - \sigma_s \tilde{S}_j$ is accretive, $j = 1,2$. This may be proved by following the argument of Madsen [24, Theorem 1] and using Lemma 2.

The usefulness of the alternating direction method depends on being able to easily solve equations of the form

$$(rI+L_1-\sigma_s S_1)v = g \quad , \tag{5.13}$$

$$(rI+L_2-\sigma_s S_2)v = g \quad . \tag{5.14}$$

To solve (5.13) in the discretized form, one may consider a subset of the equations (5.13) corresponding to a row of mesh points. To solve for these unknowns amounts to solving a one dimensional transport problem, where the scattering operator has the particular form given by $\tilde{S}_1$. Suppose the discrete ordinates are ordered in the direction of decreasing Ω_2^m, and suppose the unknowns u_{ij}^m are, at each mesh point, ordered in the same way. Then the scattering matrix $\tilde{S}_1$ is block triangular, and the problem (5.13) may be solved efficiently by direct elimination. A similar strategy may be used for the solution of (5.14), except that the mesh points must be swept along vertical columns and the ordinates must be arranged in the opposite manner. Figure 5 shows, schematically, the ordering of the mesh points for the two sweeps.

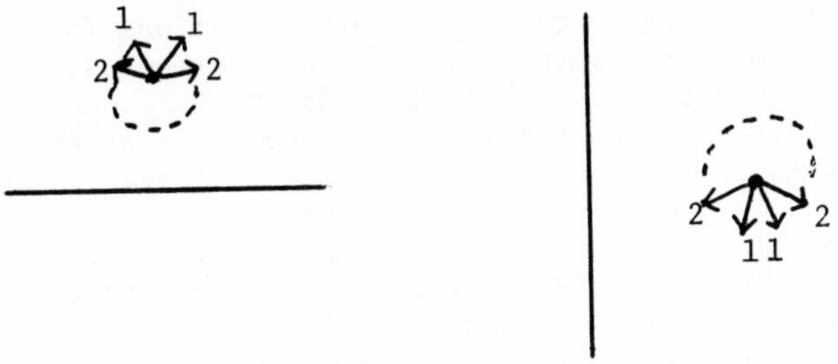

Figure 5
Ordering the ordinates

REFERENCES

1. G. I. Bell and Samuel Glasstone, Nuclear Reactor Theory, Van Nostrand Reinhold, New York, 1970.
2. V. S. Vladimirov, "Mathematical problems in the one-velocity theory of particle transport," Trudy Mat. Inst. Steklov, Vol. 61 (1961), translated by Atomic Energy of Canada Ltd., Chalk River, Ontario, 1963, AECL-1661.
3. C. Mark, "The neutron density near a plane surface," Phys. Rev. 72 (1947), 558-564.
4. B. Davison, "Angular distribution of neutrons at the interface of two adjoining media," Canadian J. Res. A28 (1950), 303-314.
5. H. G. Kaper and R. B. Kellogg, "Asymptotic behavior of the solution of the integral transport equation in slab geometry," to appear.
6. H. G. Kaper and R. B. Kellogg, "Continuity and differentiability properties of the solution of the linear transport equation," to appear.
7. E. M. Stein, Singular integrals and differentiability properties of functions, Princeton University Press, Princeton, 1970.

8. R. B. Kellogg, "First derivatives of solutions of the plane neutron transport equation," Technical Note BN-783, IFDAM, University of Maryland, 1974.
9. J. Arkuszewski, T. Kulikowska, and J. Mika, "Effect of singularities on approximation in S_n methods," Nucl. Sci. and Eng. 49 (1972), 20-26.
10. H. G. Kaper, G. K. Leaf, and A. J. Lindeman, "An approximation procedure for the neutron transport equation based on the use of surface harmonic tensors," ANL-8081, Argonne National Laboratory, Argonne, Ill., 1974.
11. S. K. Godunov and U. M. Sultangazin, "The dissipativity of V. S. Vladimirov's boundary conditions for a symmetric system of the method of spherical harmonics," Z. Vycisl. Mat. i Mat. Fiz. 11 (1971), 688-704.
12. B. Davison, "On the rate of convergence of the spherical harmonics method," Can. J. Phys. 38 (1960), 1526-1545.
13. K. O. Friedrichs, "Symmetric positive linear differential equations," Comm. Pure Appl. Math. 11 (1968), 333-418.
14. K. D. Lathrop, "Ray effects in discrete ordinate equations," Nucl. Sci. and Eng. 32 (1968), 357-369.
15. K. D. Lathrop, "Remedies for ray effects," Nucl. Sci. and Eng., 45 (1971), 255-268.
16. J. Jung, H. Chijiwa, K. Kobayashi, and N. Nishihara, "Discrete ordinate neutron transport equation equivalent to P_L approximation," Nucl. Sci. and Eng. 49 (1972), 1-9.
17. W. H. Reed, "Spherical harmonic solutions of the neutron transport equation from discrete ordinate codes," Nucl. Sci. and Eng. 49 (1972), 10-19.
18. H. B. Keller, "On the pointwise convergence of the discrete ordinate method," SIAM J. Appl. Math. 8 (1960), 560-567.
19. Paul Nelson Jr., "Convergence of the discrete ordinate method for anisotropically scattering multiplying particles in a subcritical slab," SIAM J. Numer. Anal., Vol. 10 (1973), 175-181.
20. N. K. Madsen, "Pointwise convergence of the three-dimensional discrete ordinate method," SIAM J. Numer. Anal. 8 (1971), 266-269.
21. J. H. Bennett, "Trucation errors in numerical solutions of the transport equation," Ph.D. thesis, Harvard Univ., 1962.
22. S. L. Sobolev, Introduction to the theory of quadrature formulas, Izdatel'sto Nauka, Moscow, 1974.
23. H. G. Kaper, G. K. Leaf, and A. J. Lindeman, "Formulation of a Ritz-Galekin type procedure for the approximate solution of the neutron transport equation," J. Math. Anal. Appl., 50 (1975), 42-65.

24. N. K. Madsen, "Convergent centered difference schemes for the discrete ordinate neutron transport equations," SIAM J. Numer. Anal., Vol. 12 (1975), 164-176.

25. C. Bardos, "Problemes aux limites pour les equations aux derivees partielles du premier ordre a coefficients reels; theoremes d'approximation; application a l'equation de transport," Ann. Sci. Ecole Norm. Sup. (4), Vol. 3 (1970), 185-233.

26. I. K. Abu-Shumays and E. H. Bareiss, "Adjoining appropriate singular elements to transport theory computations," J. Math. Anal. Appl. 48 (1974), 200-222.

27. T. Kulikowskaya, "Box-explicit method in one dimensional spherical geometry," Numerical Reactor Calculations, Proceedings of a Seminar on Numerical Reactor Calculations, International Atomic Energy Agency, Vienna, 1972.

28. E. M. Gelbard and L. A. Hageman, "The synthetic method as applied to the S_n equations," Nucl. Sci. and Eng., 37 (1969), 288-298.

29. W. H. Reed, "The effectiveness of acceleration techniques for iterative methods in transport theory," Nucl. Sci. and Eng., 45 (1971), 245-254.

30. G. I. Marchuk and V. I. Lebedev, Numerical methods in neutron transport theory, Atomizdat, Moscow, 1971.

31. J. A. Davis and L. A. Hageman, "An iterative method for solving the neutron transport equation in x-y geometry," SIAM J. Appl. Math. 17 (1969), 149-161.

32. R. B. Kellogg, "On the spectrum of an operator associated with the neutron transport equation," SIAM J. Appl. Math. 17 (1969), 162-171.

33. R. B. Kellogg, "An alternating direction method for operator equations," SIAM J. on Appl. Math. 12 (1964), 848-854.

INSTABILITY PHENOMENA IN FLUID MECHANICS*

Klaus Kirchgässner**

1. Introduction

Among the numerous examples of bifurcation observable in nature there are some in fluid dynamics that are most striking. Their fascination stems not only from their physical importance or their phenomenological peculiarity but also from the fact that mathematically only some very simple models are understood. The bifurcation analysis has been especially successful in describing cellular instability phenomena where the secondary motion takes place within independent bounded cells which cover the region of motion. Celebrated examples are the Taylor model and the Bénard model. For various generalizations see Chandrasekhar [3] and Joseph [4].

The mathematical analysis always takes advantage of the invariance structure inherent in the problem by cutting down the underlying function space until the eigenvalue, where bifurcation is studied, is simple. Then, general results apply, yielding existence of a nontrivial solution branch and its analytical construction. Naturally, such a procedure provides solutions of high symmetry and the question arises whether, at an eigenvalue of higher multiplicity, more exotic solutions are possible. In the linear case the answer is easy, since arbitrary superposition is possible. For the linearized Bénard model see [2], p. 51.

---

* This work was supported by the Deutsch Forschungsgemeinschaft.

** Universität Stuttgart

We shall treat this question for the nonlinear generalized Bénard problem, reviewing at the same time the now classical results on cellular instability by Iudovich [24], Busse [1], et al. It turns out that a multiple eigenvalue really is the source of noncellular solutions; we give some examples. It is well known how to reduce the question of existence to a finite-dimensional system of nonlinear equations, the dimension of which is the multiplicity of the eigenvalue (Schmidt-Liapunov). However, except for a simple equation, no general method is known to solve this system. The additional structure needed is provided by an observation made by Busse [2] in another connection, which implies that the leading nonlinear term is the gradient of some real valued functional. A recent result of Grundmann [6] gives the topological index of so-called isolated extremal fixed points of this leading term, and thus all of these fixed points can be continued to a solution of the full equations. This is the basic idea of the construction; the examples at the end demonstrate some consequences.

We treat the "generalized" Bénard model, where the density is allowed to vary quadratically with temperature, for rigid boundaries. It will be necessary to construct the eigenprojections explicitly in order to exhibit the required structure of the bifurcation equations. We shall restrict our attention to the case where the quadratic terms do not vanish identically. Similar arguments are possible when the third order term dominates the nonlinearity in the bifurcation equations (for some of the implications see [14]).

Careful experimental observations for the rigid-boundary-case are available through the work of Rossby [17] showing the dominance of roll-like motions for a wide range of parameters; however, there is no detailed information near the critical eigenvalue. No stability analysis is yet available for the noncellular solutions. Therefore, arguments about their physical relevance relative to the cellular solutions would not be on safe ground.

2. The Model

A viscous fluid of viscosity μ and density ρ is contained between two horizontal planes separated by a distance d. The lower plane is held at temperature T_1 and the upper at temperature $T_2 < T_1$. Let us introduce dimensionless quantities by the use of the following reference quantities: d for length, ν/d for velocity, T_1-T_2 for temperature $(\nu = \mu/\rho)$. The similarity parameters for this problem are the *Prandtl number* $Pr = \nu/k$, where k is the thermal conductivity coefficient, and the *Grashoff number* $Gr = agd^3(T_1-T_2)/\nu^2$, where g denotes the gravitational acceleration. The density is assumed to vary quadratically with the temperature

$$\rho = \rho_0(1-a(T-T_0) - b(T-T_0)^2)$$

with

$$T_0 = \frac{T_1 + T_2}{2} .$$

Cartesian coordinates $\underline{x} = (x,y,z)$ are defined so that the planes have the equation $z = \pm 1/2$ and gravity acts in the negative z-direction. The velocity components are denoted by $\underline{u} = (u_1,u_2,u_3)$, the pressure by p, and the temperature by θ.

The fluid, subjected to buoyancy- and viscous forces, stays at rest for small values of Gr. This purely conductive solution has the form

$$\underline{u}_o = 0 \ , \ \theta_o = -z + T_o/(T_1 - T_2) \ ,$$

$$p_o'(z) = -(1 + az(T_1 - T_2) - bz^2(T_1 - T_2)^2)d^3 g/\nu^2 \ .$$

It is the unique steady flow up to the smallest eigenvalue of the linearized problem (cf. [10]). The convective motion will be determined via the ansatz:

$$\underline{u} = \underline{u}_o + \lambda \underline{u} \ , \ p = p_o + \lambda q \ , \ \tilde{\theta} = \theta_o + \theta \ , \ \lambda = (Gr)^{1/2} \ .$$

Neglecting the temperature dependence of all material quantities except those causing buoyancy (Boussinesq-approximation), one obtains (cf. [5]):

$$(1) \qquad -\Delta \underline{u} - \lambda \underline{k} f_1(\theta) + \nabla q = -\lambda(\underline{u}\cdot\nabla)\underline{u} + \lambda \underline{k} f_2(\theta)$$

$$-(Pr)^{-1}\Delta\theta - \lambda u_3 = -\lambda(\underline{u}\cdot\nabla)\theta$$

$$\nabla\cdot\underline{u} = 0$$

$$\underline{u} = 0 \ , \ \theta = 0 \quad \text{for} \quad z = \pm 1/2 \ .$$

Here, the following notations are used

$$\underline{k} := (0,0,1) \ , \ \gamma := b(T_1 - T_2)/a \ , \quad \nabla := (\partial_x, \partial_y, \partial_z) \ ,$$

$$f_1(\theta) = \theta(1 - 2\gamma z) \ , \quad f_2(\theta) = \gamma\theta^2 \ .$$

The domain $D = \mathbb{R}^2 \times (-1/2, 1/2)$ on which (1) is considered and all differential operators in (1) are invariant under translations and orthogonal transformations of the (x,y)-plane. Precisely, if $\underline{a} = (a_1, a_2, 0)$ and

$$T_\omega = \begin{pmatrix} \tilde{T}_\omega & & 0 \\ & & 0 \\ 0 & 0 & 1 \end{pmatrix}$$

with

$$\tilde{T}_\omega = \begin{pmatrix} \cos\omega & -\sin\omega \\ \sin\omega & \cos\omega \end{pmatrix} \text{ or } \begin{pmatrix} \cos\omega & \sin\omega \\ \sin\omega & -\cos\omega \end{pmatrix}$$

then, the differential operators in (1) act in the space of functions satisfying (cf. [10]).

$$\underline{u}(T_\omega \underline{x}+\underline{a}) = T_\omega \underline{u}(\underline{x})$$

$$\theta(T_\omega \underline{x}+\underline{a}) = \theta(\underline{x}) \ , \ q(T_\omega \underline{x}+\underline{a}) = q(\underline{x}) \ .$$

If we add to (1) the requirement of double-periodicity

(1') $\underline{u}$, θ, q are invariant under $x \mapsto x + \frac{2\pi}{\alpha}$, $y \mapsto y + \frac{2\pi}{\beta}$

then the following Lemma holds.

LEMMA 0 ([11],p. 281):
Let $(\underline{u},q,\theta)$ be a nontrivial solution of (1) , (1') , which is invariant under T_ω . Then, $\omega = 2\pi/n$, and $n = 1,2,3,4$, or 6.

Actually, the values of α,β and n are not independent; e.g. $n = 3,6$ implies $\alpha = \beta\sqrt{3}$ or $\beta = \alpha\sqrt{3}$. For a more concise formulation let us define

$$\Omega := [0,\tfrac{2\pi}{\alpha}] \times [0,\tfrac{2\pi}{\beta}] \times (-\tfrac{1}{2},\tfrac{1}{2})$$

$$\underline{v} := (\underline{u},\theta) \ , \nabla := (\partial_x,\partial_y,\partial_z,0)$$

(cont.)

(cont.)

$\underline{a} := (\frac{2\pi}{\alpha}, \frac{2\pi}{\beta}, 0)$

$$H^k_{\alpha,\beta} := \{f \in L_{2,loc}(\Omega) / f|_\Omega \in H^k(\Omega) \ , \ f(\underline{x}+\underline{a}) = f(\underline{x}) \text{ a.e.}\}$$

where $H^k(\Omega)$ denotes the usual complex Sobolev space $W^k_2(\Omega)$, $k \in \mathbb{N}$, with scalar product

$$(f,g)_k := \sum_{|\alpha| \leq k} (D^\alpha f, D^\alpha g)_0 \ , \ (f,g)_0 := \int_\Omega f\bar{g}\, d\underline{x}$$

$$\mathbb{H}^k := \mathbb{H}^k_{\alpha,\beta} := (H^k_{\alpha,\beta})^4$$

$$\|\underline{v}\|_k := \{\sum_{|\alpha| \leq k} \| |D^\alpha \underline{v}| \|^2_0\}^{1/2}$$

$\mathbb{H}^k_0$ is the closure of vector functions in $\mathbb{H}^k$ that vanish in a neighborhood of $z = \pm 1/2$; $\mathbb{H}^k_\sigma$ denotes the subspace of solenoidal vectorfields $(\nabla \cdot \underline{v} = 0)$ and $\mathbb{H}^k_{0,\sigma} = \mathbb{H}^k_0 \cap \mathbb{H}^k_\sigma$. According to a Lemma of Weyl, the orthogonal complement of $\mathbb{H}^0_\sigma$ in $\mathbb{H}^0$ consists of gradient fields $(\nabla q, 0)$ where $q \in H^1_{\alpha,\beta}$ (cf. [11], p. 284). Let $W : \mathbb{H}^0 \to \mathbb{H}^0_\sigma$ denote the corresponding orthogonal projector; we define the following operators

(2) $$A := \underline{v} \to W(-\Delta\underline{u}, -\frac{1}{Pr}\Delta\theta) \ , \ D(A) = \mathbb{H}^2 \cap \mathbb{H}^1_{0,\sigma}$$

$$B := \underline{v} \to W(\underline{k}f_1(\theta), u_3)$$

$$C := \underline{v} \to W(-(\underline{v}\cdot\nabla)\underline{v} + \tilde{\underline{k}}f_2(\theta))$$

where $\tilde{\underline{k}} = (\underline{k}, 0)$. It is well known that A is positive definite and self-adjoint and that $A^{-1} : \mathbb{H}^0_\sigma \to \mathbb{H}^1_{0,\sigma}$ is compact.

It is not hard to see that $D(A^{1/2}) = \mathbb{H}^1_{0,\sigma}$ and $\|A^{1/2}\cdot\|_0$ defines an equivalent norm on $\mathbb{H}^1_{0,\sigma}$. Henceforth, we shall use $X_1 := (\mathbb{H}^1_{0,\sigma}, \|\cdot\|_1)$ with $\|\cdot\|_1 := \|A^{1/2}\cdot\|_0$ and the corresponding scalar product. Using (2) one obtains

$$(3) \qquad \left|(C(\underline{v}),\underline{w})_0\right| = \left|((\underline{v}\cdot\nabla)\underline{w},\underline{v})_0 + (\tilde{\underline{k}}f_2(\theta),\underline{w})_0\right|$$

$$\leq c_1\|\underline{v}\|^2_{L_4}\ \|\underline{w}\|_1$$

for $\underline{v},\underline{w} \in X_1$. Since X_1 is compactly imbedded in L_4, the relation

$$(4) \qquad (C(\underline{v}),\underline{w})_0 = (F(\underline{v}),\underline{w})_1$$

defines a continuous, compact quadratic operator $F: X_1 \to X_1$. (F is an extension of $A^{-1}C$). Finally, we define $L := A^{-1}B$ which is linear and compact (symmetric for $\gamma = 0$).

With the operators just defined we obtain a weak form of (1):

$$(5) \qquad \underline{v} - \lambda L\underline{v} = \lambda F(\underline{v}) \ , \ \underline{v} \in X_1 \ ,$$

which, by standard regularity results, is equivalent to (1).

3. Structure of the Bifurcation Equation

In view of the compactness of L and F (5) can be reduced to a finite-system of equations with a special structure. It will be shown that the dominant nonlinear term is the gradient of a real functional which is perturbed by "nongradient" higher order terms. The question arises which of the fixed points of the dominant term (critical points of the functional on a sphere) can be continued to a solution of the full equation. A partial answer is supplied by a recent result concerning the topological index of "extremal" fixed points of polynomial operators [6]. The special structure of the bifurcation equation, exploited here to obtain non-cellular solutions, was anticipated by Busse in another connection [2]. In order to exhibit the special structure of the bifurcation equation we have to construct the eigenprojections explicitly.

Since doubly periodic solutions of (1) are considered, every eigenfunction of L is of the form

$$\underline{\varphi}^j = \underline{\phi}^j(z)\, e^{i\underline{k}_j \cdot \underline{y}}$$

$$\underline{k}_j = (n_j\alpha, m_j\beta)\ ,\ \underline{y} = (x,y)\ ,\ \underline{\phi}^j = (\varphi_1^j,\ldots,\varphi_4^j)\ .$$

In particular, the φ_i^j satisfy [3]:

(6)

a) $$\varphi_1^j = \frac{i n_j \alpha}{\sigma^2}\partial_z \varphi_3^j\ ,\quad \varphi_2^j = \frac{i m_j \beta}{\sigma^2}\partial_z \varphi_3$$

b) $$(\partial_z^2 - \sigma^2)^2 \varphi_3^j = \sigma^2 \lambda (1-2\gamma z)\ \varphi_4^j$$

c) $$(\partial_z^2 - \sigma^2)\ \varphi_4^j = -\ \lambda \Pr\ \varphi_3^j$$

d) $$\varphi_3^j = \partial_z \varphi_3^j = \varphi_4^j = 0 \quad \text{for} \quad z = \pm\frac{1}{2}$$

$$\sigma^2 = n_j^2 \alpha^2 + m_j^2 \beta^2\ .$$

Equation (6) shows that the eigenvalue λ does not depend explicitly on n_j , m_j , α , β , but solely on σ^2 . Denote by $\lambda_1 = \lambda_1(\sigma^2)$ the eigenvalue of 6 b)-d) of smallest modulus. It is real and positive for $|\gamma| < 1$. Take

σ^2 so that λ_1 is simple (true for "almost" all σ^2 [8]). Then, $\varphi_3^j = \phi_3$, $\varphi_4^j = \phi_4$ for all j and ϕ_3, ϕ_4 can be chosen to be real and positive in $(-\frac{1}{2},\frac{1}{2})$. The multiplicity of λ_1 as an eigenvalue of L is determined by the number of grid-points (n,m) on the ellipse $n^2\alpha^2 + m^2\beta^2 = \sigma^2$. Let the geometric eigenspace be spanned by

$$\underline{\varphi}^{-s},\ldots,\underline{\varphi}^{-1}, \underline{\varphi}^{1},\ldots,\underline{\varphi}^{s}$$

with

$$\underline{\varphi}^j = \underline{\phi}^j(z)\, e^{i\underline{k}_j\cdot\underline{y}} ,$$

$$\underline{\phi}^j = \left(\frac{in\alpha}{\sigma^2}\phi_3', \frac{im\beta}{\sigma^2}\phi_3', \phi_3, \phi_4\right) .$$

The equation adjoint to $\underline{\varphi} - \lambda_1 L\underline{\varphi}$ in X_1 is given by

$$\underline{\psi} - \lambda_1 A^{-1}B^T\underline{\psi} = 0 \quad \text{where} \quad B^T\underline{\psi} = W(\underline{k}\,\psi_4, f_1(\psi_3)) .$$

Again, the adjoint eigenfunctions $\underline{\psi}^j$ are of the form

$$\underline{\psi}^j = \underline{\Psi}^j(z)\, e^{-i\underline{k}_j\cdot\underline{y}}$$

$$\underline{\Psi}^j = \left(-\frac{in\alpha}{\sigma^2}\Psi_3', -\frac{im\beta}{\sigma^2}\Psi_3', \Psi_3, \Psi_4\right)$$

and Ψ_3, Ψ_4 satisfy a system similar to 6 b)-d) where the right sides of b) and c) are replaced by $\lambda_1\sigma^2\Psi_4$ resp. $-\lambda_1 Pr(1-2\gamma z)\Psi_3$. In particular, Ψ_3 and Ψ_4 can be chosen to be real and positive in $(-\frac{1}{2},\frac{1}{2})$.

Obviously $(\underline{\varphi}^j,\underline{\psi}^k)_1 = 0$ holds for $j \neq k$. Moreover

$$(\underline{\varphi}^j,\underline{\psi}^j)_1 = (A\underline{\varphi}^j,\underline{\psi}^j)_0 = \lambda_1(B\underline{\varphi}^j,\underline{\psi}^j)_0$$

$$= \lambda_1 \frac{4\pi^2}{\alpha\beta} \int_{-1/2}^{1/2} \{(1-2\gamma z)\phi_4\Psi_3 + \phi_3\Psi_4\}\, dz > 0$$

if $|\gamma| < 1$. Thus, we can normalize $\underline{\varphi}^j,\underline{\psi}^k$ so that

$$(\underline{\varphi}^j,\underline{\psi}^k)_1 = \delta_{jk} \ , \ j,k = \pm 1,\ldots,\pm s$$

holds for $|\gamma| < 1$, and hence, λ_1 is semisimple.

The eigenprojection P , commuting with L , is defined by

$$P := \sum_{|j|=1}^{s} (\cdot,\underline{\psi}^j)\underline{\varphi}^j \ ; \ Q := \mathrm{id} - P \ .$$

$J := (\mathrm{id}-\lambda_1 L)|_{QX_1}$ is a topological isomorphism on QX_1 . Applying P and Q to (5) one obtains:

(7) (a) $$0 = \frac{\tau}{\lambda_1} P\underline{v} + \lambda PF(P\underline{v}+Q\underline{v}) \ ,$$

(b) $$Q\underline{v} = \tau J^{-1}LQ\underline{v} + \lambda J^{-1}QF(P\underline{v}+Q\underline{v}) \ , \ \tau = \lambda - \lambda_1 \ .$$

The system (7) represents an equivalent formulation to (5). Near $(0,\lambda_1)$, i.e. for small $\|P\underline{v}\|_1$ and $|\tau|$, (7b) defines $Q\underline{v}$ as a function of $P\underline{v}$ and τ :

(7c) $$Q\underline{v} = \lambda_1 J^{-1}QF(P\underline{v}) + O(\|Pv\|_1^3 + \tau\|Pv\|_1^2)$$

Insertion into (7a) yields the *bifurcation equation:*

(8) $$0 = \frac{\tau}{\lambda_1} P\underline{v} + \lambda_1 PF(P\underline{v}) + O(\|Pv\|_1^3 + \tau\|Pv\|_1^2)$$

Solutions of (8) define via (7c) solutions of (5) and hence of (1). In this sense, (8) describes the set of solutions of (1) near $\underline{v} = 0$, $\lambda = \lambda_1$. For other useful reductions of bifurcation problems in fluid dynamics see [22].

For $\tau \neq 0$, we insert $-\tau c_j := (\underline{v}, \underline{\psi}^j)_1$ into (8) to obtain the system:

$$(8')\qquad 0 = c_\ell - \lambda_1^2 \Pi_\ell(\underline{c}) + S_\ell(\tau, \underline{c}) \ , \ |\ell| = 1,\ldots,s \ ,$$

where the following notations have been used:

$$\Pi_\ell(\underline{c}) := \sum_{|i|,|j|=1}^{s} c_j c_k a_{jk\ell}$$

$$a_{jk\ell} := (F^{(2)}(\underline{\varphi}^j, \underline{\varphi}^k), \underline{\psi}^\ell)_1$$

$S_\ell(\tau,\underline{c})$ is continuous (actually analytic, a property not used subsequently), and $S(0,\underline{c}) = 0$ holds for all $\underline{c} \in \mathbb{C}^{2s}$. $F^{(2)}$ denotes the symmetric polar form of F .

<u>LEMMA 1</u>

Let be $|\gamma| < 1$, then

$$a_{jk\ell} = d\ \delta(\underline{k}_j + \underline{k}_k + \underline{k}_\ell)$$

holds, where

$$\delta(\underline{k}_j + \underline{k}_k + \underline{k}_\ell) = \begin{cases} 1 & \text{if } \underline{k}_j + \underline{k}_k + \underline{k}_\ell = 0 \\ 0 & \text{otherwise} \end{cases} \ ,$$

$$d = d(\sigma^2) := \int_{-1/2}^{1/2} \{ \frac{1}{4\sigma^2} \phi_3'^2 \Psi_3' + \frac{1}{2\sigma^2} \phi_3 \phi_3'' \Psi_3' - \frac{3}{2} \phi_3 \phi_3' \Psi_3$$

$$- \frac{3}{2} \phi_3 \phi_4' \Psi_4 - \frac{\gamma}{2} \phi_4^2 \Psi_3 \} \, dz \ .$$

PROOF: It is clear that $\underline{k}_j+\underline{k}_k+\underline{k}_\ell \neq 0$ implies $a_{jk\ell} = 0$. Observe that, if $\underline{k}_j+\underline{k}_k+\underline{k}_\ell = 0$ holds, then $|\underline{k}_j+\underline{k}_k| = \sigma^2$ and hence

$$\underline{k}_j\cdot\underline{k}_k = n_j n_k \alpha^2 + m_j m_k \beta^2 = -\sigma^2/2 \quad .$$

According to (4) and (2) we have

$$a_{ijk} = (F^{(2)}(\underline{\varphi}^j,\underline{\varphi}^k),\underline{\psi}^\ell)_1 = (C^{(2)}(\underline{\varphi}^j,\underline{\varphi}^k),\underline{\psi}^\ell)_0$$

$$= -\frac{1}{2}((\underline{\varphi}^j\cdot\nabla)\underline{\varphi}^k + (\underline{\varphi}^k\cdot\nabla)\underline{\varphi}^j \;,\; \underline{\psi}^\ell)_0 + \gamma\int_\Omega \varphi_4^j\varphi_4^k\overline{\psi_3^\ell}$$

$$= \frac{1}{2\sigma^6}\int_{-1/2}^{1/2} \phi_3'^2\Psi_3' \, dz \, \{(\underline{k}_j\cdot\underline{k}_k)((\underline{k}_j\cdot\underline{k}_\ell) + (\underline{k}_k\cdot\underline{k}_\ell))\}$$

$$- \frac{1}{2\sigma^4}\int_{-1/2}^{1/2} \phi_3\phi_3''\Psi_3' \, dz \, (\underline{k}_j\cdot\underline{k}_\ell + \underline{k}_k\cdot\underline{k}_\ell) - \frac{\gamma}{2}\int_{-1/2}^{1/2} \phi_4^2\Psi_3 \, dz$$

$$+ \int_{-1/2}^{1/2} \phi_3\phi_3'\Psi_3 \, dz \, (\frac{1}{\sigma^2}\,\underline{k}_j\cdot\underline{k}_k - 1) = d(\sigma^2) \quad , \quad \text{q.e.d.}$$

For $\gamma = 0$, (6) is invariant under $z \to -z$. Hence, ϕ_3 and Ψ_3 are even functions of z implying that $d = 0$ for $\gamma = 0$. In this case $\Pi_\ell = 0$ for all $|\ell| = 1,\ldots,s$. The term of 3rd order has to be calculated in (8). The same is true if $\underline{k}_j+\underline{k}_k+\underline{k}_\ell = 0$ for all j,k,ℓ.

The symmetry of $a_{jk\ell}$ with respect to the indices suggests that Π_2 is the gradient of a functional. To verify this, we change to real variables setting $c_\ell := \xi_\ell + i\eta_\ell$, $c_{-\ell} = \overline{c_\ell}$. Observe that $a_{jk\ell} \in \mathbb{R}$ and separate real- and imaginary parts in (8'):

$$(8'') \quad \begin{aligned} 0 &= \xi_\ell - \lambda_1^2 d p_\ell(\underline{\xi},\underline{\eta}) + s_\ell(\tau,\underline{\xi},\underline{\eta}) , \\ 0 &= \eta_\ell - \lambda_1^2 d q_\ell(\underline{\xi},\underline{\eta}) + t_\ell(\tau,\underline{\xi},\underline{\eta}) \end{aligned} \qquad \ell = 1, \ldots , s ,$$

where

$$p_\ell(\underline{\xi},\underline{\eta}) = \sum_{|j|,|k|=1}^{s} (\xi_j\xi_k - \eta_j\eta_k)\ \delta(\underline{k}_j + \underline{k}_k + \underline{k}_\ell)$$

$$q_\ell(\underline{\xi},\underline{\eta}) = \sum_{|j|,|k|=1}^{s} (\xi_j\eta_k + \xi_k\eta_j)\ \delta(\underline{k}_j + \underline{k}_k + \underline{k}_\ell)$$

and s_ℓ , t_ℓ are continuous functions of their arguments, $s_\ell(0,\underline{\xi},\underline{\eta}) = t_\ell(0,\underline{\xi},\underline{\eta}) = 0$. Define the real valued functional r by

$$(9) \quad r(\underline{\xi},\underline{\eta}) := - \frac{\lambda_1^2 d}{3} \sum_{\ell=1}^{s} \sum_{|j|,|k|=1}^{s} \{(\xi_j\xi_k - \eta_j\eta_k)\xi_\ell + (\xi_j\eta_k + \xi_k\eta_j)\eta_\ell\}$$

then, an elementary calculation yields

$$\nabla_\xi r = \underline{p} \quad , \quad \nabla_\eta r = \underline{q} \ .$$

We summarize the results obtained so far:

Let be $|\gamma| < 1$, $d \neq 0$, let λ_1 be a simple eigenvalue of 6 b)-d) and a (semisimple) eigenvalue of multiplicity $2s$ of L then, for $\lambda \neq \lambda_1$, equation (1) is equivalent to (8) and hence to (8'') via (7c) near $(0,\lambda_1)$. The leading nonlinear terms p , q in (8'') are polynomial operators homogeneous of degree 2 and gradients of the real valued functional r in (9).

4. Existence of Cellular and Noncellular Motions

The usual device to prove existence of solutions of (1) (or (5)) consists in adding various invariance requirements to the double-periodicity until λ_1 is a simple eigenvalue of L ; e.g., $\alpha = \beta\sqrt{3}$ and invariance under $T_{\pi/3}$ yields hexagonal cell - flow [8], $\alpha = \beta$ and invariance under $T_{\pi/2}$ yields square cells, and $\beta = 0$ and invariance under T_{π} yields rolls. Existence of a nontrivial solution of (5) having these invariance properties and bifurcating in $(0,\lambda_1)$ follows from a general topological result of Krasnoselskii [13] (for a global version see Rabinowitz [15]). For constructive methods cf. [5], [10], [19].

Naturally, the cutting of the function space by invariance results in solutions of high symmetry. Now, having exhibited the special structure of (8'') one is able to abandon the additional requirements. A fixed point $(\underline{\xi}_0,\underline{\eta}_0)$ of $\lambda_1^2 d(\underline{p},\underline{q})$ can be continued to a solution of (8'') if 1 is not an eigenvalue of $\lambda_1^2 dD(\underline{p},\underline{q})(\underline{\xi}_0,\underline{\eta}_0)$, where $D(\underline{p},\underline{q})$ denotes the Hessian matrix of $(\underline{p},\underline{q})$. If 1 is an eigenvalue, the following Lemma permits the calculation of the topological index of certain "extremal" fixed points.

<u>LEMMA 2, [6]</u>:

Let $\underline{\zeta}_0 = (\underline{\xi}_0,\underline{\eta}_0)$ be a fixed point of $\tilde{p} = \lambda_1^2 d(\underline{p},\underline{q})$ such that $|\underline{\zeta}| = |\underline{\zeta}_0|$ and $0 < |\underline{\zeta} - \underline{\zeta}_0| < \delta$ for δ sufficiently small, $r(\underline{\zeta}) > r(\underline{\zeta}_0)$ (+ case) or $r(\underline{\zeta}) < r(\underline{\zeta}_0)$ (- case). If 1 is an eigenvalue of $D\tilde{p}$ of multiplicity m then

$$\operatorname{ind}(\mathrm{id}-\tilde{p},\underline{\zeta}_0) = \begin{cases} (-1)^m & , \ + \text{ case} \\ -1 & , \ - \text{ case} \end{cases} .$$

Here, "ind' denotes the local Brouwer degree, which is defined since $\underline{\zeta}_0$ is an isolated fixed point. A fixed point $\underline{\zeta}_0$ satisfying the assumptions of Lemma 2 will be called *isolated* and *extremal*. Using the preceding result, the continuity of

$(\underline{s},\underline{t})$, and $\underline{s}(0,\underline{\zeta}) = \underline{t}(0,\underline{\zeta}) = 0$ one obtains

THEOREM 1:

Let $\underline{\zeta}_0 = (\underline{\xi}_0,\underline{\eta}_0)$ be a fixed point of $\tilde{p}$ such that either (i) 1 is not an eigenvalue of $D\tilde{p}$, or (ii) $\underline{\zeta}_0$ is isolated and extremal. Moreover, let λ_1 , γ satisfy the assumptions of the preceding section. Setting

$$\underline{c}_0 := \underline{\xi}_0 + i\underline{\eta}_0 \quad , \quad \underline{v}_0 := -(\lambda-\lambda_1) \sum_{|j|=1}^{s} c_j \underline{\varphi}^j$$

then, there exists a solution $\underline{v}(\lambda) \neq 0$ of (5) (or (1)) such that

$$\lim_{\lambda\to\lambda_1} \frac{\underline{v}(\lambda) - \underline{v}_0}{\lambda - \lambda_1} = 0 \quad .$$

This theorem enables us to prove existence of bifurcating solutions near multiple eigenvalues of L by verifying one of the two alternatives. Whenever (i) holds, one obtains, by using the implicit function theorem and the analyticity of all operators of (5), a solution which can be represented in X_1 by a power series in $\tau = \lambda - \lambda_1$. The condition (ii) gives existence by the use of topological arguments and is basically nonconstructive.

If $d = 0$, e.g., for $\gamma = 0$ or $\underline{k}_j+\underline{k}_k+\underline{k}_\ell \neq 0$, then the terms of 3rd order have to be constructed explicitly. Again, one obtains a homogeneous polynomial operator which is the gradient of a real valued functional, and the same methods can be applied. The details will be given in a forthcoming paper (see also [14]).

The same reasoning can be applied when all material constants depend on the temperature. It was also applied in a number of other physical problems [2]. In the case considered here the applicability of condition (ii) is closely tied to the fact that the vertical component of the vorticity vanishes in 1st order ($\partial_y\underline{\varphi}_2 - \partial_x\underline{\varphi}_1 = 0$) . Hence, it cannot be used in convection problems where rotation takes place or in the Taylor model of the flow between two rotating cylinders. Here, bifurcation phenomena can still be studied by cutting down the space through invariance properties. However, at least in the Taylor case, the invariance under translations in the axial direction only reflects the freedom in the choice of the coordinate frame.

The stability and instability of the solutions $\underline{v}(\lambda)$ constructed in Theorem 1 depend on whether they are attractors or repellers with respect to solutions of the unstationary equations corresponding to (1):

$$d\underline{v}/dt + A\underline{v} + \lambda B\underline{v} + \lambda C(\underline{v}) = 0 \quad . \tag{10}$$

Naturally these properties depend on the topology chosen. It is known that $\underline{v}(\lambda)$ is stable in $D(A)$ if the spectrum of

$$\tilde{A}(\lambda,\underline{v}(\lambda)) := A + \lambda B + \lambda C'(\underline{v}(\lambda))$$

is strictly contained in the positive complex half-plane [7]. It there are spectral points with negative real parts then, $\underline{v}(\lambda)$ is unstable in the following sense: In every L_2-neighborhood of $\underline{v}(\lambda)$ one can find initial conditions for a "strict" solution eventually leaving a fixed L_2-neighborhood of $\underline{v}(\lambda)$ [11]. Therefore, the question of stability and instability is answered when certain sign properties of the spectrum are known. Rigorously, these have been established only in the case that 0 is a simple eigenvalue of $\tilde{A}(\lambda;\underline{v}(\lambda))$ at $\lambda = \lambda_1$. Again, this assumption requires so much symmetry of the solution of (10) that $\underline{v}(\lambda)$ is shown to attract solutions of a very restrictive and physically uninteresting class only. Instability however is maintained when the underlying function space is enlarged. Therefore, instability statements obtained under highly restrictive symmetry requirements are still of physical relevance.

In this realm the stability and instability properties of cellular solutions are well known [9], [11]. For solutions bifurcating in multiple eigenvalues results are known by formal expansion methods [2]. For the examples studied in the next section these methods are certainly not applicable, since generally (ii) of Theorem 1 has to be applied. In this case not even the Puiseux-series for $\underline{v}(\lambda)$ are known.

5. Examples

1. We take $\alpha = \beta\sqrt{3}$, $4\beta^2 = \sigma^2$. Then, λ_1 is of multiplicity 6 since the grid points $(0,\pm 2)$, $(\pm 1,\pm 1)$ are on $n^2\alpha^2 + m^2\beta^2 = \sigma^2$. Setting

$$\underline{k}_1 := (-\alpha,\beta) \quad , \quad \underline{k}_2 := (\alpha,\beta) \quad , \quad \underline{k}_3 := (0,2\beta)$$

$$\underline{k}_{-j} := -\underline{k}_j$$

one obtains $\underline{k}_1+\underline{k}_{-2}+\underline{k}_3 = 0$. We assume $d \neq 0$. The functional r in (9) is given by

$$r = \rho(\xi_1\xi_2\xi_3 + \xi_3\eta_1\eta_2 - \xi_2\eta_1\eta_3 + \xi_1\eta_2\eta_3)$$

where $\rho = 2d\lambda_1^2$.

a) $\xi_j = 1/\rho$, $\eta_j = 0$,

b) $\xi_j = -1/2\rho$, $\eta_j = -\sqrt{3}/2\rho$,

c) $\xi_1 = \xi_3 = -1/\rho$, $\xi_2 = 1/\rho$, $\eta_j = 0$, $j = 1,2,3,$

determine isolated extremal fixed points of $\tilde{p} = (\nabla_\xi r,\nabla_\eta r)$. In case a) the projected streamlines satisfy the differential equation:

$$(\sin\alpha x \cos\beta y - \sin 2\beta y)dx - \sqrt{3}\cos\alpha x \sin\beta y\, dy = 0 .$$

One obtains the classical hexagonal flow (figure 1). Case b) yields the same flow translated by $(\pi/2\alpha , -\pi/2\beta)$ and case c) is congruent to a) by a translation $(0, -2\pi/3\beta)$.

2. Again we take $\alpha = \beta\sqrt{3}$, but $28\beta^2 = \sigma^2$. Then, λ_1 is of multiplicity 12. The grid points $(\pm1,\pm5)$, $(\pm2,\pm4)$, $(\pm3,\pm1)$ satisfy $n^2\alpha^2 + m^2\beta^2 = \sigma^2$. Define

$$\underline{k}_1 := (\alpha,-5\beta) , \quad \underline{k}_2 := (\alpha,5\beta) , \quad \underline{k}_3 := (2\alpha,-4\beta) ,$$

$$\underline{k}_4 := (2\alpha,4\beta) , \quad \underline{k}_5 := (3\alpha,-\beta) , \quad \underline{k}_6 :- (3\alpha,\beta) ,$$

$$\underline{k}_{-j} := -\underline{k}_j$$

One obtains

$$\underline{k}_1 + \underline{k}_4 + \underline{k}_{-5} = 0$$

$$\underline{k}_2 + \underline{k}_3 + \underline{k}_{-6} = 0$$

Hence, the functional r in (9) is of the form

$$r = \rho(\xi_1\xi_4\xi_5 + \xi_1\eta_4\eta_5 + \xi_4\eta_1\eta_5 - \xi_5\eta_1\eta_4$$

$$+ \xi_2\xi_3\xi_6 + \xi_2\eta_3\eta_6 + \xi_3\eta_2\eta_6 - \xi_6\eta_2\eta_3)$$

a) $\xi_j = 1/\rho$, $\eta_j = 0$, $j=1,2,\ldots,6$, defines an isolated extremal fixed point. The projected streamlines (in 1st order) are determined by the differential equation

$$-(5\cos\alpha x \sin 5\beta y + 4\cos 2\alpha x \sin 4\beta y - \cos 3\alpha x \sin\beta y)dx +$$

$$+ \sqrt{3}(\sin\alpha x \cos 5\beta y + 2\sin 2\alpha x \cos 4\beta y + 3\sin 3\alpha x \cos\beta y)dy = 0$$

In figure 2 the motion is visualized. There are sphere-like cells alternating with pentagons.

b) $\xi_1 = \xi_4 = \xi_5 = 1/\rho \quad , \quad \eta_1 = \eta_4 = \eta_5 = 0 \quad ,$

$$\xi_2 = \xi_3 = \xi_6 = -1/2\rho \; , \; \eta_2 = \eta_3 = -\eta_6 = \sqrt{3}/2\rho$$

is another isolated extremal fixed point of $\tilde{p}$. The projected streamlines (1st order) are given by the differential equation

$$\Big\{\tfrac{5}{2}(\sqrt{3}\ \sin\alpha x - \cos\alpha x)(\sqrt{3}\ \cos5\beta y + \sin5\beta y) + 2(\sqrt{3}\ \sin2\alpha x + - \cos2\alpha x)(-\sqrt{3}\ \cos4\beta y + \sin4\beta y) - \tfrac{1}{2}(\sqrt{3}\ \sin3\alpha x + \cos3\alpha x)(-\sqrt{3}\ \cos\beta y + \sin\beta y)\Big\}\, dx \; -$$

$$-\tfrac{\sqrt{3}}{2}\Big\{(\sin\alpha x + \sqrt{3}\ \cos\alpha x)(\sqrt{3}\ \sin5\beta y - \cos5\beta y) - 2(\sin2\alpha x + + \sqrt{3}\ \cos2\alpha x)(\sqrt{3}\ \sin4\beta y + \cos4\beta y) - 3(\sin3\alpha x - \sqrt{3}\ \cos3\alpha x)(\sqrt{3}\ \sin\beta y + \cos\beta y)\Big\}\, dy = 0$$

The result is shown in figure 3.

REFERENCES

1. F. Busse, "Das Stabilitätsverhalten der Zellularkonvektion bei endlicher Amplitude," Inauguraldissertation, Munich, 1962.
2. F. Busse, "The Stability of finite amplitude convection and its relation to an extremum principle," J. Fluid Mech., 30 (1967), 625-649.
3. S. Chandrasekhar, Hydrodynamic and hydromagnetic stability, International Series of Monographs on Physics, Clarendon Press, Oxford, England, 1961.
4. D. D. Joseph, "Nonlinear hydrodynamic stability," to appear.
5. P. C. Fife and D. D. Joseph, "Existence of convective solutions of the generalized Bénard problem which are analytic in their norm," Arch. Rational Mech. Anal., 33 (1969), 116-138.
6. A. Grundmann, "Der topologische Abbildungsgrad homogener Polynomoperatoren," Dissertation, Universität Stuttgart.

7. G. Iooss, "Théorie nonlinéaire de la stabilité des écoulements laminaires dans le cas de "l'échange des stabilités", Arch. Rational Mech. Anal., 40 (1971), 166-208.

8. V. I. Iudovich, "On the origin of convection," J. Appl. Math. Mech., 30 (1966), 1193-1199.

9. V. I. Iudovich, "Stability of convection flows," J. Appl. Math. Mech. 31 (1967), 294-303.

10. K. Kirchgässner, "Bifurcation in Nonlinear Hydrodynamic Stability," SIAM Review, to appear.

11. K. Kirchgässner and H. Kielhöfer, "Stability and bifurcation in fluid dynamics," Rocky Mountain J. Math. 3 (1973), 275-318.

12. K. Kirchgässner and P. Sorger, "Branching analysis for the Taylor problem," Quart. J. Mech. Appl. Math., 32 (1969), 183-209.

13. M. A. Krasnoselskii, <u>Topological Methods in the Theory of Nonlinear Integral Equations</u>, GITTL, Moscow, 1956; English transl., Macmillan, New York, 1964.

14. P. Maschmann, "Existenz nichtregulärer Lösungen beim Bénard Problem," Dissertation, Universität Stuttgart, 1973.

15. P. H. Rabinowitz, "Some global results for nonlinear eigenvalue problems," J. Functional Analysis, 7 (1971), 487-513.

16. P. H. Rabinowitz, "Existence and nonuniqueness of rectangular solutions of the Bénard problem," Arch. Rational Mech. Anal., 29 (1968), 32-57.

17. H. T. Rossby, "A study of Bénard convection with and without rotation," J. Fluid Mech., 36 (1969), 309-335.

18. D. Sather, "Branching of solutions of nonlinear equations," Rocky Mountain J. Math., 3 (1973), 203-250.

19. A. Schlüter, D. Lortz and F. Busse, "On the stability of steady finite amplitude convection, J. Fluid Mech., 23 (1965), 129-144.

20. L. A. Segel and J. T. Stuart, "On the question of the preferred mode in cellular thermal convection," J. Fluid Mech., 13 (1962), 289-306.

21. J. T. Stuart, "On the cellular patterns in thermal convection," Ibid., 18 (1964), 481-498.

22. E. Zeidler, "Zur Verzweigung und zur Stabilitätstheorie der Navier-Stokesschen Gleichungen," Math. Nachr., 52 (1972), 167-205.

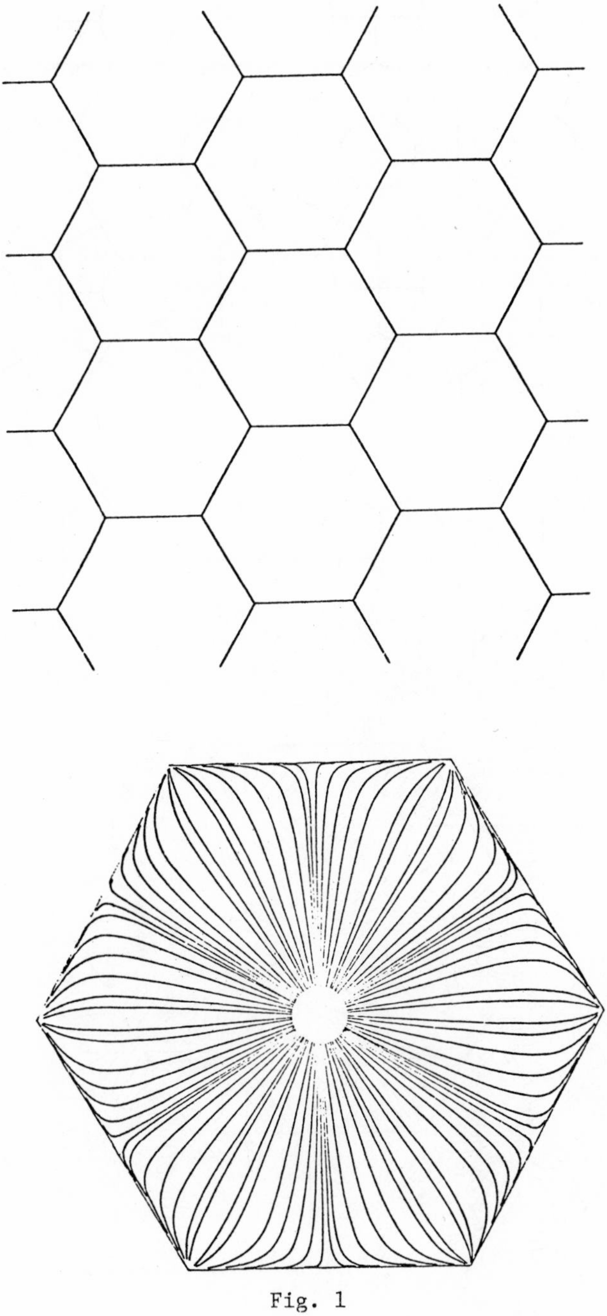

Fig. 1

Fig. 2

Fig. 3

INTRODUCTION TO SOME ASPECTS OF FREE SURFACE PROBLEMS

J. L. Lions*

Introduction

The general aim of this lecture is to give connections between some free boundary problems and some classes of Variational Inequalities[1], and to show how these connections can be useful.

Without entering, for the time being, into technical details, what we call, in a loose way, a V.I., is a set of partial differential inequalities and equalities, say:

(1) $Pu-f \geq 0 \quad , \quad u \geq 0 \quad , \quad (Pu-f)u = 0$ in a region Q ,

where P is a partial differential operator, which will be taken in this lecture a linear second order parabolic operator. In (1) u is also subject to initial and boundary conditions.

Again in a (very) loose way, that there should be some connections between V.I. as (1) and free boundary problems is obvious: indeed, we shall have in Q two regions, one, Q_1 , in which $Pu = f$, the other one, say Q_2 , where $u = 0$: the interface S between Q_1 and Q_2 is not a priori given: it is a "free surface".

In Section 1, we shall make this connection precise on a very simple example: the one phase Stefan's problem: we shall show that if θ denotes the ≥ 0 temperature, extended

---

*College De France and Iria-Laboria

[1] We shall write V.I. for Variational Inequalities.

by $\tilde{\theta}$ by 0 outside the free surface, then

$$u(x,t) = \int_0^t \tilde{\theta}(x,s)ds \tag{2}$$

is a solution of a V.I.

In Section 2 we present the penalty method which gives an approximation procedure for the solution of the V.I. and we give an error estimate.

In Section 3 we combine the techniques of V.I. together with asymptotic methods of Sanchez-Palencia [1], de Giorgi-Spagnolo [1], Babuška [1] [2] to obtain an asymptotic result for the solution of a free boundary problem in a very inhomogeneous material.

In Section 4 we briefly indicate some possible extensions and modifications of the above ideas, essentially giving bibliographical references.

V.I. of type (1) have been introduced in Lions-Stampacchia [1] for the solution of problems in mechanics with unilateral constraints: it was shown in Duvaut-Lions [1] how V.I. techniques can contribute to the solution of problems arising in plasticity, in Bingham's flow, etc.: in all these problems the physical unknown (should it be a displacement, a potential, a speed of flow, etc.) was shown to satisfy a V.I.

Later on, in Bensoussan-Lions [1]...[4], it was shown how optimal cost functions in problems of optimal stopping times or of impulse controls are also characterized by V.I. or by inequalities similar to (1) but with a non local character and that we decided to call quasi variational inequalities (Q.V.I.): again in these problems the physical unknown was directly shown to satisfy a V.I. (or a Q.V.I.).

Another important class of problems is when a transformation of the unknown is shown to satisfy a V.I. (or a Q.V.I.):

this idea (of which (2) gives a simple example) has been introduced by C. Baiocchi [1] to "reduce" a classical free boundary problem arising in infiltration into porous media to a V.I. (cf. also Baiocchi, Comincioli, Magenes and Pozzi [1]), and later on, (C. Baiocchi [2] [3]) to a Q.V.I.

This idea - which has been applied since to many other situations by many authors - has another source: it has been shown by Samuelson McKean [1], Chernoff [1] that problems of optimal stopping time could be solved by reduction to a Stefan's free boundary problem (these works being prior to Bensoussan-Lions, loc. cit. and not using V.I. or Q.V.I. techniques): since it was shown in Bensoussan and the A. that optimal stopping time problems were equivalent to V.I., it was therefore clear that V.I. arising in optimal control could be reduced to Stefan's free boundary problem (this is indeed the "reverse" idea than the one of Baiocchi!). For the "direct" approach, see Duvaut [1].

But we shall not refer anymore to optimal control theory in this lecture (we refer, for a brief review, to the Vancouver I.C.M. lectures of Bensoussan [1] and of the A.[1], [3]) nor to the numerical analysis of the V.I. or Q.V.I. (we refer to Glowinski, Lions and Trémolières [1], Comincioli [1], Lions [2] and to the bibliography therein).

At the end of this Introduction, we want to emphasize that we give here some aspects of some free surface problems by V.I. techniques. There are other free boundary problems arising in Physics for which it is not known if they are equivalent to V.I. or to Q.V.I. or if they are equivalent to some optimal control problems (and this seems to be an interesting field of research). And there are other methods to solve some of the problems considered here, let us refer among other works, to Cannon-Hill [1] [2], J. Douglas [1], A. Friedman [1], Kamenostoskaya [1], Kolodner [1],

O. A. Ladyzenskaya, V. Solonnikov and N. Uralceva [1], O. A. Oleinik [1], B. Sherman [1] and the bibliography of these works.

1. Transformation of a Free Boundary Problem into a V.I.

1.1 An example of a free boundary problem

Let us denote by θ the temperature $\theta = \theta(x,t)$ of phase 1, phase 2 being assumed to be at temperature 0 . The region of phase 1 is limited by a fixed boundary Γ' and by a free boundary $S(t)$ which depends on t : let $\mathcal{O}_1(t)$ denote this region at time t and let Q_1 be the union of these regions when t varies in, say, $[0,T]$.

In Q_1 we have $\theta(x,t) \geq 0$, which satisfies

$$(1.1) \qquad \frac{\partial\theta}{\partial t} - \Delta\theta = 0 \quad , \quad \Delta = \sum_{i=1}^{n} \frac{\partial^2}{\partial x_i^2}$$

(n = 1,2 or 3 in the applications).

We have the boundary condition

$$(1.2) \qquad \theta(x,t) = g_0(x,t) \quad , \quad x \in \Gamma' ,$$

g_0 being a given ≥ 0 function,

and on the free surface $S = \cup\, S(t)$:

$$(1.3) \qquad \left| \begin{array}{l} \theta = 0 \\ \dfrac{\partial\theta}{\partial n} = - L\, V.n \quad , \end{array} \right.$$

where n denotes a normal at $S(t)$, V=speed of $S(t)$, L given > 0 constant. The initial temperature is given

(1.4) $$\left| \begin{array}{l} \theta(x,o) = \theta_o(x) \ , \\ \theta_o \ \text{ given } \geq 0 \ \text{ in } \ \mathcal{O}_1(o) \ . \end{array} \right.$$

Remark 1.1

We emphasize again that the above problem is the simplest of the free boundary problems: we want to explain the method of V.I. on this simple example.

1.2 Another formulation

We want first to give an equivalent formulation of (1.1)..(1.4) which uses in an essential manner the function χ_1 :

(1.5) χ_1 = characteristic function of Q_1 in the set $Q = \mathcal{O}\times]0,T[$, where $\mathcal{O}$ is chosen large enough so as to contain $\mathcal{O}_1(t)$ $\forall t$ [we can take: $\mathcal{O}$ = complementary set in R^n of $\mathcal{O}_1(o)$] .

The boundary $\partial\mathcal{O}$ of $\mathcal{O}$ consists of two parts:

(1.6) $$\partial\mathcal{O} = \Gamma' \cup \Gamma'' \ .$$

We define:

(1.7) $\tilde{\theta}$ = extension of θ to Q by 0 outside Q_1

and we now compute $\frac{\partial\tilde{\theta}}{\partial t} - \Delta\tilde{\theta}$ in the sense of distributions in Q . Assuming S to be smooth, to a given function h on S we associate the distribution

$$\varphi \longrightarrow \int_S h\varphi \, dS = <\{h\}_S \ , \varphi> \ :$$

with this notation, it is an exercise to check that

$$\frac{\partial\tilde{\theta}}{\partial t} - \Delta\tilde{\theta} = - L\ \{V\cdot N_x\}_S \quad , \tag{1.8}$$

N = unitary normal to S in $R^n_x \times R_t$ directed toward the exterior of Q_1 : we also check that:

$$\frac{\partial \chi_1}{\partial t} = - \{N_t\}_S \tag{1.9}$$

and since $V\cdot N_x = -N_t$, we obtain from (1.8)(1.9) that

$$\frac{\partial\tilde{\theta}}{\partial t} - \Delta\tilde{\theta} = -L\,\frac{\partial \chi_1}{\partial t}\ \text{, in}\ \ Q = \mathcal{O}\times]0,T[\ , \tag{1.10}$$

together with

$$\left|\begin{array}{lll} \tilde{\theta}(x,t) = g_0(x,t) & \text{on} & \Sigma' = \Gamma'\times]0,T[\\ \tilde{\theta}(x,t) = 0 & \text{on} & \Sigma'' = \Gamma''\times]0,T[\\ \tilde{\theta}(x,o) = \tilde{\theta}_0(x) & \text{in} & \mathcal{O}\ . \end{array}\right. \tag{1.11}$$

If we recall that χ_1 is the <u>characteristic function of the support of</u> $\tilde{\theta}$, (1.10)(1.11) summarize all the information.

1.3 <u>Transformation of (1.10) (1.11) into a V.I.</u>

By looking at equation (1.10), it is natural to introduce a new function

$$u(x,t) = \int_0^t \theta(x,s)ds\ . \tag{1.12}$$

By integrating (1.10) in t , we obtain:

$$\tilde{\theta}(x,t) - \tilde{\theta}(x,o) - \Delta u(x,t) = -L\chi_1(x,t) + L\chi_1(x,o)\quad . \tag{1.13}$$

Let us define:

(1.14) $\chi_{o2}(x)$ = characteristic function in $\mathcal{O}$ of the complementary set of $\mathcal{O}_1(o)$:

then $\chi_1(x,o) = 1 - \chi_{o2}(x)$ and (1.13) becomes

(1.15) $$\frac{\partial u}{\partial t}(x,t) - \Delta u(x,t) = \tilde{\theta}_o(x) - L\chi_{o2}(x) + L(1-\chi_1) \quad .$$

If we set

(1.16) $$f(x) = \tilde{\theta}_o(x) - L\chi_{o2}(x)$$

we have, after noticing that $u(1-\chi_1) = 0$

(1.17) $$\left|\begin{array}{l} \frac{\partial u}{\partial t} - \Delta u - f \geq 0 \quad , \\ u \geq 0 \quad , \quad \left(\frac{\partial u}{\partial t} - \Delta u - f\right) u = 0 \quad \text{in} \quad Q \end{array}\right.$$

(1.18) $$\left|\begin{array}{l} u(x,t) = \int_0^t g_o(x,s)ds = g(x,t) \quad \text{on} \quad \Sigma' \quad , \\ u(x,t) = 0 \quad \text{on} \quad \Sigma'' \quad , \end{array}\right.$$

(1.19) $$u(x,o) = 0 \quad \text{in} \quad \mathcal{O} \quad .$$

Consequently, <u>the "transformed" function</u> u , <u>defined by</u> (1.12), <u>satisfies the V.I.</u> (in the sense of the introduction) (1.17) (1.18) (1.19).

Reciprocally, if u is a (smooth) solution of the V.I., then $\tilde{\theta} = \frac{\partial u}{\partial t}$ is solution of the free boundary problem in its first formulation.

1.4 <u>Variational formulation</u> (I)

We now introduce:

(1.20) $$V = \{v | v \in L^2(\mathcal{O}) \ , \ \frac{\partial v}{\partial x_i} \in L^2(\mathcal{O}) \quad , \quad v = 0 \ \text{on} \ \Gamma''\} \quad ,$$

(1.21) $a(u,v) = \Sigma \int_{\mathcal{O}} \frac{\partial u}{\partial x_i} \frac{\partial v}{\partial x_i} dx$, $(u,v) = \int_{\mathcal{O}} uv\, dx$,

(1.22) $K(t) = \{v | v \in V \text{ , } v \geq 0 \text{ a.e. in } \mathcal{O} \text{ , } v(x) = g(x,t) \text{ on } \Gamma'\}$

The set $K(t)$ is a <u>closed convex subset</u> of V, which is not empty (since $g_o \geq 0$).

It is a simple exercise to verify that - provided all functions are smooth enough - $u(x,t)$ satisfies

(1.23)
$$\left| \begin{array}{l} (\frac{\partial u}{\partial t}, v-u) + a(u,v-u) - (f,v-u) \geq 0 \quad \forall v \in K(t) \text{ ,} \\ u(t) \in K(t) \text{ ,} \\ u(o) = 0 \text{ .} \end{array} \right.$$

We shall say that u is a "<u>strong solution</u>" of (1.23) if

(1.24) $u \in L^2(0,T;V)$, $\frac{\partial u}{\partial t} \in L^2(0,T;H)$, $(H = L^2(\mathcal{O}))$

and satisfies (1.23).

We now introduce the notion of "<u>weak solution</u>".

1.5 <u>Variational formulation</u> (II)

We define the set κ as follows:

(1.25) $\kappa = \{v | v \in L^2(0,T;V) \text{ , } \frac{\partial v}{\partial t} \in L^2(0,T;H) \text{ ,}$

$v \geq 0 \text{ in } Q \text{ , } v = g \text{ on } \Sigma'\}$

and we make the <u>only assumption</u> on g that

(1.26) $$\kappa \neq \emptyset \text{ .}$$

Then, it is again a simple exercise to verify that, if u satisfies (1.23), then

$$(1.27)\quad \left| \begin{array}{l} \displaystyle\int_0^T \left[\left(\frac{\partial v}{\partial t}, v-u\right) + a(u,v-u) - (f,v-u) \right] dt + \frac{1}{2}|u(o)|^2 \geq 0 \\ \forall\, v \in K \end{array} \right.$$

and we shall say that u is a "weak solution" of the V.I. if

$$(1.28)\quad \left| \begin{array}{l} u \in L^2(0,T;V)\ ,\ u \geq 0 \ \text{ in } \ Q\ , \\ u = g \text{ on } \Sigma' \text{ a.e.} \end{array} \right.$$

and if it satisfies (1.27).

1.6 Application of general results

The advantage of the formulation as a V.I., in a strong form or in a weak form, are:

(i) one can apply general results and general methods: we give examples below and in Section 2:

(ii) one can apply the general numerical methods for V.I.: we refer to Glowinski, Lions and Trémolières [1] and to the bibliography therein:

(iii) one can use the V.I. to obtain properties of the solution: we give an Example in Section 3.

For the problem in its weak form, one can apply a result of Mignot-Puel [1]: under the hypothesis (1.26), there exists a minimum solution of (1.27)(1.28) - that is a solution u which is such that, if w is any weak solution, then $u \leq w$. Moreover, u is the limit (in $L^2(0,T;V)$) of the solutions of the penalized approximations (defined below in Section 2), so that the minimum solution is the "physical solution" of the problem.

2. The Penalty Method

2.1 Introduction of the method

In what follows, we shall set

(2.1) $$-\Delta = A$$

and, for every $\eta > 0$, we introduce the penalized problem: find u_η such that

(2.2) $$\left| \begin{array}{l} \frac{\partial u_\eta}{\partial t} + Au_\eta - \frac{1}{\eta}u_\eta^- = f \quad \text{in } Q ,^{(1)} \\ u_\eta = \left| \begin{array}{ll} g & \text{on } \Sigma' \\ o & \text{on } \Sigma'' \end{array} \right. \\ u_\eta(x,o) = 0 \quad . \end{array} \right.$$

This problem admits a unique solution (by application of the general theory of monotone operators - see for instance Lions [4] and the bibliography therein) such that

(2.3) $$u_\eta \in L^2(0,T;V) \ , \ \frac{\partial u_\eta}{\partial t} \in L^2(0,T;H^{-1}(\mathcal{O})) \ ,$$

provided g satisfies (for instance) (1.26).

In (2.2) the term $-\frac{1}{\eta}u_\eta^-$ is the so-called penalty term.

2.2 A priori estimates

By taking the scalar product of (2.2) with $\frac{\partial u_\eta}{\partial t} - \frac{\partial \hat{g}}{\partial t}$, where $\hat{g}$ satisfies:

(2.4) $$\left| \begin{array}{l} \hat{g} \in L^2(0,T;V) \quad , \quad \frac{\partial \hat{g}}{\partial t} \in L^2(0,T;V) \ , \\ \hat{g} = g \quad \text{on } \Sigma' \ : \ \hat{g} \geq 0 \quad , \quad \frac{\partial \hat{g}}{\partial t} \geq 0 \quad \text{a.e. in } Q \end{array} \right.$$

(this is a rather harmless hypothesis, since we assumed that $g_0 \geq 0$ and we have $\frac{\partial g}{\partial t} = g_0$), we obtain, after some computations:

(1) $\varphi^- = \sup(-\varphi,o)$.

$$\|u_\eta\|_{L^\infty(0,T;V)} + \left\|\frac{\partial u_\eta}{\partial t}\right\|_{L^2(0,T;H)} \leq C \tag{2.5}$$

where the C's denote constants which do not depend on η, and by taking the scalar product of (2.2) with u_η^-, we obtain

$$\|u_\eta^-\|_{L^2(0,T;H)} \leq C\eta \quad , \tag{2.6}$$

$$\|u_\eta^-\|_{L^2(0,T;V)\cap L^\infty(0,T;H)} \leq C\sqrt{\eta} \quad . \tag{2.7}$$

2.3 Strong solutions

From the estimates (2.5) (2.6) (2.7) we obtain, by letting $\eta \to 0$: under hypothesis (2.4) there exists a unique solution u of (1.23) (1.24). Moreover

$$u \in L^\infty(0,T;V) \tag{2.8}$$

$$u \in L^2(0,T;H^2(\mathcal{O})) \quad . \tag{2.9}$$

2.4 Error estimate in the penalty method

One can show the following: under hypothesis (2.4), if u (resp u_η) denotes the solution of the V.I. (1.23) (1.24) (resp. of the penalized equation (2.2)) then

$$\|u-u_\eta\|_{L^2(0,T;V)\cap L^\infty(0,T;H)} \leq C\eta^{\frac{1}{4}} \quad . \tag{2.10}$$

2.5 Penalty on the Stefan's problem

By analogy with what has been done in Section 1, it is natural to introduce:

$$\theta_\eta = \frac{\partial u_\eta}{\partial t} \quad . \tag{2.11}$$

It follows from (2.2) that

$$\frac{\partial\theta_\eta}{\partial t} + A\theta_\eta - \frac{1}{\eta}\frac{\partial}{\partial t}(u_\eta^-) = 0 \quad , \tag{2.12}$$

$$\left|\begin{array}{l} \theta_\eta = g_0 \ \text{ on } \ \Sigma' \ , \ 0 \ \text{ on } \ \Sigma'' \ , \\ \theta_\eta(x,o) = f(x) \ \text{ on } \ \mathcal{O} \ . \end{array}\right. \tag{2.13}$$

If we denote by $\chi_{\psi<0}$ the characteristic function of the set where $\psi < 0$, we observe that (2.12) can be written:

$$\frac{\partial\theta_\eta}{\partial t} + A\theta_\eta + \frac{1}{\eta}\theta_\eta \chi_{\int_0^t \theta_\eta(x,s)ds<o} = 0 \ ; \tag{2.14}$$

this equation, together with (2.13), is the penalized equation independently introduced by H. Kawarada [1] (cf. also H. Kawarada and M. Natori [1] for numerical applications of this method).

3. An Asymptotic Result

3.1 Free boundary problem in a composite material

We now consider a problem similar to the one of Section 1 but with a very inhomogeneous material. We assume that the problem can be modelled as follows. We introduce an operator A^ε given by

$$A^\varepsilon v = - \Sigma \frac{\partial}{\partial x_i}\left(a_{ij}\left(\frac{x}{\varepsilon}\right) \frac{\partial v}{\partial x_j}\right) \tag{3.1}$$

where

$$\left|\begin{array}{l} a_{ij}(y) \in L^\infty(\pi^n) \ \ \pi^n = \text{n-dimensional torus,} \\ a_{ij} = a_{ji} \quad \forall i,j \\ \Sigma\, a_{ij}(y)\, \xi_i\xi_j \geq \alpha\, \Sigma\, \xi_i^2 \ , \ \alpha > 0 \ \text{ a.e. in } \ \pi^n \end{array}\right. \tag{3.2}$$

and where $\varepsilon > 0$.

We are looking for a function $\theta^\varepsilon = \theta^\varepsilon(x,t)$ such that, with similar notations to Section 1.1:

(3.3) $$\frac{\partial\theta^\varepsilon}{\partial t} + A^\varepsilon \theta^\varepsilon = 0 \text{ in } Q_1^\varepsilon ,$$

(3.4) $$\theta^\varepsilon(x,t) = g_o(x,t) , \; x \in \Gamma' ,$$

(3.5) $$\left| \begin{array}{l} \theta^\varepsilon = 0 , \; \Sigma\, a_{ij}(\frac{x}{\varepsilon}) \frac{\partial\theta^\varepsilon}{\partial x_j} n_i = -L\, V\cdot n \\ \text{on the free surface } S^\varepsilon(t) \end{array} \right.$$

(3.6) $$\theta^\varepsilon(x,o) = \theta_o(x) \text{ in } \mathcal{O}_1(o) .$$

Of course what has been said above immediately extends to the present situation (for ε fixed). If we introduce $\tilde{\theta}^\varepsilon$ and

(3.7) $$u^\varepsilon(x,t) = \int_o^t \tilde{\theta}^\varepsilon(x,s)ds ,$$

then u^ε is <u>the solution of the</u> V.I.

(3.8) $$\left| \begin{array}{l} (\frac{\partial u^\varepsilon}{\partial t}, v-u^\varepsilon) + a^\varepsilon(u^\varepsilon, v-u^\varepsilon) - (f, v-u^\varepsilon) \geq 0 \\ \forall\, v \in K(t) \;\; (K(t) \text{ defined as in (1.22)}),\; u^\varepsilon(t) \in K(t) , \\ u^\varepsilon(o) = 0 \end{array} \right.$$

where

(3.9) $$a^\varepsilon(u,v) = \sum_{i,j=1}^n \int_{\mathcal{O}} a_{ij}\, (\frac{x}{\varepsilon}) \frac{\partial u}{\partial x_j} \frac{\partial v}{\partial x_i} dx .$$

The problem we want to study is the behaviour of u^{ε} (and of θ^{ε}) as $\varepsilon \to 0$.

3.2 The "homogenized" free boundary problem

We recall a known construction (cf. Sanchez Palencia [1], de Giorgi-Spagnolo [1], Babuška [1]) of an "homogenized" operator $\mathcal{a}$ associated to A^{ε}.

Let us set:

$$W = \{\varphi | \ \varphi \in L^2(\pi^n), \ \frac{\partial \varphi}{\partial y_i} \in L^2(\pi^n) ,$$

$$\varphi \text{ "periodic" i.e. } \varphi(y_1,\dots,y_{i-1}, o, y_{i+1},\dots,y_n) =$$

$$= \varphi(y_1,\dots,y_{i-1},1,y_{i+1},\dots,y_n) \ \forall \ i \ \} ,$$

$$\alpha(\varphi,\psi) = \Sigma \int_{\pi n} a_{ij}(y) \frac{\partial \varphi}{\partial y_j}(y) \frac{\partial \varphi}{\partial y_i}(y)\, dy :$$

there exists a unique element $\chi_i \in W/R$ such that

$$\alpha(\chi_i,\psi) = \alpha(y_i,\psi) \quad \forall \ \psi \in W \tag{3.10}$$

so that we uniquely define $q_{ij} \in R$ by

$$q_{ij} = \alpha(\chi_i - y_i, \ \chi_j - y_j) . \tag{3.11}$$

We define in this manner a symmetric elliptic operator $\mathcal{a}$ by

$$\mathcal{a}v = - \sum_{i,j-1}^{n} q_{ij} \frac{\partial^2 v}{\partial x_i \partial x_j} . \tag{3.12}$$

We now consider the "homogeneized" free boundary problem as follows: to find a function θ such that

(3.13) $$\frac{\partial\theta}{\partial t} + a\theta = 0 \text{ in } Q_1 ,$$

(3.14) $$\theta(x,t) = g_0(x,t) , \; x \in \Gamma' ,$$

(3.15) $$\theta = 0 , \; \Sigma\, q_{ij} \frac{\partial\theta}{\partial x_j} n_i = -L\, v.n \text{ on the free surface } S(t)$$

(3.16) $$\theta(x,o) = \theta_0(x) \text{ in } \mathcal{O}_1(o) .$$

As before, if we introduce $\tilde{\theta}$ and

(3.17) $$u(x,t) = \int_0^t \tilde{\theta}(x,s)\, ds ,$$

then u is the solution of the V.I.

(3.18) $$\left| \begin{array}{l} (\frac{\partial u}{\partial t}, v-u) + a(u,v-u) - (f,v-u) \geq 0 , \; \forall\, v \in K(t) , \\ u(t) \in K(t) , \; u(o) = 0 \end{array} \right.$$

where

(3.19) $$a(u,v) = \Sigma \int_{\mathcal{O}} q_{ij} \frac{\partial u}{\partial x_j} \frac{\partial v}{\partial x_i}\, dx .$$

3.3 Asymptotic result

One can prove the following:

(3.20) $$\left| \begin{array}{l} \text{under the hypothesis (2.4), one has, as } \varepsilon \to 0 : \\ u^\varepsilon \to u \text{ in } L^2(0,T;V) \text{ weakly and in } L^\infty(0,T;H) \text{ weak star} \end{array} \right.$$

(where u^ε (resp u) is the solution of (3.8) (resp. (3.18)).

Of course it follows from (3.20) that

(3.21) $$\tilde{\theta}^\varepsilon = \frac{\partial u^\varepsilon}{\partial t} \to \tilde{\theta} = \frac{\partial u}{\partial t} \text{ in } H^{-1}(0,T;V) \text{ weakly.}$$

This is indeed a rather weak convergence which naturally leads to:

Open question 1: Is it possible to improve (3.21)?
(so as to obtain at least weak convergence in $L^2(0,T;V)$).

Another natural open question is

Open question 2: Is it possible to show that, in a suitable sense, $S^\varepsilon(t)$ (the free boundary for θ^ε) "converges" to the free boundary for θ ?

3.4 Sketch of proof of (3.20)

1) One uses the analogous of the error estimate (2.10) for u^ε , u^ε_η instead of u , u_η : one shows that the corresponding constant (in (2.10)) can be chosen independently of ε .

2) One has then to show the analogous of (3.20) for the penalized equation, with fixed η (penalty weight). One does that by using the techniques of Spagnolo [1] [2], Sbordone [1].

Remark 3.1

We refer to Bensoussan, Lions and Papanicolaou [1] for other results and complete proofs.

4. Extensions and Modifications of the Methods

4.1 Other problems which can be reduced to V.I.

Multiphase Stefan's problems can be reduced to V.I. by a transformation somewhat similar to (1.12). We refer to Duvaut [2] (cf. also Frémond [1] who does not use the approach by derivatives of characteristic functions of various phases; this A. remarks that the new unknown u has been introduced in the Engineering literature under the name of "freezing index").

Various problems arising in hydrodynamics have been reduced to V.I. by H. Brezis and G. Duvaut [1], H. Brezis and G. Stampacchia [1] and for infiltration in porous media in a non stationary situation, Torelli [1].

4.2 Problems which can be reduced to Q.V.I.

The method of Q.V.I. introduced in Bensoussan-Lions [4] for the solution of impulse control problems has been shown by C. Baiocchi [2] [3] to be useful for infiltration problems in porous media; he showed how a suitable transformation of the problem reduced the question to a stationary Q.V.I.

A. Friedman and D. Kinderlehrer [1] showed how problems of the Stefan's type could also be reduced to some kind of Q.V.I.

A much more complete report will be given in Duvaut-Lions [2].

4.3 Asymptotic analysis

The asymptotic analysis presented in Section 3 can be extended to all of the above problems - and to many others. We refer to Bensoussan, Papanicolaou and the A. [1].

REFERENCES

C. Baiocchi [1] C.R.A.S. 273 (1971), pp. 1215-1217.
[2] C.R.A.S. 278 (1974),
[3] Lecture at the I.C.M. Vancouver, August 1974.

C. Baiocchi, V. Comincioli, E. Magenes and G. A. Pozzi [1] Annali di Mat. 96 (1972) pp. 1-82.

I. Babuška [1] "Solution of problems with interfaces and singularities," Inst. Fluid Dyn. Applied Math. April 1974.
[2] "Solution of the interface problem by homogenization II., Inst. Fluid Dyn. Applied Math. March 1974.

A. Bensoussan [1] Lecture at I.C.M. Vancouver, August 1974.

A. Bensoussan and J. L. Lions [1] "Problèmes de temps d'arrêt optimal et I.V. paraboliques," Applicable Analysis (1973) (3), pp. 267-294.

A. Bensoussan and J. L. Lions [2] "I.V. non linéaires du premier et du second ordre," C.R.A.S., Paris, 176 (1973), pp. 1411-1415.

[3] Book to appear, Hermann, Paris, Vol. 1 (1976).

[4] "Nouvelle formulation de problèmes de contrôle impulsionnel et applications," C.R.A.S. Paris, 276, pp. 1189-1192; pp. 1333-1338.

A. Bensoussan, J. L. Lions and G. Papanicolaou [1] Notes in the C.R.A.S. 1975 and book to appear, North Holland.

H. Brezis and G. Duvaut [1] C.R.A.S. 276 (1973), pp. 875-878.

H. Brezis and G. Stampacchia [1] C.R.A.S. 276, (1973), pp. 129-132.

J. R. Cannon and C. D. Hill [1] "Existence, uniqueness ...," J. Math. Mech. 17 (1967), pp. 1-19.

[2] "Remarks on a Stefan problem," 17 (1967), pp. 433-449.

H. Chernoff [1] "Optimal stochastic control," Sankhya 30 (1968), pp. 221-252.

V. Comincioli [1] "A theoretical and numerical approach to some free boundary problems," Ann. Mat. Pura Appl. 1975.

J. Douglas [1] Proc. Amer. Math. Society 8 (1957), pp. 402-408.

G. Duvaut [1] C.R.A.S. Paris (1973).

[2] Lecture Rio-de-Janeiro, January 1975.

G. Duvaut and J. L. Lions [1] Inéquations en Mécanique et en Physique, Dunod, Paris, (1972) English translation by Mrs. John, Springer, 1976.

[2] Book in preparation.

M. Fremond [1] Lecture in Computational Methods in non-linear Mechanics, The Univ. of Texas at Austin, Austin, Texas, September 1974.

A. Friedman [1] "Free boundary problems for parabolic equations," I - J. Math. Mech. 8 (1959), pp. 483-498.
II - ibid. 9 (1960), pp. 19-66.

A. Friedman and D. Kinderlehrer [1] "A class of parabolic quasi-variational inequalities," to appear.

E. de Giorgi and S. Spagnolo [1] "Sulla convergenza degli integrali dell' energia per operatori ellittici del secondo ordine," Boll. UMI I (4) 8 (1973), pp. 391-411.

R. Glowinski, J. L. Lions and R. Tremolieres, [1] Analyse numérique des Inéquations Variationnelles, Paris, Dunod, 1976.

S. Kamenostoskaya [1] "On Stefan's problem," Mat. Sbornik, 53 (95), pp. 489-514.

H. Kawarada [1] "Stefan-type free boundary problems for heat equations," R.I.M.S. Kyoto University 9 (1974), pp. 517-533.

K. Kawarada and M. Natori [1] To appear.

I. I. Kolodner [1] "Free boundary problem for the heat equation with applications to problems of change of phase," C.P.A.M. 9 (1956), pp. 1-31.

O. A. Ladyzenskaya, V. Solonnikov, and N. Uralceva, Linear and quasi-linear equations of parabolic type, Moscow, 1967,

J. L. Lions [1] Sur le contrôle optimal, I.C.M. Vancouver, August 1974.

[2] On free surface problems: Methods of V.I. and Q.V.I., Austin, Int. Conference on Computational Methods in non linear Mechanics, September 1974.

[3] Variational problems and free boundary problems, Int. Symp. on Math. Problems in Theoretical Physics. Kyoto, January 1975.

[4] Quelques méthodes de résolution des problèmes aux limites non linéaires, Dunod 1969.

J. L. Lions and G. Stampacchia [1] "Variational Inequalities," C.P.A.M. 20 (1967), pp. 493-519.

F. Mignot and J. P. Puel [1] "Solution maximum ...," C.R.A.S. Paris, 280 (1974), pp. 259-262.

O. A. Oleinik [1] "On Stefan type free boundary problems for parabolic equation," Sem. dell'Ist. Naz. di Alta Mat. 1962, 63, pp. 388-403.

Samuelson and McKean [1] "Rational theory of warrant pricing," Ind. Man. Rev. 6 (1974), pp. 331-351.

E. Sanchez-Palencia [1] "Comportement local et macroscopique d'un type de milieux physiques hétérogènes," Int. J. Eng. Sc. 12 (1974), pp. 331-351.

G. Sbordone [1] Sulla G-convergenza di equazioni ellittiche e paraboliche, Ric. di Mat. 1975.

B. Sherman [1] "General one phase Stefan problems ...," SIAM J. Applied Math. 20 (1971), pp. 555-570.

S. Spagnolo [1] "Sulla convergenza di soluzioni di equazioni paraboliche et ellittiche," Ann. Scuola Norm. Sup. Pisa, XXII, (1968), pp. 571-597.

[2] Lecture at this Conference.

Torelli, [1] C.R.A.S. (1975) To appear.

REGULARITY OF SOLUTIONS OF THE STOKES PROBLEM IN A POLYGONAL DOMAIN*

John E. Osborn**

1. Introduction

It is the purpose of this paper to discuss several regularity results for the Stokes problem in a polygonal domain. Analogous regularity results for solutions of a single $2m^{th}$ order elliptic equation in a polygonal domain have been extensively developed [1,2,4,5,6,7,10,12,13,16,18]. The paper by Professor Grisvard in these proceedings surveys many of these results.

Regularity results are of fundamental importance in the analysis of numerical methods for the Stokes problem (see, e.g., Crouzeix and Raviart [3], Jamet and Raviart [9], Osborn [22]). Also, regularity results are used in analyzing the stability of stationary solutions of the Navier-Stokes equations (see, e.g., Iudovich [8], Kirchgässner and Kielhöffer [15], Prodi [23]).

The results presented here are all based on the method used by Kondratév [16] to study the regularity of a single $2m^{th}$ order elliptic equation. In our context this method must be applied to the system of equations given by the Stokes problem. (See Sovin [25] for a sketch of some related results.) The results depend in an essential way on the spectral properties

*This work was partially supporeted by NSF Grant MPS71-03498-A02.

**Department of Mathematics, University of Maryland.

of a system of ordinary differential equations which is associated with the Stokes equations; this dependence is fully discussed in Section 4.

In this paper we merely state the presented results. In Section 3 we discuss two low order regularity results; these results are proved in [14]. In Section 4 we discuss two higher order results; the proofs of these results will appear elsewhere. Kondratév [17] has obtained results related to those in Section 4 (see Remark 1 in Section 4).

Throughout the paper we will use various Sobolev and weighted Sobolev spaces. $W^m(D)$, $m = 0,1,2,\cdots$, will denote the usual m^{th} Sobolev space on a domain D in the plane (R^2). On this space we use the norm given by

$$\|w\|^2_{W^m(D)} = \sum_{|\alpha|\leq m} \int_D |D^\alpha w|^2 dx\, dy .$$

We will also use the local versions of these spaces. Note that $W^0(D) = L_2(D)$. $W_0^1(D)$ will denote the subspace of $W^1(D)$ consisting of functions which vanish on ∂D. We will also use the Sobolev spaces $W^m[0,\omega]$ on a closed interval $[0,\omega]$.

For D a polygonal domain in the plane, the weighted Sobolev space $\mathring{W}^m_\alpha(D)$ is defined to be the class of functions for which the following norm is finite:

$$\|w\|^2_{\mathring{W}^m_\alpha(D)} = \sum_{j=0}^{m} \int_D \delta(x,y)^{\alpha-2(m-j)} |D^j w|^2 dx\, dy ,$$

where

$$\delta(x,y) = \min\{\text{dist}((x,y),P) : P \text{ a vertex of } D\}$$

and

$$D^j w(x,y) = \Big(\sum_{|\alpha|=j} |D^\alpha w(x,y)|^2 \Big)^{1/2} \quad .$$

I would like to thank R. B. Kellogg for many illuminating discussions on the material in this paper.

2. Formulation of the problem

Let $D \subset R^2$ be a bounded domain with boundary ∂D and consider the generalized Stokes problem

$$(2.1) \qquad \begin{cases} -\Delta u + p_x = f_1 \quad \text{in} \quad D \ , \\ -\Delta v + p_y = f_2 \quad \text{in} \quad D \ , \\ u_x + v_y = g \quad \text{in} \quad D \ , \\ u = v = 0 \quad \text{on} \quad \partial D \ . \end{cases}$$

These equations are obtained by suppressing the nonlinear terms in the two dimensional stationary Navier-Stokes equations. The viscosity ν has been set equal to 1 for convenience. We suppose f_1 , $f_2 \in L_2(D)$ and $g \in w^1(D)$. If $g = 0$ we refer to the Stokes problem, as opposed to the generalized Stokes problem. A triple of functions $u \in w_0^1(D)$, $v \in w_0^1(D)$, $p \in L_2(D)$ is a generalized solution of (2.1) if, for any smooth φ with compact support in D ,

$$\int_D (\varphi_x u_x + \varphi_y u_y - \varphi_x p) dx\, dy = \int_D f_1 \, \varphi \, dx\, dy \quad ,$$

$$\int_D (\varphi_x v_x + \varphi_y v_y - \varphi_y p) dx\, dy = \int_D f_2 \, \varphi \, dx\, dy \quad ,$$

and if $u_x + v_y = g$ a.e. in D . Note that the latter equation plus the boundary conditions imply that g satisfies the constraint

$$\int_D g \, dx \, dy = 0 \quad . \tag{2.2}$$

If $g = 0$ one can prove the existence of a generalized solution of (2.1) by variational techniques. If $g \neq 0$ but satisfies (2.2) and ∂D is a smooth curve one can also show the existence of a generalized solution of (2.1) (see, e.g., Temam [27]). u and v , the velocity components of the solution of (2.1), are unique, and p , the pressure, is unique up to a constant.

Regularily results for the solution of (2.1) are well-known for the case in which ∂D is a smooth curve (see Ladyzhenskaya [19]). In particular, one obtains local regularity of the solution, and global regularity of the solution if the boundary is smooth. More precisely, the following result may be obtained from Theorem 2, Chapter 3, in [19].

<u>Theorem</u> Let $m \geq 0$ be an integer, let $f_1 \in W^m(D)$, $f_2 \in W^m(D)$, $g \in W^{m+1}(D)$, and let u,v,p be a generalized solution of (2.1). Then if D_1 is any domain with $\overline{D_1} \subset D$, we have $u \in W^{m+2}(D_1)$, $v \in W^{m+2}(D_1)$ and $p \in W^{m+1}(D_1)$. Furthermore, if ∂D is smooth, then $u \in W^{m+2}(D)$, $v \in W^{m+2}(D)$, $p \in W^{m+1}(D)$, and there is a constant $c > 0$ depending only on D such that

$$\|u\|_{W^{m+2}(D)} + \|v\|_{W^{m+2}(D)} + \|p\|_{W^{m+1}(D)}$$

$$\leq c\left[\|f_1\|_{W^m(D)} + \|f_2\|_{W^m(D)} + \|g\|_{W^{m+1}(D)} + \|p\|_{L_2(D)}\right] .$$

The purpose of this paper is to discuss several related regularity results for the case in which ∂D is not smooth;

more specifically, we consider the case in which D is a polygon.

3. Lower order regularity

It is well-known (Birman and Skvorcov [2], Kadlec [10]) that the solution u of the problem

$$\Delta u = f \quad \text{in} \quad D \quad ,$$

$$u = 0 \quad \text{on} \quad \partial D$$

has square integrable second derivatives provided $D \subset R^2$ is a convex polygon and $f \in L_2(D)$. A similar result was proved by Kellogg and Osborn [14] for the Stokes problem. The precise statement is given in the following

Theorem 1. Let D be a convex polygon, let $f_1 \in L_2(D)$, $f_2 \in L_2(D)$, $g \in \overset{\circ}{W}{}^1_0(D)$, and let u, v, p be a generalized solution of (2.1). Then $u \in w^2(D)$, $v \in w^2(D)$ and $p \in w^2(D)$. Furthermore, there is a constant $c > 0$ depending only on D such that

$$\|u\|_{w^2(D)} + \|v\|_{w^2(D)} + \|D^1 p\|_{L_2(D)}$$

$$\leq c\left[\|f_1\|_{L_2(D)} + \|f_2\|_{L_2(D)} + \|g\|_{\overset{\circ}{W}{}^1_0(D)}\right] \quad .$$

Theorem 1, as well as the analogous result for Laplace's equation, is of central importance in the analysis of certain numerical methods (see, e.g., [3,9,20,22]).

It was pointed out to us by R. Temam that Theorem 1 implies a corresponding result for solutions of the Navier-Stokes equations

$$(3.1) \quad \begin{cases} - \Delta u + p_x + uu_x + vu_y = f_1 & \text{in } D, \\ - \Delta v + p_y + uv_x + vv_y = f_2 & \text{in } D, \\ u_x + v_y = 0 & \text{on } \partial D \end{cases}$$

in a convex polygon D . In fact we have

Theorem 2. Suppose $u \in w_0^1(D)$, $v \in w_0^1(D)$ and $p \in L_2(D)$, and suppose (3.1) is satisfied in the generalized sense. Then $u \in w^2(D)$, $v \in w^2(D)$ and $p \in w^1(D)$.

For complete proofs of Theorems 1 and 2 see [14].

4. Higher order regularity results

Let D be a polygon in R^2 . Using the smooth domain regularity theorems in Section 2 one can study the regularity of the solution of (2.1) in any subdomain of D with smooth boundary not containing a vertex of D . The study of the regularity of the solution near the vertices of D can be reduced to the study of the Stokes problem in a sector. This can be seen as follows.

Let P be one of the vertices of D and let $\zeta(x,y)$ be a smooth function which is identically 1 in a neighborhood of P and which satisfies $\zeta(x,y) = 0$ for $\text{dist}((x,y),P) \geq r_0$, where $2r_0$ is the length of the smallest side of D . We let Ω denote the infinite sector whose vertex is placed at P and whose sides are the extensions to infinity of the two sides of ∂D which meet at P . From (2.1) we formally obtain

$$(4.1)\quad \begin{cases} -\Delta(\zeta u) + (\zeta p)_x = \zeta f_1 - 2\nabla\zeta\cdot\nabla u - u\Delta\zeta + p\zeta_x & \text{in } \Omega , \\ -\Delta(\zeta u) + (\zeta p)_y = \zeta f_2 - 2\nabla\zeta\cdot\nabla v - v\Delta\zeta + p\zeta_y & \text{in } \Omega , \\ (\zeta u)_x + (\zeta v)_y = \zeta g + u\zeta_x + v\zeta_y & \text{in } \Omega , \\ \zeta u = \zeta v = 0 \text{ on } \partial\Omega \text{ and for } r \geq r_0 . \end{cases}$$

It is easily seen that ζu and ζv are in $W^2_{loc}(\Omega)$ and ζp is in $W^1_{loc}(\Omega)$, and that (4.1) is satisfied a.e. in Ω . Also, the right hand sides of the first three equations of (4.1) are in $L_2(\Omega)$, $L_2(\Omega)$, and $W^1(\Omega)$, respectively, and ζu , ζv , ζp is a generalized solution of the generalized Stokes equations (4.1). If this construction is made for each vertex P_i of D , $1 \leq i \leq I$, we see that the original solution u,v,p may be written in the form

$$u = \sum_0^I u_i \ , \quad v = \sum_0^I v_i \ , \quad p = \sum_0^I p_i \ ,$$

where the functions u_i,v_i,p_i, $1 \leq i \leq I$, are the generalized solutions of a problem of the form (4.1) corresponding to the vertex P_i , and the remaining triple u_0,v_0,p_0 is a generalized solution of the generalized Stokes problem which vanishes in a neighborhood of the vertices of D . Since (u,v,p) equals (u_i,v_i,p_i) in a neighborhood of P_i we see that the study of (u,v,p) near the vertices of D is reduced to the study of the Stokes problem in a sector. The remaining results in the paper will all be stated for the Stokes equations in a sector; it is a simple matter to translate the results back to the polygon D .

Let the vertex of the sector Ω be placed at the origin 0 and suppose one of the sides of Ω lies on the positive x-axis. Letting the angle of Ω be $\omega < 2\pi$, the sector Ω is given, in polar coordinates, by the inequality

$0 < \theta < \omega$. We exclude the uninteresting case $\omega = \pi$. For the remainder of the paper we assume u,v and p are functions defined on Ω such that

$$u \in W^1(\Omega) \cap W^2_{loc}(\Omega) \ , \ v \in W^1(\Omega) \cap W^2_{loc}(\Omega) \ , \ p \in L_2(\Omega) \cap W^1_{loc}(\Omega),$$

$$u \equiv 0 \ , \ v \equiv 0 \ , \ p \equiv 0 \quad \text{for} \quad r > r_0 \ ,$$

and, in both the generalized and pointwise sense, u,v and p satisfy

$$(4.2) \qquad \begin{cases} -\Delta u + p_x = f_1 & \text{in } \Omega \ , \\ -\Delta v + p_y = f_2 & \text{in } \Omega \ , \\ u_x + v_y = g & \text{in } \Omega \ , \\ u = v = 0 & \text{on } \partial\Omega \ . \end{cases}$$

There is associated with the Stokes equations, the sector Ω , and the boundary conditions a sequence of complex numbers λ_j in the complex $\zeta = \xi + i\eta$ plane together with multiplicities m_j , in terms of which we will state our higher order regularity results. Before stating the first result we will briefly discuss the definitions of the numbers λ_j , m_j .

We begin by writing (4.2) in terms of polar coordinates $x = r \cos \theta$, $y = r \sin \theta$. Toward this end we introduce the functions

$$U = u \cos \theta + v \sin \theta \ ,$$
$$V = -u \sin \theta + v \cos \theta \ ,$$

representing the components of the velocity in the radial and targential directions, respectively, and the functions

$$F_1 = f_1 \cos\theta + f_2 \sin\theta ,$$
$$F_2 = -f_1 \sin\theta + f_2 \cos\theta ,$$

representing the components of the force in the radial and targential directions, respectively. Using these functions, (4.2) may be rewritten as

$$(4.3)\quad \begin{cases} -\frac{1}{r}(rU_r)_r - \frac{1}{r^2}U_{\theta\theta} + \frac{1}{r^2}U + \frac{2}{r^2}V_\theta + p_r = F_1 & \text{in } \Omega , \\ -\frac{1}{r}(rV_r)_r - \frac{1}{r^2}V_{\theta\theta} + \frac{1}{r^2}V - \frac{2}{r^2}U_\theta + \frac{1}{r}p_\theta = F_2 & \text{in } \Omega , \\ U_r + \frac{1}{r}U + \frac{1}{r}V_\theta = g & \text{in } \Omega , \\ V = V = 0 \quad \text{for} & \theta = 0 , \theta = \omega . \end{cases}$$

Next we introduce the new variable τ by $r = e^{-\tau}$. In the $\tau\theta$ plane the sector Ω becomes the strip $S = \{(\tau,\theta) : 0 < \theta < \omega , -\infty < \tau < +\infty\}$. We set $q = e^{-\tau}p$ and transform (4.3) into $\tau\theta$ coordinates, obtaining

$$(4.4)\quad \begin{cases} -U_{\tau\tau} - U_{\theta\theta} + U + 2V_\theta - q_\tau - q = e^{-2\tau}F_1 & \text{in } S , \\ -V_{\tau\tau} - V_{\theta\theta} + V - 2U_\theta + q_\theta = e^{-2\tau}F_2 & \text{in } S , \\ -U_\tau + U + V_\theta = e^{-\tau}g & \text{in } S , \\ U = V = 0 \quad \text{for} & \theta = 0 , \omega . \end{cases}$$

We now introduce the Fourier transform

$$(Fw)(\zeta,\theta) = \hat{w}(\zeta,\theta) = \frac{1}{\sqrt{2\pi}} \int_{-\infty}^{\infty} e^{-i\zeta\tau} w(\tau,\theta)\, d\tau , \quad (\zeta = \xi + i\eta) .$$

Applying the Fourier transforms we readily see that (4.4) is transformed into

$$(4.5)\begin{cases} -\hat{U}_{\theta\theta} + (\zeta^2+1)\hat{U} + 2\hat{V}_\theta - (i\zeta+1)\hat{q} = \hat{F}_1(\zeta-2i,\theta) \ , \ 0 \le \theta \le \omega, \\ -\hat{V}_{\theta\theta} + (\zeta^2+1)\hat{V} - 2\hat{U}_\theta + \hat{q}_\theta = F_2(\zeta-2i,\theta) \ , \quad 0 \le \theta \le \omega \ , \\ (1-i\zeta)\hat{U} + \hat{V}_\theta = \hat{g}(\zeta-i,\theta) \ , \ 0 \le \theta \le \omega \ , \\ \hat{U} = \hat{V} = 0 \quad , \quad \theta = 0 \ , \omega \ . \end{cases}$$

Consider the system (4.5) with arbitrary right hand sides h_1 , h_2 , h_3 :

$$(4.6)\begin{cases} -\hat{U}_{\theta\theta} + (\zeta^2+1)\hat{U} + 2\hat{V}_\theta - (i\zeta+1)\hat{q} = h_1(\theta) \in L_2[0,w] \ , \\ -\hat{V}_{\theta\theta} + (\zeta^2+1)\hat{V} - 2\hat{U}_\theta + \hat{q}_\theta = h_2(\theta) \in L_2[0,w] \ , \\ (1-i\zeta)\hat{U} + \hat{V}_\theta = h_3(\theta) \in W^1[0,w] \ , \\ \hat{U} = \hat{V} = 0 \quad , \quad \theta = 0,\omega \ . \end{cases}$$

We will refer to this system of ordinary differential equations containing the parameter ζ as the auxiliary system corresponding to the Stokes equations (4.2). (4.6) is uniquely solvable for arbitrary h_1 , h_2 , h_3 for all ζ satisfying $\eta > 0$ except for a sequence of values λ_j ; these values are the eigenvalues of the homogeneous system corresponding to (4.6) and can be characterized as the roots of the transcendental equations

$$(4.7) \qquad \nu(\zeta) \equiv \sinh^2\zeta\omega - \zeta^2\sin^2\omega = 0$$

which lie in the upper half plane. The numbers λ_j are also the poles of the resolvent operator $R(\zeta)$ associated with

(4.6). m_j is defined to be the order of λ_j as a zero of $\nu(\zeta)$ (or as a pole of $R(\zeta)$) . It is in terms of the λ_j and m_j that we can now state the higher order regularity results.

Theorem 3. Suppose f_1 , $f_2 \in \overset{\circ}{W}{}_0^m(\Omega)$, $g \in \overset{\circ}{W}{}_0^{m+1}(\Omega)$, $m = 1,2,\cdots$, and that no λ_j lies on the line $\operatorname{Im} \zeta = \eta = m+1$ in the complex $\zeta = \xi + i\eta$ plane. Then

$$u = \tilde{u} + \sum_{0 < \operatorname{Im} \lambda_j < m+1} \sum_{k=0}^{m_j - 1} c^u_{j,k}(\theta) r^{-i\lambda_j} \ell n^k r ,$$

$$v = \tilde{v} + \sum_{0 < \operatorname{Im} \lambda_j < m+1} \sum_{k=0}^{m_j - 1} c^v_{j,k}(\theta) r^{-i\lambda_j} \ell n^k r ,$$

$$p = \tilde{p} + \sum_{0 < \operatorname{Im} \lambda_j < m+1} \sum_{k=0}^{m_j - 1} c^p_{j,k}(\theta) r^{-i\lambda_j - 1} \ell n^k r ,$$

where

$$\tilde{u}, \tilde{v} \in \overset{\circ}{W}{}_0^{m+2}(\Omega) \quad , \quad \tilde{p} \in \overset{\circ}{W}{}_0^{m+1}(\Omega)$$

and

$$c^u_{j,k}(\theta) \quad , \quad c^v_{j,k}(\theta) \quad , \quad c^p_{j,k}(\theta)$$

are C^∞ functions of θ . In addition there is a constant $c = c(m,\Omega)$ such that

$$\|\tilde{u}\|_{\overset{\circ}{W}{}_0^{m+2}(\Omega)} + \|\tilde{v}\|_{\overset{\circ}{W}{}_0^{m+2}(\Omega)} + \|\tilde{p}\|_{\overset{\circ}{W}{}_0^{m+1}(\Omega)}$$

$$c\left[\|f_1\|_{\overset{\circ}{W}{}_0^{m}(\Omega)} + \|f_2\|_{\overset{\circ}{W}{}_0^{m}(\Omega)} + \|g\|_{\overset{\circ}{W}{}_0^{m+1}(\Omega)}\right] .$$

This theorem shows that $u(v,p)$ can be written as the sum of a smooth function $\tilde{u}(\tilde{v},\tilde{p})$ and several singular functions which correspond to the poles of $R(\zeta)$.

The functions $C^u_{j,k}(\theta)$, $C^v_{j,k}(\theta)$, $C^p_{j,k}(\theta)$ are eigenfunctions and generalized eigenfunctions of (4.6) corresponding to the eigenvalue λ_j . The infinite differentiability of these functions follows from this. Thus we see that the regularity of the solution is completely determined by the data f_1 , f_2 , g and the spectral properties of the auxiliary system (4.6). If a fixed basis is chosen for the space of generalized eigenfunctions of (4.6) corresponding to λ_j , then the functions $C^u_{j,k}(\theta)$, $C^v_{j,k}(\theta)$, $C^p_{j,k}(\theta)$ can be expressed as linear combinations of these basis functions (which do not depend on the data f_1 , f_2 , and g) with coefficients which are linear functionals of f_1 , f_2 and g .

Remarks: 1) Our proof of Theorem 3 consists of the application of the method of Kondratév to the Stokes system. A different approach (at least for studying the regularity of u and v) would be to introduce the stream function ψ , reduce the study of the Stokes system to the study of the biharmonic equation in ψ , and apply the results of Kondratév [16] directly to the biharmonic equation to obtain an expansion formula for ψ . In [17] Kondratév obtains such an expansion for the stream function corresponding to the stationary Navier-Stokes equations (cf. also Ogonesjan [21]). This approach does not appear to be possible in the case of Theorem 1.

2) Theorem 1 is not a consequence of Theorem 3 since $R(\zeta)$ has a pole (at i) on the line $\eta = 1$. (There are no poles in $0 < \eta < 1$ if $\omega < \pi$.) For this reason the method of Kondratév must be modified to obtain the proof of Theorem 1 (see [14]).

It is of interest to have regularity results for the case in which the data (i.e., f_1, f_2, g) are in the usual Sobolev spaces, as opposed to the weighted Sobolev spaces. We present next a partial result of this type.

Theorem 4. Suppose $\omega \neq \pi$, $f_1, f_2 \in W^m(\Omega)$, $g = 0$, and in addition suppose that $\int_\Omega r^\alpha |D^m f_i|^2 dx\, dy < \infty$ for some $\alpha < 0$. Then for small r we have the expansions

$$u = \tilde{u} + \sum_{0 < \operatorname{Im} \lambda_j < 2} \sum_{k=0}^{m_j - 1} c^u_{j,k}(\theta) r^{-i\lambda_j} \ell n^k r + \sum_{\ell=0}^{m-1} r^{\ell+2} d^u_\ell(\theta) ,$$

$$v = \tilde{v} + \sum_{0 < \operatorname{Im} \lambda_j < 2} \sum_{k=0}^{m_j - 1} c^v_{j,k}(\theta) r^{-i\lambda_j} \ell n^k r + \sum_{\ell=0}^{m-1} r^{\ell+2} d^v_\ell(\theta) ,$$

$$p = \tilde{p} + \sum_{0 < \operatorname{Im} \lambda_j < 2} \sum_{k=0}^{m_j - 1} c^p_{j,k}(\theta) r^{-i\lambda_j - 1} \ell n^k r + \sum_{\ell=0}^{m-1} r^{\ell+1} d^p_\ell(\theta) .$$

where

$$\tilde{u}, \tilde{v} \in \mathring{W}_0^{m+2}(\Omega) \quad , \quad \tilde{p} \in \mathring{W}_0^{m+1}(\Omega)$$

and

$$c^u_{j,k} , c^v_{j,k} , c^p_{j,k} , d^u_\ell , d^v_\ell , d^p_\ell$$

are C^∞ functions of θ.

5. Some special cases

In this section we discuss the theorems presented in Section 4 in certain special cases. In addition we make several remarks about the results.

a) Recall that the λ_j are the poles of the resolvent operator $R(\zeta)$ associated with (4.6). Although $\zeta = i$ is a pole of $R(\zeta)$ for all values of ω, it is a removable singularity of the u and v components of $R(\zeta)(h_1, h_2, h_3)$ unless $\tan\omega = \omega$. For $0 < \omega < 2\pi$ this equation has a unique solution $\omega_1 \simeq 1.43\pi$ radians $\simeq 257.40$ degrees. Let us suppose $\omega \neq \omega_1$; we will discuss the case $\omega = \omega_1$ in Subsection e of this section.

The q component of $R(\zeta)(h_1, h_2, h_3)$ has a simple pole at i for all $\omega \neq \omega_1$. We denote this pole by λ_1. Since the u and v components of $R(\zeta)(h_1, h_2, h_3)$ are analytic at i, the expansion formulas for u and v given in Theorem 3 do not contain terms corresponding to λ_1. It can be shown that the singular term in the expansion for p is just a constant. Thus the expansion formulas in Theorem 3 take the form

$$u = \tilde{u} + \sum_{\substack{0 < \operatorname{Im}\lambda_j < m+1 \\ j \neq 1}} \sum_{k=0}^{m_j - 1} C^u_{j,k}(\theta) r^{-i\lambda_j} \ell n^k r .$$

$$v = \tilde{v} + \sum_{\substack{0 < \operatorname{Im}\lambda_j < m+1 \\ j \neq 1}} \sum_{k=0}^{m_j - 1} C^v_{j,k}(\theta) r^{-i\lambda_j} \ell n^k r ,$$

$$p = \tilde{p} + \sum_{\substack{0 < \operatorname{Im}\lambda_j < m+1 \\ j \neq 1}} \sum_{k=0}^{m_j - 1} C^p_{j,k}(\theta) r^{-i\lambda_j - 1} \ell n^k r + C .$$

Note that the constant in the formula for p corresponds to the fact that p is only unique up to a constant. We also note that all of the poles of $R(\zeta)$ have order less than or equal to two, i.e., $m_j \leq 2$ for all j.

b) Consider now the case $m = 1$. We can apply Theorem 3, provided no λ_j lies on the line $\eta = 2$, and we find that

$$u = \tilde{u} + \sum_{\substack{0< \operatorname{Im} \lambda_j<2 \\ j \neq 1}} \sum_{k=0}^{m_j-1} C^u_{j,k}(\theta) r^{-i\lambda_j} \ell n^k r \quad ,$$

$$v = \tilde{v} + \sum_{\substack{0< \operatorname{Im} \lambda_j<2 \\ j \neq 1}} \sum_{k=0}^{m_j-1} C^v_{j,k}(\theta) r^{-i\lambda_j} \ell n^k r \quad ,$$

$$p = \tilde{p} + \sum_{\substack{0< \operatorname{Im} \lambda_j<2 \\ j \neq 1}} \sum_{k=0}^{m_j-1} C^p_{j,k}(\theta) r^{-i\lambda_j-1} \ell n^k r + C \quad ,$$

where

$$\tilde{u},\tilde{v} \in \overset{\circ}{W}{}^3_0(\Omega) \quad , \quad \tilde{p} \in \overset{\circ}{W}{}^2_0(\Omega) \quad .$$

We conclude this subsection with a remark about the condition that no λ_j lies on $\eta = 2$. For $0 < \omega < \pi$, there is a pole on $\eta = 2$ if and only if $\pi/2 < \omega < {}^{3\pi}/4$ and

$$\cosh\left(\frac{\omega \cos 2\omega}{\cos \omega}\right) = \frac{-1}{\cos \omega} \quad .$$

This equation has at least one solution in the interval $\pi/2 < \omega < {}^{3\pi}/4$: $\omega_2 \simeq 2.20$ radians $\simeq 126.28$ degrees. It is interesting to note that in this regard $\omega = \pi/2$ is an exceptional angle for the Dirichlet problem for Laplace's equation.

c) Suppose now $m = 1$ and $\omega = \pi/2$. In this case there are no poles in $0 < \eta < 2$ (except for the pole at i of the q component of $R(\zeta)$ (h_1, h_2, h_3)). Thus we have

$$u,v \in W_0^3(\Omega) \quad , \quad p = \tilde{p} + C$$

where

$$\tilde{p} \in \overset{\circ}{W}{}_0^2(\Omega) \quad .$$

d) Consider now the case in which $\omega = \pi/2$ and $m \geq 1$. In this case there are no poles on $\eta = m+1$ and there are no double poles. Thus we see that

$$u = \tilde{u} + \sum_{\substack{0< \operatorname{Im}\lambda_j<m+1 \\ j \neq 1}} C_j^u(\theta) r^{-\lambda_j i}$$

$$v = \tilde{v} + \sum_{\substack{0< \operatorname{Im}\lambda_j<m+1 \\ j \neq 1}} C_j^v(\theta) r^{-\lambda_j i}$$

$$p = \tilde{p} + \sum_{\substack{0< \operatorname{Im}\lambda_j<m+1 \\ j \neq 1}} C_j^p(\theta) r^{-\lambda_j i-1} + C \quad ,$$

where

$$\tilde{u},\tilde{v} \in \overset{\circ}{W}{}_0^{m+2}(\Omega) \quad , \quad \tilde{p} \in \overset{\circ}{W}{}_0^{m+1}(\Omega) \quad .$$

e) We conclude this section with a remark about the angle $\omega_1 \simeq 257.40$ degrees mentioned in Subsection a . For $\omega = \omega_1$, the u and v components of $R(\zeta)(h_1, h_2, h_3)$ have a pole at $\zeta = i$ and the q component has a pole of (possibly) higher order than one at i (recall that i is a simple pole of the q component for $\omega \neq \omega_1$). To show the exceptional nature of ω_1 we point out that one can prove an analogue of Theorem 1 for the case where D is not convex (this corresponds to $\omega > \pi$) provided $\omega \neq \omega_1$. This result would show that the solution u (or v) which corresponds to $f_1, f_2 \in L_2(\Omega)$, $g \in \overset{\circ}{W}{}^1_0(\Omega)$ can be written as the sum of a function in $\overset{\circ}{W}{}^2_0(\Omega)$ and several singular functions (corresponding to poles of $R(\zeta)$ in $0 < \eta < 1$).

The exceptional angle ω_1 arises in a different context (one involving the equations of elasticity) in the work of Sternberg and Koiter [26].

The transcendental equations (4.7) occurs in other work on problems in a polygon or a sector (see, e.g., [11,24]). Also, various exceptional angles arise in these studies.

REFERENCES

1. A. Avantaggiati and M. Troisi, "Spazi di Sobolev con peso e problemi ellipttici in an angolo I, II, III," Annali di Matem. Pura ed Applicato, 95 (1973), 361-348, 97 (1973), 207-252, 99 (1974), 1-64.
2. M. Š. Birman and G. E. Skvorcov, "On the square summability of highest derivatives of the solutions of the Dirichlet problem in a domain with piecewise smooth boundary," Izv. Vysš. Učebn. Zaved. Mathematika 1962, no. 5 (30), 11-21.
3. M. Crouzeix and P.-A. Raviart, "Conforming and nonconforming finite element methods for solving the stationary Stokes equations. I," R.A.I.R.O., Série mathématiques, 7 annee, R-3, 33-76.
4. P. Grisvard, "Problème de Dirichlet dans un cone," Ricerche Mat. 20 (1971), 175-192.
5. P. Grisvard, "Alternative de Fredholm relative au probleme de Dirichlet dans un polygone on un polyedre," Bollettino U.M.I. 4 (1972), 132-164.
6. P. Grisvard, "Alternative de Fredholm relative au problème de Dirichlet dons un polygone ou un polyedre, seconde partie," Ann. Scuola Norm. Sup. Pisa (to appear).
7. P. Grisvard, "Behavior of the solutions of an elliptic boundary value problem in a polygon or polyhedral domain," these proceedings.

8. V. Iudovich, "On the stability of stationary flows of a viscous incompressible fluid," Dokl. Akad. Nauk. SSSR 161 (1965), 1037-1040, Soviet Physics Dokl. 10 (1965), 293-295.

9. P. Jamet and P.-A. Raviart, "Numerical solution of the stationary Navier-Stokes equations by finite element methods," (to appear).

10. J. Kadlec, "The regularity of the solution of the Poisson problem in a domain whose boundary is similar to that of a convex domain," Czechoslovak Math. J. 89 (1964), 386-393.

11. S. N. Karp and F. C. Karol, "The elastic-field behavior in the neighorbood of a crack of arbitrary angle," Comm. Pure Appl. Math., 15 (1962), 413-421.

12. R. B. Kellogg, "Singularities in interface problems," Numerical Solutions of Partial Differential Equations, (B. Hubbard, editor), Academic Press, New York, 1971, 351-400.

13. R. B. Kellogg, "Higher order singularities for interface problems," The Mathematical Foundations of the Finite Element Method with Applications to Partial Differential Equations, (A. K. Aziz, editor), Academic Press, New York, 1973, 589-602.

14. R. B. Kellogg and J. E. Osborn, "A regularity result for the Stokes problem in a convex polygon," J. Functional Analysis, (to appear).

15. K. Kirchgässner and H. Kielhöfer, "Stability and bifurcation in fluid dynamics," Rocky Mountain J. Math. 3 (1973), 275-318.

16. V. A. Kondratév, "Boundary Problems for elliptic equations with conical or angular points," Trans. Moscow Math. Soc. 16 (1967), 209-292, translated by Am. Math. Soc., 1968.

17. V. A. Kondratév, "Asymptotic of a solution of the Navier-Stokes equation near the angular part of the boundary," Prikl. Mat. Meh. 31 (1967), 119-123, J. Appl. Math. Mech. 31 (1967), 125-129.

18. V. A. Kondratév, "The smoothness of a solution of Dirichlet's problem for 2nd order elliptic equations in a region with a piecewise smooth domain," Differencialńye Uravnenija 6 (1970), 1831-1843.

19. O. A. Ladyzhenskaya, The Mathematical Theory of Viscous Incompressible Flow, Gordon and Breach, New York, 1962.

20. J. Nitsche, "Ein Kriterion für die Quasi-Optimalität des Ritzschen Verfahrens," Numer. Math. 11 (1968), 346-348.

21. L. A. Oganesjan, "Singularities at corners of the solution of the Navier-Stokes equation," Zap. Naučn. Sem. Leningrad. Otdel Mat. Inst. Steklov. (LOMI) 27 (1972), 131-144.

22. J. E. Osborn, "Approximation of the eigenvalues of a non-selfadjoint operator arising in the study of the stability of stationary solutions of the Navier-Stokes equations," SIAM J. Numer. Anal. (to appear).

23. G. Prodi, "Teoremi di tipo locale per il sistema di Navier-Stokes e stabilitá delle soluzioni stazionaire," Padua Universita Seminario Matematico Rendiconti 32 (1962), 374-397.

24. J. B. Seif, "On the Green's function for the biharmonic equation in an infinite wedge," Trans. Amer. Math. Soc. 182 (1973), 241-260.

25. A. Sovin, "Boundary value problems in the domains with conic points and plane boundary value problems for the systems with gaps," Dopovidi Akad. Nauk. Ukrain RSR, Ser. A (1970), 426-429.

26. E. Sternberg and W. T. Koiter, "The wedge under a concentrated couple: a paradox in the two-dimensional theory of elasticity," J. Appl. Mech. 25 (1958), 575-581.

27. R. Temam, "On the Theory and Numerical Analysis of the Navier-Stokes Equations," Lecture Notes #9, Department of Mathematics, University of Maryland, College Park, 1973.

HYPERBOLIC EQUATIONS IN REGIONS WITH CHARACTERISTIC BOUNDARIES OR WITH CORNERS

Stanley Osher*

1. Introduction

In the last decade a general theory of hyperbolic mixed problems for first order systems has been developed. (See [9], [11], [29], [30], [32].) Moreover, a parallel theory has been partially worked out for the more complicated finite difference analogue ([6], [12], [14], [23], [24].) Unfortunately, the hypotheses required by this theory have been restrictive on two counts:

1) The boundary must be smooth.
2) The boundary must be non-characteristic.

We consider a general hyperbolic system of the form:

$$(1.1)\quad Lu = E(t,x)u_t - \sum_{j=1}^{m} A_j(t,x)\partial x_j u - C(t,x)u = F(t,x)$$

where F,u are complex valued k vectors and E, A_j, and C are uniformly bounded smoothly varying complex-valued $k\times k$ matrices defined for $(t,x) = (t,x_1,\ldots,x_n)$ in $(-\infty,\infty)\times\overline{\Omega}$, and they are constant for $|t| + |x|$ sufficiently large with $E > cI > 0$.

We require that the system is hyperbolic in the sense that the Cauchy problem is well posed in the L^2 norm. Usually strict hyperbolicity, symmetry, or some variant, is assumed.

* State University of New York at Stony Brook

At this point the hypotheses 1) and 2) are usually made. In particular 2) may be written:

2') $\det\left[\sum_{j=1}^{m} A_j(t,x)N_j(x)\right] \neq 0 \quad \text{for} \quad x \in \partial\Omega ,$

with $\overline{N} = (N_1,\ldots,N_m)$, the inward unit normal.

A simple calculation shows that the symmetric hyperbolic systems arising from Maxwell's equations violates 2'). In particular any real non-trivial linear combination of the 6×6 matrices A_j is always of rank four.

The linearized 3×3 systems arising from the shallow-water equations and the equations of gas dynamics in two dimensions have the property that some linear combination of the form 2') will be of rank two. Thus if the boundary of Ω is a smooth, simple, closed curve, then it must be characteristic at some points, but not everywhere.

This last situation presents special difficulties for the numerical analyst. If we wish to calculate solutions numerically over a bounded region of space, then artifical boundaries must be put in. The only way that the boundary can avoid being characteristic at some points is if it is non-smooth. Boundary instabilities were first noticed here by Elvius and Sundstrom [5].

We shall begin with a brief review of the theory in the smooth boundary non-characteristic case (section II). In section III we shall describe the results for non-characteristic corner problems and we give several examples and counter-examples. In section IV we describe the work done jointly with A. Majda concerning uniformly characteristic smooth boundary problems. In V we discuss the difficulties in the non-uniformly characteristic smooth boundary case, and also describe some work in progress with A. Majda on the shallow water equations. Finally, in VI we contrast these results with those of the analogous elliptic and parabolic boundary value problems and their numerical analogues.

The work we describe here will be mostly concerned with first order systems. A single higher order equation can be reduced to a system in a somewhat standard fashion.

2. Non Characteristic Smooth Boundaries

We assume for simplicity that the operator L is strictly hyperbolic, i.e. $E^{-1}(\sum_{j=1}^{m} A_j\omega_j)$ has k distinct real eigenvalues for any $\omega = (\omega_1,\ldots,\omega_m) \in R^m$.

In particular, we shall take $E \equiv I$, the identity matrix.

The mixed problem we will study is

(2.1)a) $Lu = F$ in $[0,T] \times \overline{\Omega}$

b) $u|_{t=0} = f$ in $\overline{\Omega}$ (initial condition)

c) $u^I - S(t,x)u^{II} = g$ for $x \in \partial\Omega$ (boundary condition).

Condition c) requires some explanation.

$$\sum_{j=1}^{m} A_j(t,x)N_j(x) = A^N(t,x) \tag{2.2}$$

defined above, in this case is non-singular, smoothly varying and with distinct real eigenvalues. Without loss of generality we may assume

$$A^N(t,x) = \begin{pmatrix} A^I & 0 \\ 0 & A^{II} \end{pmatrix} \text{ with } A^I < 0\ ,\ A^{II} > 0\ , \tag{2.3}$$

A^I is a diagonal $\ell\times\ell$ matrix, A^{II} is a diagonal $(k-\ell)\times(k-\ell)$ matrix.

We define

$$u^{I} = (u_1,\ldots,u_\ell)^T \ .$$

$$u^{II} = (u_{\ell+1},\ldots,u_k)^T \ .$$

Finally we assume that S is a smoothly varying rectangular matrix which becomes constant for $|t| + |x|$ large.

In order to check whether the problem is well posed, we proceed as follows:

For each $x \in \partial\Omega$ we make a local change of coordinates which maps the boundary into $x_1 = 0$ and Ω into $x_1 > 0$. We then freeze the resulting A_j and S coefficients to be constant at their values at this boundary point, throw away the lower order term C, and consider the new homogeneous constant coefficient problem:

(2.4)

a) $u_t - A_1 u_{x_1} - \sum_{j=2}^{m} A_j u_{x_j} = 0$ for $t, x_1 \geq 0, \{x_1,\ldots,x_m\} \in R^m$

b) $u^I - Su^{II} = 0$ at $x_1 = 0$.

By a growing exponential function is meant a function of the form

$$(2.5) \qquad \psi(t,x) = \sum_j P_j(x_1) e^{i(x'\zeta' + x_1\eta_j + t\tau)}$$

with $x', \zeta' \in R^{m-1}$, $\text{Im}\ \eta_j > 0$, $\text{Im}\ \tau < 0$, $V_j \in C^k$, and P_j are polynomials. For fixed t, such a function is bounded and decays exponentially as $x_1 \to \infty$ but it grows exponentially in time. If $\psi(t,x)$ satisfies (2.4), then so does $\psi_\lambda = \psi(\lambda t, \lambda x)$. As $\lambda \to \infty$, the values of ψ_λ and its derivatives at $t = 0$ grow like a polynomial in λ but the values at $t = T$ grow exponentially with λ. Thus there can be no estimate of the form

$$(2.6) \qquad |||\psi_\lambda(T,\cdot)||| \le C(\|L_0\psi_\lambda\|_{(0,T]\times\Omega} + \|\psi_\lambda(0,\cdot)\|)$$

where L_0 is the frozen operator defined in (2.4)a) and the norms are deliberately vague.

It is clear that such an exponential eigenfunction would be disastrous for stability of difference approximations to such problems.

It was shown by Hersh [7], after the early work of Agmon [1], that in order for the frozen mixed problem with smooth data to have a classical solution, it is necessary and sufficient that there be no non-trivial growing exponential solution for $|\tau|^2 + |\zeta'|^2 = 1$, $\operatorname{Im} \tau < 0$.

In order to check this, we may choose a basis for the ℓ-dimensional space of growing exponential functions for τ,ζ fixed, orthonormalized at $x_1 = 0$, and apply the boundary operator c). This will lead us to an $\ell\times\ell$ matrix $N(\tau,\zeta')$, and for a classical solution to exist, it is necessary and sufficient that $\det N(\tau,\zeta') \neq 0$.

However, it is possible that $\det N(\tau,\zeta') \neq 0$ for these values, but it approaches 0 as $\operatorname{Im} \tau \to 0$ for some τ_0,ζ_0' . Such a problem is said to have a generalized eigenvalue. If this determinant vanishes for $\operatorname{Im} \tau > 0$ it is said to have an eigenvalue.

We can now state the main theorem in this field (mostly due to Kreiss).

2.1) Theorem (Kajitani, Kreiss, Ralston, Rauch, Sakamoto).

If for each $x \in \partial\Omega, t$, the frozen mixed problem has no eigenvalues or generalized eigenvalues, then $\forall T > 0$ and for $F \in L^2([0,T]\times\Omega)$, $f \in L^2(\Omega)$, $g \in L^2([0,T]\times\partial\Omega)$ there is a unique strong solution of 2.1). In addition, the solution has the following properties:

$$(2.7)\quad \|u\|_{[0,T]\times\Omega} + \|u\|_{[0,T]\times\partial\Omega} + \|u(T)\|_{\Omega}$$

$$\leq e^{KT}C_K[\|F\|_{[0,T]\times\Omega} + \|g\|_{[0,T]\times\partial\Omega} + \|f\|_{\Omega}]$$

where the norms are L^2 norms and $K > K_0 > 0$ but is otherwise arbitrary.

In addition, Massey and Rauch [31], have shown that under suitable compatibility conditions at the space time corner the L^2 norms above may be replaced by H^s norms for positive integers s. This is impossible in general in both the characteristic and non-smooth boundary value case as we shall show below.

Many important physical problems such as the Neumann problem for the wave equation and various natural boundary value problems for the shallow water equations have generalized eigenvalues (but of course, they have no eigenvalues). Under such conditions, it is sometimes (but not always) possible to obtain existence, uniqueness, and an estimate for the problem (2.1) where $g \equiv 0$ and the term $\|u\|_{[0,T]\times\partial\Omega}$ is omitted from (2.7). See Kreiss [13], and Miyatake [21]. Differentiability results can then be obtained with no difficulty in these cases. Such generalized eigenvalues are called elliptic generalized eigenvalues. The algebraic condition for the ellipticity of a generalized eigenvalue is not yet obtained in the fullest generality, hence we omit it here. However, the maximal dissipative boundary value problems of Lax-Phillips in [18] are often in this category if their solutions do not satisfy (2.7).

The following simple example due to Kreiss [13] illustrates all the phenomena:

Suppose the frozen coefficient half-space problem is of the form:

$$(2.8)\qquad \begin{pmatrix} u_1 \\ u_2 \end{pmatrix}_t = \begin{pmatrix} -1 & 0 \\ 0 & 1 \end{pmatrix}\begin{pmatrix} u_1 \\ u_2 \end{pmatrix}_{x_1} + \begin{pmatrix} 0 & 1 \\ 1 & 0 \end{pmatrix}\begin{pmatrix} u_1 \\ u_2 \end{pmatrix}_{x_2}$$

with $u_1 = \alpha u_2 + f$ prescribed at $x_1 = 0$.

Then there exist no eigenvalues or generalized eigenvalues for $|\alpha| < 1$, there exists no eigenvalue for $|\alpha| = 1$ or $|\alpha| > 1$ and α real, and the generalized eigenvalue is elliptic if and only if $|\alpha| = 1$.

3. Corners

We shall begin by considering the hyperbolic problem.

$$(3.1)\text{a)}\qquad Lu = Eu_t - \sum_{j=1}^{m} A_j \partial_{x_j} u - C(x,t)u = F(t,x)$$

where the constant $k\times k$ matrices A_j and $E > 0$ are complex valued, $C(x,t)$ is as before. We wish to solve this in the region $x_1, x_2, t \geq 0$, $\{x_3, \ldots, x_m\} = x' \in R^{m-2}$. The boundary is non-characteristic i.e. $\det A_1 \neq 0 \neq \det A_2$. [An angle of less than 180° can be transformed to this 90° angle by a linear change of variables].

We again impose initial and boundary conditions of the form:

$$(3.1)\text{b)}\qquad u(x,0) = f(x) \qquad \text{(initial conditions)}$$

$$\text{c)}\quad u^{I} - Su^{II} = g(x_2, x', t) \text{ at } x_1 = 0 \text{ (boundary condition 1)}$$

$$u^{III} - Ru^{IV} = h(x_1, x', t) \text{ at } x_2 = 0 \text{ (boundary condition 2)}$$

where u^{I} and u^{II} span to the negative and positive eigenspaces of A_1, u^{III} and u^{IV} are defined analogously for A_2 and S and R are rectangular constant complex valued matrices.

It is clear from finite speed of propagation considerations that in order for this problem to be well posed it is necessary that both half space problems have no eigenvalues or generalized eigenvalues. (The latter is required if we desire an analogue of the estimate in (2.7).)

The half-space problems are of course defined to be (3.1)a), b) to be solved for $x_1, t \geq 0$ and all $\{x_2, x'\}$ with the first boundary condition in (3.1)c) and the analogous problem to be solved for $x_2, t \geq 0$ and all $\{x_1, x'\}$ using the second boundary condition.

It might be conjectured that this condition suffices for well-posedness of the corner problem. We present here four different examples which prove that this conjecture is in general false. The first three examples come from Osher [27], the last one comes from Sarason and Smoller [35]. They each violate well-posedness for different reasons.

Consider:

$$\text{(3.2)a)}\quad \begin{pmatrix} u_1 \\ u_2 \end{pmatrix}_t = \begin{pmatrix} -2 & 0 \\ 0 & 1 \end{pmatrix}\begin{pmatrix} u_1 \\ u_2 \end{pmatrix}_{x_1} + \begin{pmatrix} 1 & 0 \\ 0 & -2 \end{pmatrix}\begin{pmatrix} u_1 \\ u_2 \end{pmatrix}_{x_2}$$

with boundary conditions:

$$\text{c)}\qquad \begin{aligned} u_1 &= u_2 \quad \text{at} \quad x_1 = 0 \\ u_2 &= u_1 \quad \text{at} \quad x_2 = 0 . \end{aligned}$$

It is clear that the half space problems are well posed. Then the function:

$$u_1 = u_2 = f(t-x_1-x_2) \tag{3.3}$$

where $f(z)$ is an arbitrary smooth function which vanishes identically for $z \le 0$ but not for $z > 0$, is a solution to (3.2)a),c), with zero initial data. Thus we have non-uniqueness of solutions because a wave of the type (3.3) can pass through the corner and enter the region while identically satisfying the boundary conditions.

It is clear that if we use $f_\lambda(z) = e^{\lambda z}$ for $\lambda > 0$, then we can build a growing exponential solution which will violate (2.6). However, a first attempt at a numerical approximation to this ill-posed problem may lead to a stable difference scheme because the numerical analyst may unwittingly impose the condition $u_1(0,0,t) \equiv u_2(0,0,t) \equiv 0$ which will stabilize the scheme (and make the problem well-posed in some modified fashion).

Consider next:

(3.4)a) $$\begin{pmatrix} u_1 \\ u_2 \end{pmatrix}_t = \begin{pmatrix} -1 & 0 \\ 0 & 2 \end{pmatrix} \begin{pmatrix} u_1 \\ u_2 \end{pmatrix}_{x_1} + \begin{pmatrix} 2 & 0 \\ 0 & -1 \end{pmatrix} \begin{pmatrix} u_1 \\ u_2 \end{pmatrix}_{x_2}$$

with boundary conditions:

b) $u_1 = au_2$ at $x_1 = 0$ $\quad a \neq 0$

$u_2 = \frac{4}{a} u_1$ at $x_2 = 0$.

Then any vector function of the form:

$$(3.5)\qquad \begin{pmatrix} u_1 \\ u_2 \end{pmatrix} = \begin{pmatrix} \left(\frac{1}{2x_1+x_2}\right) f(t+x_1+x_2) \\ \frac{2}{a}\left(\frac{1}{(x_1+2x_2)}\right) f(t+x_1+x_2) \end{pmatrix}$$

with $f(z)$ smooth and having support in the region $0 < \varepsilon_1 < z < \varepsilon_2$ will satisfy the differential equation and boundary conditions and will have smooth initial data with compact support. However, as t increases the solution will blow up at the corner, hence the problem is not well posed in the L^2 norm.

This well-posedness fails, because, for certain initial data the solution will not exist. Moreover, we can see the influence of the boundary condition in terms of increasing energy along trapped rays (described below). Each reflection on a ray moving towards the corner increases the energy of the solution too much for it to stay in L^2 ultimately. However, weakening the norm in an as yet unknown manner involving a power of $r = \sqrt{x_1^2+x_2^2}$ as a weight, would probably lead us to a well posed problem.

Now we consider:

$$(3.5)a)\qquad \begin{pmatrix} u_1 \\ u_2 \end{pmatrix}_t = \begin{pmatrix} -1 & 0 \\ 0 & 1 \end{pmatrix} \begin{pmatrix} u_1 \\ u_2 \end{pmatrix}_{x_1} + \begin{pmatrix} 1 & 0 \\ 0 & -1 \end{pmatrix} \begin{pmatrix} u_1 \\ u_2 \end{pmatrix}_{x_2}$$

c) $u_1 = au_2$ at $x_1 = 0$, $u_2 = bu_1$ at $x_2 = 0$ with $|ab| > 1$.

Then any vector function of the form:

$$(3.6)\qquad \begin{pmatrix} u_1 \\ \\ u_2 \end{pmatrix} = \begin{pmatrix} (ab)^{\frac{t-x_1}{2(x_1+x_2)}} f(x_1+x_2) \\ \\ \sqrt{\frac{b}{a}}\,(ab)^{\frac{t-x_2}{2(x_1+x_2)}} f(x_1+x_2) \end{pmatrix},$$

with $f(z)$ defined as in the last example, will be a solution of the differential equations and boundary conditions and will have smooth initial data with compact support. As t increases, the initial data is multiplied by a factor $(ab)^{\frac{t}{2(x_1+x_2)}}$ and if the support of the initial data approaches the corner, this factor becomes exponentially unbounded. Thus no reasonable weakening of the norm at the corner, and no additional boundary condition at the corner, can possibly lead to a well posed problem here.

This last example is an illustration of what can happen if a periodic trapped ray increases in energy as it reflects off the boundaries.

Sarason and Smoller [35] have pointed out that if we considered a 2×2 strictly hyperbolic system, then half space well posedness $\Leftrightarrow$ corner well posedness. They prove this using a result of Strang [36] concerning 2×2 hyperbolic systems. (We presented an example of a 2×2 strictly hyperbolic system with non-local, but pseudodifferential, boundary conditions which was half space well-posed but not corner well-posed in [22].)

They also discuss the question of when periodic trapped rays can destroy well-posedness of corner problems. In particular, they construct a strictly hyperbolic 4×4 system having the desired properties.

By a trapped ray is meant the following. First approximate solutions of $Lu = 0$ are constructed by the methods of geometric optics and they are chosen so that the leading term is a sum of plane waves. The solution of the transport equations involves differentiating along directions normal to the plane wave which is taken in this case to be transversal to the boundary. When such a ray hits a boundary, it is reflected according to geometrical optics, but the amplitude changes according to the boundary conditions. Given this setup, Sarason and Smoller construct such a non-symmetric system for which the geometric optics approximation is ill-posed, hence so is the original problem.

It should be pointed out that other things can cause ill-posedness besides trapped rays, e.g. our example (3.2) above has none.

Before describing the (complicated) condition which is equivalent to well-posedness, and the main positive theorem in this field, we give one more simple example to show that even in the most well-posed of corner problems, with given C^∞ initial data supported away from the corner, that derivatives of the solution will in general blow up at the corner for later time. We do have a positive differentiability result discussed below.

The vector function

$$(3.7)\qquad \begin{pmatrix} u_1 \\ u_2 \end{pmatrix} = \begin{pmatrix} (2x_1+x_2)^{1/4} \\ 2^{-1/2}(x_1+2x_2)^{1/4} \end{pmatrix} f(t+x_1+x_2)$$

has the desired properties, i.e. for f chosen as in (3.4) it obeys (3.4)a) with boundary conditions $u_1 = 2^{1/4}u_2$ at $x_1 = 0$, $u_2 = 2^{-3/4}u_1$ at $x_2 = 0$ (which gives a well posed problem from integration by parts) but differentiability will indeed fail at the corner as time progresses.

In order to check for well-posedness of the corner problems (3.1), given that both half-space problems have no eigenvalues or generalized eigenvalues, we proceed as follows.

We assume that $f \equiv 0 \equiv h \equiv C \equiv F$ and proceed to "solve" this problem by applying a Fourier transform in the variables $\{x_3,\cdots,x_m\} = x'$, letting the dual variable be $\{\omega_3,\cdots,\omega_m\} = \omega$, and multiplying the equation (3.1)a) by $e^{-\eta t}$ for $\eta > 0$, and Fourier transforming in time, letting the dual variable (unfortunately) be ξ. Let $s = \eta + i\xi$ with $\eta > 0$.

The resulting partial differential equation in two variables is

$$(3.8)\text{a)} \qquad A_1 u_{x_1} + A_2 u_{x_2} + [-sE + iB(\omega)]u = 0$$

to be solved for $x_1, x_2 \geq 0$ with the induced boundary conditions at $x_1 = 0$ and $x_2 = 0$ and defining

$$iB(\omega) = (i\sum_{j=3}^{m} A_j\omega_j)$$

Suppose we define a function $V(x_1,x_2)$ whose domain is the half space $x_2 \geq 0$, which equals u in the quarter plane, and is identically zero for $x_1 < 0$. We then Fourier transform in x_1 and call the dual variable ω_1.

$$(3.9) \qquad A_2 V_{x_2} + [A_1 i\omega_1 - Es + iB(\omega)]V = A_1\, u(0,x_2)$$

where $u(0,x_2)$ belongs to the space defined by $u^I = Su^{II} + g$.

Thus we may solve this half space problem uniquely for V using boundary condition 2 in (3.1)c) and view the solution V as depending linearly on $u^{II}(0,x_2)$ which is to be determined.

We then restrict the resulting solution $V(x_1,x_2)$ to the half line $x_2 = 0$, $x_1 \geq 0$. Next we define $W(x_1,x_2)$ in the half space $x_1 \geq 0$ to equal V in the quarter plane and to be zero for negative x_2 . We solve the resulting Fourier transformed ordinary differential equation for W

$$A_1 W_{x_1} + [A_2 i\omega_2 - Es + iB(\omega)]V = A_2 V(x_1,0) \tag{3.10}$$

where W satisfies boundary condition 1 in (3.1)c).

Finally we restrict the resulting $W(x_1,x_2)$ to the half line $x_1 = 0$, $x_2 \geq 0$. The resulting function satisfies the boundary conditions $W^I(0,x_2) = SW^{II}(0,x_2) + g$.

Moreover we have

$$W^{II}(0,x_2) = T_{\omega,s}\, u^{II}(o,x_2) + R_{\omega,s}\, g \tag{3.11}$$

where $T_{\omega,s}$ and $R_{\omega,s}$ are the induced linear operators which we shall discuss below.

If a solution exists to this problem we must have:

$$(I - T_{\omega,s})\, u^{II}(0,x_2) = (I + R_{\omega,s})\, g \, . \tag{3.12}$$

$T_{\omega,s}$ is easily seen to be densely defined in $L_2[0,\infty)$ for any ω,s .

The condition that the corner problem be well-posed is thus the following:

Corner Condition

The operator $(I - T_{\omega,s})$ whould have an inverse which is a bounded map from $L_2[0,\infty)$ into itself for ω real, $\operatorname{Re} s > 0$, for all ω,s with $|\omega| + |s| = 1$. Moreover, the norm $\|(I-T_{\omega,s})^{-1}\| \le C$, with C independent of ω and s .

Unfortunately for the theory, $T_{\omega,s}$ is neither compact, nor local, nor pseudo-local, in general. We pause here for an example [27].

Given the system

$$\text{(3.13) a)} \quad \begin{pmatrix} u_1 \\ u_2 \end{pmatrix}_t = \begin{pmatrix} -c_1 & 0 \\ 0 & c_2 \end{pmatrix} \begin{pmatrix} u_1 \\ u_2 \end{pmatrix}_{x_1} + \begin{pmatrix} d_1 & 0 \\ 0 & -d_2 \end{pmatrix} \begin{pmatrix} u_1 \\ u_2 \end{pmatrix}_{x_2}$$

with boundary conditions

$$\text{c)} \quad u_1(0,x_2,t) = au_2(0,x_2,t) + g(x_2,t)$$

$$u_2(x_1,0,t) = bu_1(x_1,0,t) + h(x_1,t)$$

with $c_1, c_2, d_1, d_2 > 0$.

T_s (there is no ω) is defined by

$$\text{(3.16)} \quad T_s u = ab \exp\{-sx_2[\frac{1}{c_1 d_2} + \frac{c_2}{c_1^2 d_2}]\}\ u(\frac{c_2 d_1}{c_1 d_2}x_2) .$$

Thus we can show uniform invertibility if and only if $|ab| < \sqrt{\frac{c_2 d_1}{c_1 d_2}}$, and our counter-examples above come from analyzing this operator when it fails to be unifromly invertible.

We can now state our main Theorem which is an improvement of our earlier result [26] made possible by some recent results of Sarason [33] and Beals [3] (but still only valid here for constant coefficients). We shall prove it in detail in a forthcoming paper.

(3.1) <u>Theorem</u>. If L has constant coefficient principal part, is a) strictly hyperbolic, or b) symmetric, obeying a certain additional technical hypothesis (assumption 2.3 of [26]) and c) every real eigenvalue of $A_1^{-1}A_2$ is negative, and d) each half space problem has no eigenvalues or generalized eigenvalues, then the corner problem (3.1) is well-posed in the sense defined below if and only if the corner condition is valid.

The corner problem is said to be well posed if for $\forall T > 0$ and for $F \in L^2([0,T]\times R_+^2\times R^{m-2})$, $f \in L^2[R_+^2\times R^{m-2}]$, $g \in L^2([0,T]\times R_+^1\times R^{m-1})$, $h \in L^2([0,T]\times R_+^1\times R^{m-1})$ there is a unique strong solution of (3.1). In addition, the solution has the property enjoyed by the analogous smooth-boundary solution in equation (2.7) with obvious modifications.

We also have a modified differentiability theorem: Define the operator $D = (r\partial_r, \partial_\theta, \partial_t, \partial_{x_3}, \cdots, \partial_{x_m})$, where $r = \sqrt{x_1^2+x_2^2}$, $\theta = \tan^{-1}\left(\frac{x_2}{x_1}\right)$. Then we have

(3.2) <u>Theorem</u>. Under the previous hypothesis, the solution obeys the estimate as above with L^2 norms replaced by H^s norms with respect to the above differentiation (where the convention applies that only internal derivatives appear in any norm which involve an inhomogeneous function .

The main job now seems to be to make the corner condition tractable. Sarason [33] has shown that if $m = 2$, i.e. no tangential variables, that T_s is a continuous map of $L_2[0,\infty)$ into $L_2[0,\infty)$ and depends continuously on the boundary matrix. Thus, if a problem is known to be well-posed in this case then, for some nearby open set of boundary conditions, it is still well posed.

For well posedness of problems with variable coefficients, we conjecture that well-posedness of frozen problems at the corner should suffice. We constructed a positivizer with this end in mind in [26], however, the perennial problem of "grazing frequencies" destroys its smoothness in general.

For additional counterexamples to well posedness involving the shallow water equations and the wave equation, see [27]. All these examples, however, have non-elliptic generalized eigenvalues in at least one of their half space problems.

For an example of a closed form solution to a physically oriented multidimensional corner problem, see Kupka and Osher [16].

IV. Uniformly Characteristic Smooth Boundaries

The results recently obtained jointly with Andrew Majda [9] are more satisfying here than in the corner case in that they work for variable coefficients, and they require us only to check for growing exponential solutions. However, there are some interesting pathologies involving both non-symmetric systems, and non-differentiability of solutions of well-posed problems.

We study problems of the form:

(4.1)a) $$Lu = E(t,x)u_t + \sum_{j=1}^{m} A_j(t,x)\partial_j u - C(t,x)u \quad \text{in } [0,T]X\Omega .$$

b) $$u(0,x) = f(x) \quad \text{in } \Omega .$$

c) $$u^{I}(t,x) - S(t,x)u^{II}(t,x) = g(t,x) \quad \text{in } [0,T]X\partial\Omega .$$

This time, however, we must assume that the A_j are symmetric. It is an interesting fact that this hypothesis is necessary in the following sense: We shall present an example of a strictly hyperbolic, but non-symmetric system, which has no eigenvalues or generalized eigenvalues in the sense made precise below, but for which the basic theorem below is false.

The last condition will be explained below.

We consider characteristic boundary value problems which satisfy the constant rank hypothesis,

$$\text{rank}\left(\sum_{j=1}^{m} A_j(t,x)N_j(x)\right) = \text{constant} < k \tag{4.2}$$

for x in a neighborhood of $\partial\Omega$. This means that if we draw a straight line perpendicular to $[0,\infty)X\partial\Omega$ through (t_0,x_0) and into Ω the rank of the matrix $\sum_{j=1}^{m} A_j(t,x)N_j(x_0)$ is constant for (t,x) on this line and close to (t_0,x_0). Without loss of generality, we may assume the matrix $A^N(t,x)$ defined above is of the form

$$A^N(t,x) = \begin{pmatrix} 0 & 0 & 0 \\ 0 & A^I & 0 \\ 0 & 0 & A^{II} \end{pmatrix} \text{ with } A^I < 0 < A^{II} \tag{4.3}$$

The dimension of the null space is ℓ_0, the rank of A^I is ℓ_1, and the rank of A^{II} is ℓ_2, with $k = \ell_0 + \ell_1 + \ell_2$. We denote

$$u^0 = [u_1,\cdots,u_{\ell_0}]^T ,$$

$$u^I = [u_{\ell_0+1},\cdots,u_{\ell_0+\ell_1}]^T ,$$

$$u^{II} = [u_{\ell_0+\ell_1+1},\cdots,u_k]^T .$$

Then the boundary conditions are written in the form (4.1)c). It is easy to construct counter-examples to our main theorem in the case when the boundary conditions involve u^0. This was done in [19].

We make various technical hypotheses involving the symbol of L. These are valid for all reasonable problems, e.g. Maxwell's equations, equations of magneto-hydrodynamics.

Next we perform the by now familiar ritual of examing the associated frozen coefficient half-space problems as in (2.4). We again check for growing exponentials and the space of such functions for fixed τ, ζ' is again ℓ_1 dimensional. This time, at $x_1 = 0$, we orthonormalize this basis with respect to the <u>last</u> $\ell_1+\ell_2$ components, apply the boundary operator, and proceed as in section II. We may now define eigenvalues, generalized eigenvalues, and elliptic generalized eigenvalues, as in that section.

(4.1) <u>Theorem</u>. If, for each $x \in \partial\Omega, t$, the frozen mixed problem has no eigenvalues or generalized eigenvalues, then $\forall T > 0$, and for $F \in L^2([0,T]X\Omega)$, $f \in L^2(\Omega)$, $g \in L^2([0,T]X\partial\Omega)$, there is a unique strong solution of (4.1). In addition, the solution has the following property:

$$\begin{aligned} &\|u\|_{[0,T]X\Omega} + \|v\|_{[0,T]X\partial\Omega} + \|u(T)\|_{\Omega} \\ &\le C_k e^{kT}[\|F\|_{[0,T]X\Omega} + \|g\|_{[0,T]X\partial\Omega} + \|f\|_{\Omega}] \end{aligned} \tag{4.4}$$

where the norms are L^2 norms and $k > k_0 \ge 0$, but is otherwise arbitrary.

An important fact about the above estimate is that the quantity which appears in the above estimate on the boundary is $v = [u_{\ell_0+1},\cdots,u_k]^T$, i.e. just those components of u which are not annihilated by the boundary matrix.

We have in [19] applied these results to Maxwell's equations defined by

$$\frac{\partial}{\partial t}\begin{bmatrix} E \\ H \end{bmatrix} = \begin{bmatrix} 0 & \text{curl} \\ -\text{curl} & 0 \end{bmatrix}\begin{bmatrix} E \\ H \end{bmatrix} + F, \tag{4.5}$$

E and H are 3 vectors. We then show that all local boundary conditions for Maxwell's equations may be expressed invariantly as

$$(4.6)\qquad \Pi_+ \begin{bmatrix} E \\ H \end{bmatrix} = S(x,t)\Pi_- \begin{bmatrix} E \\ H \end{bmatrix}, \quad (x,t) \in \partial\Omega X[0,T]$$

where S is any smoothly varying 6×6 matrix of rank 2 satisfying

$$(4.7)\qquad \begin{aligned} \Pi_- S \Pi_- &= 0 \\ (\bar{N},0)\cdot S\Pi_- &= 0 \\ (0,\bar{N})\cdot S\Pi_- &= 0 \end{aligned}$$

with

$$\Pi_+ \begin{bmatrix} E \\ H \end{bmatrix} = \begin{bmatrix} E-(E\cdot\bar{N})\bar{N} - (\bar{N}\times H) \\ H-(H\cdot\bar{N})\bar{N} + (\bar{N}\times E) \end{bmatrix}$$

and

$$\Pi_- \begin{bmatrix} E \\ H \end{bmatrix} = \begin{bmatrix} E-(E\cdot\bar{N})\bar{N} + (\bar{N}\times H) \\ H-(H\cdot\bar{N})\bar{N} - (\bar{N}\times E) \end{bmatrix}$$

The basic result is true if and only if the matrix S has spectral radius (not necessarily norm) less than 1. Moreover under these circumstances, higher energy estimates are also valid. There is also an energy conserving class where $\|S\| = 1$, e.g. the perfectly reflecting conductor. It turns out that these problems have generalized eigenvalues of the elliptic type and the basic estimate above (4.4) is true when $g \equiv 0$ and the boundary term $\|v\|$ does not appear on the left.

We present here a simple example to show that non-symmetric, but strictly hyperbolic, systems have a certain loss of derivative property in general.

We take

$$(4.8)\quad \begin{pmatrix} u_0 \\ u_1 \\ u_2 \end{pmatrix}_t = \begin{pmatrix} 0 & 0 & 0 \\ 0 & -1 & 0 \\ 0 & 0 & 1 \end{pmatrix} \begin{pmatrix} u_0 \\ u_1 \\ u_2 \end{pmatrix}_{x_1} + \begin{pmatrix} 0 & 1 & 0 \\ 0 & 0 & 1 \\ 0 & 1 & 0 \end{pmatrix} \begin{pmatrix} u_0 \\ u_1 \\ u_2 \end{pmatrix}_{x_2}$$

for $t, x_1 \geq 0$, $-\infty < x_2 < \infty$ with $u_1 = au_2 + g$ at $x_1 = 0$ for $|a| < 1$ and $u(0,x) = 0$ initially.

We apply a Fourier-Laplace transform as usual and arrive at

$$(4.9)\quad \hat{u}_0(x,s,\omega) = \frac{i\omega}{s}\,\hat{g}(s,\omega)\,\frac{(s+\sqrt{s^2+\omega^2})e^{-\sqrt{s^2+\omega^2}\,x_1}}{s+\sqrt{s^2+\omega^2}-ai\omega} ;$$

hence

$$\eta \int_0^\infty (\hat{u}_o(x,s,\omega))^2 dx_1$$

$$= \eta\left|\frac{\omega}{s}\right|^2 |\hat{g}(s,\omega)|^2 \frac{\left|s+\sqrt{s^2+\omega^2}\right|^2}{2\mathrm{Re}\sqrt{s^2+\omega^2}\left|s+\sqrt{s^2+\omega^2} - ai\omega\right|^2} ,$$

we merely let $s = \eta$ fixed and take ω/η large. The right hand side is greater than a constant times $(\frac{\omega}{\eta})\cdot(\hat{g}(s,\omega))^2$, thus no estimate of the type (4.4) is possible.

We also give here an example to show that the "best" differentiability results, i.e. those which replace L^2 norms by H^s norms in (4.4), are false in general in this case. However, if the inhomogeneous data is C_0^∞ and satisfies reasonable compatibility conditions, then the solution is indeed C^∞ . We have a precise differentiability theorem in [19]. Suffice to say that the example we give here is the "worst" that can happen, and that the physical examples which are well posed do have the best differentiability properties.

Consider

$$(4.10)\qquad \begin{bmatrix} u_0 \\ u_1 \end{bmatrix}_t = \begin{bmatrix} 0 & 0 \\ 0 & -1 \end{bmatrix} \begin{bmatrix} u_0 \\ u_1 \end{bmatrix}_{x_1} + \begin{bmatrix} 0 & -1 \\ -1 & 0 \end{bmatrix} \begin{bmatrix} u_0 \\ u_1 \end{bmatrix}_{x_2}$$

to be solved in the usual region with $u_1 = g$ at $x_1 = 0$ and zero initial data. A simple calculation on the Fourier-Laplace side leads to:

$$(4.11)\qquad \begin{bmatrix} \hat{u}_0 \\ \hat{u}_1 \end{bmatrix} = e^{-(s + \frac{w^2}{s})x_1} \hat{g} \begin{bmatrix} \frac{iw}{s} \\ 1 \end{bmatrix} ,$$

thus

$$\frac{\partial^k \hat{u}}{\partial x_1^k} = -(s + \frac{w^2}{s})^k \hat{u}$$

and it is easy to see that we need $2k$ derivatives on the inhomogeneous data to estimate k derivatives of the solution. It is easy to modify this example to the case where the boundary data is zero and the initial data is not.

V. Non-Uniformly Characteristic Smooth Boundaries

We are concerned here with the system in (2.1)a),b) where the boundary matrix $A^N(t,x)$ changes rank at isolated parts on $\partial\Omega$. To our knowledge, the results in this field are extremely fragmented. However, it is clear that it will not be sufficient to analyze the resulting frozen coefficient problems as above. Probably a subelliptic condition of the type invented by Hormander in [37] and used in many places, e.g. [20] in reflection of singularities, arises here.

We present here an example done with Bjorn Engquist which modifies our corner example (3.5).

Consider

$$\begin{pmatrix} u_1 \\ u_2 \end{pmatrix}_t = \begin{pmatrix} -1 & 0 \\ 0 & 1 \end{pmatrix} \begin{pmatrix} u_1 \\ u_2 \end{pmatrix}_{x_1} \tag{5.1)a}$$

to be solved in the region $t \geq 0$, $x_2 \geq x_1^2$.

Thus the boundary matrix is

$$A_N(x,t) = \frac{2x_1}{\sqrt{1+4x_1^2}} \begin{pmatrix} 1 & 0 \\ 0 & -1 \end{pmatrix}, \tag{5.2}$$

which is singular only at $x_1 = 0$ and has rank zero there. Thus the boundary conditions should be of the form:

$$\begin{aligned} u_1 &= a(x_1,t)u_2 \quad \text{for} \quad x_1 < 0 \\ u_2 &= b(x_1,t)u_1 \quad \text{for} \quad x_1 > 0 \end{aligned} \tag{5.1)b}$$

All the frozen problems are well posed for $(x_1,x_2,t) \neq (0,0,t)$. We take a,b constant, $|ab| > 1$, and notice that

$$\begin{pmatrix} u_1 \\ u_2 \end{pmatrix} = \begin{pmatrix} (ab)^{\frac{t-x_1}{4\sqrt{x_2}}} \\ \left(\frac{b}{a}\right)^{\frac{1}{2}}(ab)^{\frac{t+x_1}{4\sqrt{x_2}}} \end{pmatrix} f(x_2) \tag{5.3}$$

where $f(z)$ is defined as in section III, has all the appropriate properties for a counter-example to the supposition that if the frozen problem is well posed at all non-characteristic points, then the full problem is well posed.

[A trapped periodic ray does us in in this case.]

Moreover, integration by parts shows us that the problem is well posed if $|ab| < 1$.

A method of attack for this type of problem is the following: By using the map: $x = x_1$, $y = x_2 - x_1^2$, the equation transforms to (after a Laplace transform in time, assuming homogeneous initial date)

$$(5.4) \qquad s\begin{pmatrix} u \\ v \end{pmatrix} = \begin{pmatrix} -1 & 0 \\ -0 & 1 \end{pmatrix}\begin{pmatrix} u \\ v \end{pmatrix}_x - 2x\begin{pmatrix} -1 & 0 \\ 0 & 1 \end{pmatrix}\begin{pmatrix} u \\ v \end{pmatrix}_y ,$$

to be solved for boundary conditions

$$-\infty < x < \infty , \quad 0 \le y , \text{ with}$$

$$u = av + \Phi \quad \text{if} \quad x \le 0 , \; y = 0$$

$$v = bu \quad \text{if} \quad x > 0 , \; y = 0 .$$

The general solution to (5.4) is

$$\begin{pmatrix} u \\ v \end{pmatrix} = \begin{pmatrix} e^{-sx}h(x^2+y) \\ e^{sx}k(x^2+y) \end{pmatrix} .$$

Plugging in the boundary conditions, we see

$$e^{-sx}h(x^2) = ae^{sx}k(x^2) + \Phi(x) \qquad \text{if} \quad x < 0$$

$$e^{sx}k(x^2) = be^{-sx}h(x^2) \qquad \text{if} \quad x > 0$$

or

$$h(x^2) = ae^{-2sx}k(x^2) + \Phi(-x)$$

$$= abe^{-4sx}h(x^2) = \Phi(-x) , \text{ for } \quad x > 0 .$$

We can invert this smoothly for h if and only if $|ab| < 1$.

We are presently working with A. Majda on such initial

boundary value problems for the shallow water equation.

These are:

$$(5.5)\qquad u_t = \begin{pmatrix} a & 0 & c \\ 0 & a & 0 \\ c & 0 & a \end{pmatrix} u_{x_1} + \begin{pmatrix} b & 0 & 0 \\ 0 & b & c \\ 0 & c & b \end{pmatrix} u_{x_2} ,$$

where a, b, c are smooth functions with $0 < a^2 + b^2 < c^2$. A curve of the form $(x(s),y(s))$ will be characteristic if $(x'(s),y'(s)) = k(s)(a(x(s),y(s),t),\ b(x(s),y(s),t)),\ k(s) \neq 0$. We note that by mixing space and time and using a transformation of Elvius [4], we may change coordinates in a smooth fashion to map this equation into

$$V_t = if(x,y,t)\ \text{curl}\ V + CV$$

where C is pseudo differential of order 0 , f is a smooth function, and C and ∇f vanish if a, b, c are constant. This was carried out in [19].

For the non-uniformly characteristic boundary value problem we are using techniques described in (5.4), those of geometric optics, and reducing the problem to the boundary as in Hormander [37]. We expect a "clean" result to arise here because the characteristics of this system are those of the wave equation and a single scalar hyperbolic factor.

VI. The Elliptic, Parabolic and Numerical Analogues

We first consider an elliptic first order system in a region with smooth compact boundary. Well posedness of the problem is verified as in section II except that it is now simple - there is no time variable. We check for bounded exponential functions in a region $x_1 \geq 0$, $(x_2,\cdots,x_m) \in R^{m-1}$ of the type (2.5) with t independence. It is necessary to check for these functions only for ζ' real, with $|\zeta'| = 1$. Moreover, if the problem obeys this Lopatinskii condition, then an elliptic gain of derivative up to the boundary in all L_p norms $1 \leq p \leq \infty$ is valid. (We have left out some technical restrictions on the symbol in the two dimensional case.) This fundamental work is due to Agmon-Douglis-Nirenberg [2]. The discrete version of this in a half space has been done recently by C. Johnson [8].

Next we consider an elliptic system in a region whose compact boundary has a finite number of conical points. (This includes "wedge" type problems in two dimensions.) At all the other points of the boundary, we first verify that the problem obeys the condition above. If this is valid, then the problem is well-posed and has a gain of derivative even up to the conical points, with respect to a weighted L^2 norm, involving powers of r-distance to the concical point- and $r\partial_r$ and $\partial\theta_j$, $j = 1,\cdots,m-1$ are the derivatives in question. This is due to Kondrat'ev [10]. The discrete version in a rectangle has been recently done by us [28].

The multi-dimensional case is more complicated. Even nothing worse than translation invariance causes trouble. Our last example shows this.

Consider the question

$$(6.1) \quad \sum_{i=1}^{3} u_{x_i x_i} = 0 \quad \text{to be solved for} \quad 0 \leq x_1, x_2, \ -\infty < x_3 < \infty$$

with boundary conditions:

$$u_{x_1} = u_{x_2} \quad \text{at} \quad x_1 = 0 \ , \text{ and } \ x_2 = 0 \ .$$

If there were no x_3 dependence, then Kondrat'ev's work assures us of well-posedness in this modified sense. However we can get an infinite collection of linearly independent eigenfunctions of the type

$$u_\lambda = e^{-\lambda(x_1+x_2+i\sqrt{2}x_3)}$$

for any $\lambda \to +\infty$.

Thus something more than the check for bounded exponential functions, as in the 2 dimensional case, is needed.

The parabolic smooth boundary value problem has been completely studied by Ladyzenskaya [17]. A condition on growing exponential solutions of the type in section II is again the right one although it is modified in a simple fashion which puts it, as expected, "between" the elliptic and hyperbolic cases.

The one dimensional discrete analogue was done by us in [25]. Only hard work and perserverance is needed to generalize it to a multi-dimensional half space. Earlier work on it was done by Varah [38].

REFERENCES

1. Agmon, S., Problèmes mixtes pour les equations hyperbolique d'order supérior Colloques internationeux du Centre National de la Recherche Scientifique, 117, Paris (1962).
2. Agmon, S., Douglis, A., Nirenberg, L., Estimates near the boundary for solutions of elliptic partial differential equations satisfying general boundary conditions, Comm. Pure. Appl. Math. 12 (1959) 623-727.
3. Beals, R., Hyperbolic equations and systems with multiple characteristics, Arch. Rat. Mech. and Anal., 48 (1972) 123-152.
4. Elvius, T., Thesis, Department of Computer Science, University of Uppsala, Uppsala, Sweden.
5. Elvius, T., and Sundstrom, A., Computationally efficient schemes and boundary conditions for a fine mesh barotropic model based on shallow water equations, Tellus, 25 (1973) 132-156.
6. Gustavsson, B., Kreiss, H. O., and Sundstrom, A., Stability theory of Difference Approximations for Mixed initial boundary Value Problems, II., Math. Comp. 26 (1972) 649-686.
7. Hersh, R., Mixed problems in several variables, J. Math. Mech. 12 (1963) 317-334.

8. Johnson, C. G. L., Estimates near plane portions of the boundary for discrete elliptic problems, Math. Comp. 28 (1974) 909-935.
9. Kajitani, K., First order hyperbolic mixed problems, J. Math. Kyoto U., 11 (1971) 449-484.
10. Kondrat'ev, V. A., Boundary value problems for elliptic equations with conical or angular points, Trans. Moscow Math. Soc. V. 16 (1967) translated by Amer. Math. Soc. (1968).
11. Kreiss, H. O., Initial boundary value problems for hyperbolic systems, Comm. Pure Appl. Math. 23 (1970) 277-298.
12. Kreiss, H. O., Stability theory for different approximation of mixed initial boundary value problems, I., Math. Comp. 22 (1968) 1-12.
13. Kreiss, H. O., Generalized eigenvalues for mixed boundary value problems (to appear).
14. Kreiss, H. O., Difference approximations for initial boundary value problems, Proc. Ray. Soc. London, A, 323 (1971) 255-261.
15. Kreiss, H. O., and Oliger, J., Methods for the approximate solution of time dependent problems, Global Atmosphere Research Programme (GARP) Publication series #10 (1973).
16. Kupka, I. A. K., and Osher, S., On the wave equation in a multi-dimensional corner, Comm. Pure Appl. Math., 24 (1971) 381-393.
17. Ladyzenskaya, O. A., Boundary value problems of mathematical physics, III, Proc. Steklov Inst. of Math. #83 (1965).
18. Lax, P. D., and Phillips, R. S., Local boundary conditions for dissipative symmetric linear differential operators, Comm. Pure and Appl. Math., 13 (1960) 427-455.
19. Majda, A., and Osher, S., Initial Boundary Value Problems for Hyperbolic Equations with Uniformly Characteristic Boundary, Comm. Pure and Appl. Math (to appear).
20. Majda, A., and Osher, S., Reflections of singularities at the boundary, Comm. Pure and Appl. Math (to appear).
21. Miyatake, S., Mixed problem for hyperbolic equations of second order, J. Math., Kyoto U., 13 (1973) 455-487.
22. Osher S., An ill posed problem for a strictly hyperbolic equation in two unknowns near a corner, A.M.S. Bull., 80 (1974) 705-708.
23. Osher, S., Systems of difference equations with general homogeneous boundary conditions, Trans. Amer. Math. Soc., 137 (1969) 177-201.
24. Osher, S., Stability of difference approximations of dissipative type for mixed initial boundary value problems, I., Math. Comp., 23 (1969) 335-340.

25. Osher, S., Stability of parabolic difference approximations to certain mixed initial boundary value problems, Math. Comp. 26 (1972) 13-39.
26. Osher, S., Initial boundary value problems for hyperbolic equations in regions with corners, I, Trans. Amer. Math. Soc. 176 (1973) 141-164.
27. Osher, S., Initial boundary value problems for hyperbolic equations in regions with corners, II, Trans. Amer. Math. Soc. 198 (1974) 155-175.
28. Osher, S., Elliptic difference equations in rectangles (in preparation).
29. Ralston, J., Note on a paper of Kreiss, Comm. Pure Appl. Math., 74 (1971) 759-762.
30. Rauch, J., L_2 is a continuable initial condition for Kreiss' mixed problems, Comm. Pure Appl. Math., 15 (1972) 265-285.
31. Rauch, J., and Massey, F. J., Differentiability of solutions to hyperbolic initial boundary value problems, Trans. Amer. Math. Soc., 189 (1974) 303-318.
32. Sakamoto, R., Mixed problems for hyperbolic equations, I and II, J. Math., Kyoto Univ. 10 (1970) 349-373, 403-417.
33. Sarason, L., Hyperbolic and other symmetrizable systems in regions with corners and edges (to appear).
34. Sarason, L., On weak and strong solutions of boundary value problems, Comm. Pure Appl. Math., 15 (1962) 237-288.
35. Sarason, L., and Smoller, J. A., Geometrical optics and the corner problem, Arch. Rat. Mech. and Anal., 56 (1975) 34-69.
36. Strang, G., Hyperbolic initial boundary value problems in two unknowns, J. Diff. Eq., 6 (1969) 161-171.
37. Hormander, L., Pseudo differential operators and non-elliptic boundary value problems, Ann. of Math., 83 (1966) 129-209.
38. Varah, J., Maximum norm stability of difference approximations to the mixed initial boundary value problem for the heat equation, Math. Comp., 24 (1970) 31-44.

NUMERICAL BIFURCATION AND SECONDARY BIFURCATION: A CASE HISTORY*

Louis Bauer and Edward L. Reiss**

1. Introduction

Secondary transitions or secondary bifurcations occur frequently in many nonlinear stability problems. In hydrodynamic stability, secondary bifurcations are called secondary transitions and in elastic stability they are called secondary bucklings. Secondary bifurcation phenomena were known at least since the time of Poincaré. In fact, it was Poincaré who invented the term secondary bifurcation in his classical studies of the equilibrium states of rotating and self-gravitating liquid masses [1]. In this paper we describe some recent progress that has been made in the theory and numerical computation of secondary bifurcation by considering two problems for the nonlinear buckling of elastic plates.

2. The Buckled Circular Plate

The edge of a circular elastic plate is clamped and compressed by a uniform, radial edge thrust. For sufficiently small thrusts, the plate deforms by uniformly contracting in its plane to a circle of smaller radius. This equilibrium

* This research was supported by the National Science Foundation under Grant GP-27223 and the Office of Naval Research under Contract N00014-67-A-0467-0006. The computations were performed at the ERDA Computing and Applied Mathematics Center of the Courant Institute of Mathematical Sciences.

** Courant Institute of Mathematical Sciences, New York University.

state is called the unbuckled state. When the thrust exceeds a critical value, which is called the buckling load, the unbuckled state is unstable. Then the plate deforms by deflecting out of its plane.

We assume that the von Kármán plate theory [2] adequately describes the nonlinear deformations of the plate. In dimensionless variables this theory is equivalent to the boundary value problem,

$$\left.\begin{aligned} &(2.1a)\quad \Delta^2 f = -1/2[w,w] \\ &(2.1b)\quad \Delta^2 w + \lambda w = [f,w] \end{aligned}\right\} \text{for } 0 \le x < 1 \ , \ \text{and } 0 \le \theta \le 2\pi \ ,$$

$$(2.1c)\quad w(1,\theta) = w_x(1,\theta) = f(1,\theta) = f_x(1,\theta) = 0 \ ,$$

for the dimensionless lateral deflection $w(x,\theta)$ of the plate, and a dimensionless excess stress function $f(x,\theta)$. In (2.1), x is the dimensionless radial variable, $x = 1$ is the edge of the plate, θ is the polar angle, λ is a parameter that is proportional to the edge thrust ($\lambda > 0$ corresponds to compressive thrusts), Δ is the Laplacian in polar coordinates, and the nonlinear operator $[f,w]$ is defined by

(2.1d)

$$[f,w] = \frac{1}{x}\left\{f_{xx}\left(w_x + \frac{1}{x}w_{\theta\theta}\right) + w_{xx}\left(f_x + \frac{1}{x}f_{\theta\theta}\right) - 2x\left(\frac{w_\theta}{x}\right)_x\left(\frac{f_\theta}{x}\right)_x\right\} .$$

Axisymmetric buckled states are nontrivial solutions $\{W_0(x),F_0(x)\}$ of (2.1) that are independent of θ. Then if we define the new dependent variables $\alpha(x)$ and $\psi(x)$ by

$$\alpha(x) \equiv W_0'(x) \quad , \quad \psi(x) \equiv F_0'(x) \quad ,$$

where a prime denotes differentiation with respect to x ,

(2.1) is reduced to the ordinary differential equation boundary value problem,

$$L\psi = -(1/2)\alpha^2 \quad , \quad L\alpha + \lambda x\alpha = \alpha\psi \tag{2.2a}$$

$$\alpha(0) = \psi(0) = \alpha(1) = \psi(1) = 0 \quad . \tag{2.2b}$$

The differential operator L is defined by

$$Lv \equiv x[(xv)'/x]' \quad . \tag{2.2c}$$

The conditions in (2.2b) at $x = 0$ are a consequence of the axisymmetry of the solutions.

The unbuckled state is given by, $w = f \equiv 0$. It is a solution of (2.1) for all values of λ . The classical linearized theory of buckling is obtained by linearizing (2.1) about the unbuckled state. This is equivalent to deleting the nonlinear terms [w,w] and [f,w] in (2.1a,b). Then it follows [see (2.1a)] that f is a biharmonic function that satisfies the boundary conditions in (2.1c). Consequently, we have $f \equiv 0$. Then the linear buckling theory is obtained from (2.1b) as the eigenvalue problem,

$$\Delta^2 w + \lambda\Delta w = 0 \; , \; w = w_x = 0 \quad \text{on} \quad x = 1 \quad . \tag{2.3}$$

The eigenvalues and eigenfunctions of (2.3) are denoted by $\lambda = \lambda_{mn}$ and $w = w_{mn}(x,\theta)$, respectively. They have been determined explicitly in terms of Bessel functions, see [3]. Here m and n correspond to the number of waves in the plate in the radial and circumferential directions respectively. Consequently, for axisymmetric states we have $n = 0$ and the axisymmetric eigenvalues are $\lambda_{m,0} \equiv \lambda_m$. The buckling load of the plate, λ_c , is the minimum eigenvalue. It can be shown [3] that this corresponds to an axisymmetric eigenfunction.

That is

$$\lambda_c \equiv \min_{m,n} \lambda_{mn} = \min_{m} \lambda_{m,0} = \lambda_1 = j_{1,1}^2 \approx 14.7 \quad ,$$

where $j_{1,1}$ is the smallest root of $J_1(z) = 0$ and $J_1(z)$ is the Bessel function. All the eigenvalues of (2.3) are simple, as we can easily demonstrate.

We shall now describe some properties of the solutions of (2.1) that are related to bifurcation and multiplicity. They have been rigorously established in a variety of publications, see e.g. [4-8].

1. It follows directly from (2.1), that solutions occur in pairs. That is, if for some λ , $\{w,f\}$ is a solution of (2.1), then $\{-w,f\}$ is a solution for the same value of λ . Physically, this means that the plate can buckle in either lateral direction.

2. The unbuckled state is the only solution of (2.1) for $\lambda \leq \lambda_c$.

3. A one parameter branch of non-trivial solutions*, $\{w(\varepsilon),f(\varepsilon),\lambda(\varepsilon)\}$ of the nonlinear problem (2.1), depending continuously on the parameter ε , bifurcates from each eigenvalue of the linearized theory (2.3). That is, the eigenvalues are bifurcation points of the solutions of (2.1). We shall call them primary points. The corresponding solutions are called primary states. The primary states exist in a small neighborhood of each eigenvalue, $\lambda_{mn} < \lambda < \lambda_{mn} + \delta$, where $\delta > 0$ is small. Since $\delta > 0$, the solutions exist near λ_{mn} only for λ in excess of λ_{mn} . This is called supercritical bifurcation. Presumably, the primary states exist for all $\lambda > \lambda_{mn}$, but this result has not been completely

* We have suppressed the dependence of w and f on x and θ .

demonstrated. The primary states can be determined near each λ_{mn} by a perturbation expansion in the amplitude parameter ε .

4. The primary states that branch from axisymmetric eigenvalues are axisymmetric. Since they are determined by the solutions of ordinary differential equations (2.2), more detailed properties are known about them, than unsymmetric states. For example, the axisymmetric states exist for all $\lambda > \lambda_m$. Furthermore, the potential energy of the axisymmetric state that branches from λ_c is less than the potential energy of all other axisymmetric states* for the same value of $\lambda > \lambda_c$.

5. The linear theory of dynamic stability shows that the unbuckled state is stable for $\lambda < \lambda_c$ and unstable for $\lambda \geq \lambda_c$. The primary state that branches from λ_c is stable for λ near λ_{mn} . Thus as λ increases through λ_c , the plate buckles axisymmetrically.

We shall now study the response of the plate as λ increases above λ_c . The results of this discussion are summarized in the "bifurcation diagram" of Figure 1.

3. Numerical Results

Approximate representations of the axisymmetric, primary state that branches from λ_c have been obtained by perturbation expansions in ε [4,5,7], power series in x , and asymptotic series in λ^{-1} in [4,7] for the simply-supported plate and in [9] for the clamped plate. Difference methods are employed in [10] and the shooting method in [11,12] to obtain numerical approximations of this primary state. The details of the shooting method are described in [13]. Thus

---
* This result was proved in [4] for the simply supported circular plate. Presumably, the same analysis can be extended to the clamped plate.

accurate numerical approximations of the axisymmetric primary states are relatively easily obtained for a large range of values of λ , with the availability of high-speed computers.

A graph of the circumferential stresses $\tau(x)$, that were obtained from these numerical solutions, is shown in Figure 2 for a sequence of increasing values of λ . As λ increases, a strip of large circumferential compressive stress develops adjacent to the edge of the plate. The "width" of the strip narrows and the compressive stress intensity increases as λ increases. This suggests that for sufficiently large λ the strip may buckle unsymmetrically like a ring. That is, the plate would experience a second buckling by the edge wrinkling circumferentially about the axisymmetric state.

Morozov [14] showed, for the simply supported plate, that for sufficiently large values of λ , there are unsymmetric functions for which the potential energy of the plate is less than for all axisymmetric equilibrium states. Since it can be shown that the potential energy has a minimum, this demonstrates the existence of unsymmetric equilibrium states.* However it does not show that there are unsymmetric equilibrium states that arise by secondary buckling from axisymmetric states. This was subsequently demonstrated numerically for the clamped plate by Cheo and Reiss [11,12]. We shall now briefly describe their analysis.

Secondary buckling occurs when there is a bifurcation or branching of solutions (other than the unbuckled state) from a primary state. The values of λ at which this happens are called secondary buckling loads. In more general contexts they

---

* Presumably a similar result can be proved for the clamped plate using the same techniques. Previously, Yanowitch [15] had obtained related results for other boundary and loading conditions.

are called secondary bifurcation points. For simplicity, we shall call them secondary points. The equilibrium states that branch from the primary states at the secondary points are called secondary states. The secondary points are determined from the boundary value problem that is the linearization of (2.1) about a primary state. As a result of the discussion at the end of Section 2, we shall only consider the primary state that branches from λ_c. We denote it by $\{W_0(x;\lambda), F_0(x;\lambda)\}$ and linearize (2.1) about this solution. This gives,

$$\Delta^2 f + [W_0, w] = 0 \tag{3.1}$$

$$\Delta^2 w + \lambda \Delta w + [W_0, f] + [F_0, w] = 0$$

$$w = w_x = f = f_x = 0 \quad , \quad \text{for} \quad x = 1 \quad .$$

We wish to determine the values of $\lambda = \lambda^s$ for which the homogeneous boundary value problem (3.1) has non-trivial solutions. The "eigenvalue" parameter λ appears nonlinearly because the axisymmetric primary state $\{W_0(x;\lambda), F_0(x;\lambda)\}$, which is contained in the coefficients of (3.1), depends nonlinearly on λ.

Since the coefficients in (3.1) are independent of θ, we seek solutions in the form

$$f = y_n(x)\sin\theta \quad , \quad w = z_n(x)\sin\theta \quad .$$

This leads to a homogeneous boundary value problem for a system of two, fourth order, ordinary differential equations. By introducing the new dependent variables $u_n = y_n''$ and $v_n = z_n''$, where primes indicate derivatives with respect to x, we obtain an equivalent boundary value problem for a system of four, second order, ordinary differential equations. This problem is solved numerically by first replacing it by a system of difference equations. This results in the homogeneous system of algebraic equations

$$(3.2) \qquad Q(n,\lambda)\underset{\sim}{S} = 0$$

where $Q = [C_i, B_i, A_i]$ is a block tridiagonal matrix. The 4×4 component submatrices A_i, B_i and C_i , $i = i,\ldots,N$ are functions of λ . Here $N + 1$ is the number of mesh points on the interval, $0 \leq x \leq 1$. In addition, $\underset{\sim}{S}$ is a mesh vector which approximates the solution at the mesh points. The values of λ for which the algebraic system (3.2) has non-trivial solutions are the roots of

$$(3.3) \qquad \det Q(n,\lambda) = 0 \ .$$

These roots are then approximations to the secondary points λ^s.

To determine these roots, we first factor Q into the product of a lower block triangular matrix $L \equiv [\ell_i, I, 0]$, and an upper block triangular matrix $U \equiv [0, U_i, A_i]$, where ℓ_i and U_i are 4×4 scalar matrices. The factoring is possible if and only if $U_1,\ldots,U_{N-1}$ are nonsingular matrices. For all the calculations in [11,12] the factoring was always possible. Since $\det L = 1$, $\det U_i \neq 0$, $i = 1,2,\ldots,N-1$, and $Q = LU$, (3.3) implies that

$$(3.4) \qquad \phi(n,\lambda) \equiv \det U_N(n,\lambda) = 0 \ .$$

Thus the secondary points are approximated by the roots of the 4×4 determinant in (3.4).

The roots of (3.4) were evaluated by determining the sign changes in $\phi(n,\lambda)$ for fixed n and varying λ , and then using a chord method. To evaluate the determinant for any λ , the axisymmetric states were calculated by applying the shooting method to (2.2). These results were used to evaluate the submatrices A_i , B_i and C_i and then finally, by the factoring, the matrix U_n . For most calculations a mesh width of .02 was employed. The three lowest secondary points and the corresponding values of n at which they occurred are listed in Table 1. In Table 1, $\lambda_c \approx 14.7$ is the buckling load, and n is the number of circumferential waves in the

Table 1

n	7	10	12
λ^s/λ_c	7.5	8.1	10.4

plate. The corresponding eigenfunctions are almost zero except near the edge of the plate where they vary rapidly. Thus the plate is deformed, essentially, axisymmetrically in its interior and is circumferentially wrinkled near the edge.

The secondary states, which are unsymmetric, are determined by a perturbation expansion of the solution of (2.1) about the axisymmetric state at the secondary point. This leads to a system of linear boundary value problems for the coefficients in the expansions. They are solved numerically by the shooting and parallel shooting methods. The results of this calculation show that near the secondary points, the secondary states are subcritical, as we show in Figure 3. That is, they occur for $\lambda < \lambda^s$. The implications of these results are discussed in [12].

4. The Buckled Rectangular Plate

We consider a simply supported, rectangular plate, that is deformed by uniform compressive thrusts applied to the two edges normal to the x-axis. The other two edges are free to expand in the y direction. In dimensionless variables and rectangular coordinates, the von Kármán plate boundary value problem consists of solving the differential equations*,

$$\Delta^2 f = -\frac{1}{2}[w,w] , \tag{4.1a}$$

$$\Delta^2 w + \lambda w_{xx} = [f,w]$$

on the rectangle $0 \le x \le \ell$, $0 \le y \le 1$, subject to the conditions

$$w = \Delta w = f = \Delta f = 0 , \tag{4.1b}$$

on the boundary B of the rectangle. Here Δ is the two dimensional Cartesian Laplacian, $\ell > 0$ is the aspect ratio of the plate, λ is a parameter proportional to the thrust, and the nonlinear operator [f,w] is now defined by

$$[f,w] = f_{yy}w_{xx} + f_{xx}w_{yy} - 2f_{xy}w_{xy} . \tag{4.1c}$$

The formulation (4.1) and physical interpretations of the boundary conditions (4.1b) are discussed in [16].

Solutions of (4.1) occur in pairs, as for the circular plate. The unbuckled state $w = f \equiv 0$ is clearly a solution of (4.1) for all values of λ and ℓ . The classical, linearized theory of buckling is obtained by linearizing (4.1) about the unbuckled state. This gives $f \equiv 0$ and w satisfies,

$$\Delta^2 w + \lambda w_{xx} = 0 \tag{4.2}$$

$$w = \Delta w = 0 , \text{ on } B .$$

The eigenvalues and eigenfunctions of this problem are,

---

*cf. (2.1) .

$$(4.3a) \qquad \lambda = \lambda_{mn}(\ell) \equiv \left(\frac{\pi}{\ell}\right)^2 \left[m + \frac{n^2\ell^2}{m} \right]^2 ,$$

$$(4.3b) \quad w = w_{mn}(x,y;\ell) \equiv A_{mn} \sin \frac{m\pi x}{\ell} \sin n\pi y \ , \ m,n = 1,2,\ldots \ .$$

The constants A_{mn} are determined by the normalizing conditions $\|w_{mn}\| = 1$. The minimum eigenvalue $\lambda_c(\ell)$ is clearly achieved for $n = 1$. The corresponding value of m depends on ℓ.

For $\lambda < \lambda_c(\ell)$, it has been proved [16] that the unbuckled state is the unique solution of (4.1). As for the circular plate, the eigenvalues (4.3a) of the linearized theory are primary points of the solutions of (4.1). If λ_{mn} is a simple eigenvalue, then a unique pair of primary states branch from λ_{mn}. It is believed that these states exist for all $\lambda > \lambda_{mn}$, but this result has never been proved. Several pairs of primary states may branch from multiple eigenvalues. For example, if λ_{mn} is a double eigenvalue, then four pairs of primary states may branch from λ_{mn} [17,18].

It has been observed experimentally [19-21] that as λ increases from λ_c, the primary state may become unstable and the plate jumps to another equilibrium state. This is called mode-jumping by some experimenters. Mode-jumping may occur several times as λ increases before the plate finally fails. The experiments suggest that mode-jumping is related to secondary buckling. Investigations of various aspects of secondary buckling are described in [16,22-25].

The primary states can be determined formally near λ_{mn} by perturbation expansions, see e.g. [17,21]. These expansions may give inaccurate approximations for λ sufficiently in excess of λ_{mn}. In particular, they may be inaccurate when secondary buckling occurs. Other approximate methods such as, the Ritz and Galerkin methods, and Fourier series expansions, have been employed to obtain the primary states for a variety of boundary conditions. Some of these investigations are discussed in [26]. A comprehensive numerical study of the solutions of (4.1) was given in [16], including the first numerical determination of secondary points and secondary states. We shall now summarize these numerical techniques and results.

5. Numerical Methods

To obtain numerical solutions of (4.1), we first define the new variables $\Omega(x,y)$ and $\Phi(x,y)$ by

$$\Omega \equiv \Delta w \ , \ \Phi \equiv \Delta f \ .$$

Then (4.1) is written as

$$\text{(5.1a)} \qquad \Delta\Phi = -1/2[w,w] \ , \quad \Delta f = \Phi \ ,$$

$$\Delta\Omega = -\lambda w_{xx} + [f,w] \ , \quad \Delta w = \Omega \ ,$$

and

$$\text{(5.1b)} \qquad \Phi = f = \Omega = w = 0 \ , \quad \text{on } B \ .$$

Thus the system of two coupled fourth order differential equations in (4.1) have been transformed to a system of four coupled second-order equations. This is particularly fruitful for numerical computations, as we shall see.

Approximate solutions of (5.1) were obtained in [16] by an "accelerated" iteration procedure. Thus, for fixed values of the parameters ℓ and λ, and starting with an initial estimate $w^{(0)}(x,y)$ of the solution, a sequence of iterates $\{\Phi^{(n)},f^{(n)},\Omega^{(n)},w^{(n)}\}$ was defined by the recursions,

$$(5.2)\quad \Delta\Phi^{(n)} = -1/2[w^{(n)},w^{(n)}] \qquad \Phi^{(n)} = 0 \text{ on } B ,$$

$$\Delta f^{(n)} = \Phi^{(n)} \qquad f^{(n)} = 0 \text{ on } B ,$$

$$\Delta\Omega^{(n)} = -\lambda w_{xx}^{(n)} + [f^{(n)},w^{(n)}] , \quad \Omega^{(n)} = 0 \text{ on } B ,$$

$$\Delta\bar{w}^{(n+1)} = \Omega^{(n)} \qquad \bar{w}^{(n+1)} = 0 \text{ on } B ,$$

$$w^{n+1} = \theta\bar{w}^{(n+1)} + (1-\theta)w^{(n)} ,$$

where θ is the acceleration parameter.

Thus to evaluate each iterate, we must solve, four times at each step of the iteration, a Poisson boundary value problem

$$(5.3)\qquad \Delta u(x,y) = c(x,y) , \quad u = 0 \text{ on } B ,$$

where $c(x,y)$ is a prescribed function. This was done numerically by covering the rectangular region by a uniform mesh with mesh width δ . Then the Laplacian in (5.3) was replaced by the nine-point difference approximation on the mesh. This gives a system of algebraic equations

$$(5.4)\qquad Mu = c$$

where $\underset{\sim}{u}$ is a vector of the values of u at mesh points, $\underset{\sim}{c}$ is a vector of $c(x,y)$ at the mesh points, and the coefficient matrix M is of block tridiagonal form. The solution of (5.4) is an approximation to the solution of (5.1).

Since many iterations (5.2) may be required for convergence, particularly when λ is large, we employed the block factoring method [27] to solve (5.4) rapidly. All storage for the method was iternally, in core. This limited the permissible mesh sizes. For example, for the rectangular plate with $\ell = 2$, we used 741 interior mesh points. The calculations reported in [16] were performed on an IBM 7090 computer. Somewhat finer meshes can be used for larger machines such as the CDC 6600. For sufficiently fine meshes, it is not possible to maintain all storage internally. It is then necessary to use disc or tape storage and consequently to slow the computational speed. Then Fast-Fourier, and block reduction methods are superior procedures to solve (5.4); see the review paper [28] for a discussion of these methods.

The iterations (5.2) were terminated when

$$\max|\bar{w}^{n+1} - w^n| < \varepsilon$$

where $\varepsilon > 0$ is a prescribed small number and the maximum is taken over all mesh points. For most of the calculations $\varepsilon = 10^{-7}$ was used.

When the acceleration parameter is $\theta = 1$, the iterations (5.2) are Picard iterations. The Picard iterates converged only for a limited range of values of λ . By choosing $\theta \neq 1$, the iterations converged for a wide range of values of λ . Test calculations suggested that for each fixed ℓ and λ there is an optimal value of θ .

The calculation of a primary buckled state proceeded by first selecting a value of λ slightly greater than λ_{mn} . Then the corresponding eigenfunction $w_{mn}(x,y)$ was used as the initial iterate $w^{(0)}$. The iterations were performed until they converged to a buckled solution. This solution was then used as the initial iterate for a slightly different value of λ . In this way, solutions were obtained for increasing and decreasing sequences of values of λ on a specific branch.

6. Discussion of the Results

Some of the results of these calculations are summarized in the sketch in Figure 4 for the rectangular plate with $\ell = 2$. The four lowest primary points are,

$$\lambda_c = \lambda_{2,1} = 4\pi^2 \quad , \qquad \lambda_{3,1} = 169\pi^2/36 \quad ,$$

$$\lambda_{4,1} = \lambda_{1,1} = 25\pi^2/4 \quad , \quad \lambda_{5,1} = 8.41\pi^2 \quad .$$

We label the corresponding primary states as States 1-5, as shown in the figure. The third eigenvalue is double. The analysis in [17,18] suggests that four (pairs) of primary states should branch from this double eigenvalue. However, we were able to determine numerically only two of these branches. The remaining two branches are "combination modes."

For State 1 and at $\lambda = 120$, we used the solution for a slightly smaller value of λ as an initial iterate. Then the iterations suddenly converged to another solution. We called this solution the A-mode because of its lack of symmetry, as we show in the sketch in the figure. The A-mode branch was numerically traced by using increasing and decreasing sequences of values of λ , as we described in the previous section. At $\lambda = 450$, the iterations converged to primary State 3. As λ decreased along the A-mode, the corresponding solutions approached those of State 1. At $\lambda = 110$, the iterations converged from the A-mode to a solution of State 1. Thus $\lambda = 110$ is a numerical upper bound of a secondary point of State 1. The A-mode is a secondary state. A least square extrapolation of the A-mode solutions give $\lambda^s \approx 107 = 2.64\ \lambda_c$. The A-mode and State 1 exist numerically for λ in the interval $110 < \lambda < 120$.

7. Secondary Buckling

We shall appeal to the theory of secondary bifurcation presented in [29] to discuss a mechanism of secondary buckling. This theory relates the creation of secondary points and secondary states to multiple primary points as follows. The eigenvalues (primary points) $\lambda_{mn}(\ell)$ depend on the parameter ℓ ; see (4.3a). For special values of $\ell = \ell_0$, two or more eigenvalues coalesce to form a multiple eigenvalue $\bar{\lambda}$. For example, for $\ell_0 = 2$, the eigenvalues $\lambda_{1,1}$ and $\lambda_{4,1}$ have coalesced to form a double eigenvalue. The multiplicity M of $\bar{\lambda}$ as an eigenvalue is the dimension of the null space of (4.2). The multiplicity μ of $\bar{\lambda}$ as a primary point is the number of pairs of primary states that branch from $\bar{\lambda}$. We observe that μ and M need not be equal. In fact, we have for $\ell_0 = 2$ and $\bar{\lambda} = \lambda_{1,1} = \lambda_{4,1}$ that $M = 2$ and $\mu = 4$.

The theory in [29] states that as ℓ varies from ℓ_0 , the multiple eigenvalue splits into its component eigenvalues and, in addition, one or more secondary points, as we show in Figure 5. These secondary points move along the primary states as $|\ell-\ell_0|$ increases. Thus as $\ell \to \ell_0$, the primary and secondary points coalesce into a single, multiple primary point $\bar{\lambda}$. Figure 5 shows that the number of states near $\bar{\lambda}$ is invariant as ℓ deviates from ℓ_0 . That is, there are four pairs of states branching from $\bar{\lambda}$. After splitting there are still four pairs of states, namely the two pairs of states branching from the primary points λ_1 and λ_2 and the two pairs of secondary states.

However secondary bifurcation can also occur by splitting of $\bar{\lambda}$ even when $M = \mu$, as we have shown in [29]. Then as $|\ell-\ell_0|$ increases, $\bar{\lambda}$ splits into its primary points and several secondary points. The secondary states now connect all the secondary points. That is, the secondary states are of finite extent. Thus as $\ell \to \ell_0$ the primary and secondary points coalesce into $\bar{\lambda}$ and the secondary states disappear.

These observations suggested a perturbation method in the small parameter $\ell - \ell_0$ to calculate the secondary points and secondary states [29]. Application of this method to the present plate problem will appear in [25].

The computations reported in [29] show that there may be other secondary points which do not arise by the splitting of multiple primary points. Indeed, in the circular plate problem discussed in Sections 2 and 3 of this paper, there is no auxiliary parameter such as ℓ, and all the eigenvalues of the linearized theory are simple. Thus the secondary points obtained cannot arise by the mechanism we have discussed. Of course, it may be possible to imbed the circular plate problem in a more general family of problems containing auxiliary parameters for which the mechanism is applicable. However, we shall not pursue this idea in the present paper.

For problems containing additional parameters, there is then the possibility of multiple secondary points, which are themselves multiple eigenvalues, and their subsequent splitting. This results in tertiary and higher order bifurcations. This subject is pursued in [30].

To ascertain whether the numerically determined secondary point and secondary states satisfy this mechanism, we consider the variation of the primary State 1 and the secondary point, as ℓ varies. The eigenvalue $\lambda_{2,1}(\ell)$ forms a sequence of multiple eigenvalues $\bar{\lambda}_1, \bar{\lambda}_2, \ldots,$ as ℓ varies. That is,

$$\bar{\lambda}_k = \lambda_{2,1}(\sqrt{4k+2}) = \lambda_{2k+1,1}(\sqrt{4k+2}) = \frac{(2k+3)^2}{4k+2}\pi^2 , \quad k = 0,1,\ldots .$$

As ℓ varies through the special values $\ell_0 = \sqrt{4k+2}$, $k = 0,1,\ldots,$ new secondary points are created and they move along primary State 1. Thus, for any value of ℓ (say $\ell = 2$), the primary state may contain many secondary points, including those that do not arise by the splitting process.

We observe that State 1 is antisymmetric in x ($m = 2$, $n = 1$) and the modes $m = 2k+1$, $n = 1$ are symmetric in x. A "combination" of such modes can give rise to asymmetric modes such as the numerically determined A-mode.

We have performed a series of computations to determine $\lambda^s(\ell)$. For fixed ℓ we obtained secondary states for a decreasing sequence of values of λ. For some $\lambda = \lambda_u^s$ the iterations converged to a State 1 solution. This gave an upper bound for λ^s. At least square approximation was then made of the data $(\lambda, \|w(\lambda)\|)$ for the secondary states, where $\|w\|$ is the L_2 norm of w. The intersection of this approximation and State 1 is called the numerical approximation of λ^s. The results are shown in Figure 6. For $\ell \leq 2.0$ the values of $\lambda_u^s - \lambda^s$ are small. However, for $\ell > \sqrt{6}$ this difference is substantial. This suggests that the estimates of λ^s may be inaccurate.

In order for $\lambda^s(\ell)$ to arise from splitting, we must have $\lambda^s(\ell_0) = \bar{\lambda}_j$ for some j and the corresponding value of ℓ_0. As we see from the figure, the numerical approximations deviate considerally from any of the $\bar{\lambda}_j$ shown in the figure. Because of the possible inaccuracies in the numerical values of λ^s for the larger values of ℓ and the restricted range of ℓ-values considered, a conclusion concerning the origin of λ^s is not possible at present.

REFERENCES

1. H. Poincaré, "Sur l'equilbre d'une masse fluide animée d'un mouvement de rotation," Acta Math., 1 (1885), 259-380.
2. T. von Kármán, "Festigkeitsprobleme im Maschinenbau," Encyk. de Math. Wissen., Vol. IV, Teubner, Leipzig, 1910, 348-352.

3. C. B. Biezeno and R. Grammel, Engineering Dynamics, Blackie & Son, Ltd., London, 1956, 477-481.
4. K. O. Friedrichs and J. J. Stoker, "The non-linear boundary value problem of the buckled plate," Amer. J. Math., 63 (1941), 839-888.
5. H. B. Keller, J. B. Keller and E. L. Reiss, "Buckled states of circular plates," Q. Appl. Math., 20 (1962), 55-65.
6. J. Wolkowsky, "Existence of buckled states of circular plates," Comm. Pure Appl. Math., 20 (1967), 549-560.
7. K. O. Friedrichs and J. J. Stoker, "Buckling of the circular plate beyond the critical thrust," J. Appl. Mech., 9 (1942), A7-A14.
8. M. S. Berger and P. Fife, "On von Kármán's equations and the buckling of a thin elastic plate," Comm. on Pure and Appl. Math. 20 (1967), 687-719; 21 (1968) 227-241.
9. S. R. Bodner, "The post buckling behavior of a clamped circular plate," Q. Appl. Math., 12 (1955), 397-401.
10. H. B. Keller and E. L. Reiss, "Non-linear bending and buckling of circular plates," Proc. 3rd U. S. Nat. Cong. of Appl. Math., 1958, 375-385.
11. L. S. Cheo and E. L. Reiss, "Unsymmetric wrinkling of circular plates," Quart. Appl. Math., 31 (1973), 75-91.
12. L. S. Cheo and E. L. Reiss, "Secondary buckling of circular plates," SIAM J. on Appl. Math., 26 (1974), 490-495.
13. L. Bauer, E. L. Reiss and H. B. Keller, "Axisymmetric buckling of hollow spheres and hemispheres," Comm. Pure and Appl. Math., 23 (1970), 529-568.
14. N. F. Morozov, "Investigation of a circular symmetric compressible plate with a large boundary load," Izv. Vyssh. Uchebn. Zaved., 34 (1963), 95-97.
15. M. Yanowitch, "Non-linear buckling of circular plates," Comm. Pure Appl. Math., 9 (1956), 661-672.
16. L. Bauer and E. L. Reiss, "Nonlinear buckling of rectangular plates," J. Soc. Indust. Appl. Math., 13 (1965), 603-626.
17. B. J. Matkowsky and L. J. Putnick, "Multiple buckled states of rectangular plates," Internat. J. Non-Linear Mech., 9 (1974), 89-103.
18. G. H. Knightly and D. Sather, "Nonlinear buckled states of rectangular plates," Arch. Rational Mech. Anal., 54 (1974), 356-372.
19. L. Schuman and G. Back, "Strength of rectangular flat plates under edge compression," NASA TR 356, 1930.
20. M. Ojalvo and F. H. Hull, "Effective width of thin rectangular plates," J. Eng. Mech. Div., ASCE 84, EM 3 (1958), 1718-1 to 1718-20.

21. M. Stein, "Loads and deformations of buckled rectangular plates," NASA TR R-40, 1959.
22. T. von Kármán, E. E. Sechler and L. H. Donnell, "The strength of thin plates in compression," Trans. ASME, APM 54-5, 54 (1932), 53-57.
23. Y. W. Chang and E. L. Masur, "Vibration and stability of buckled panels," J. Eng. Mech. Div., ASCE, EM 5 (1965), 1-26.
24. G. J. Stroebel and W. H. Warner, "Stability and secondary bifurcation for von Kármán plates," J. Elast., 3 (1973), 185-202.
25. B. J. Matkowsky, L. J. Putnik and E. L. Reiss, "Secondary states of rectangular plates," to appear.
26. S. Timoshenko, Theory of Elastic Stability, McGraw-Hill, New York, 1936.
27. S. Schechter, "Quasi-tridiagonal matrices and type-insensitive difference equations," Q. Appl. Math., 18 (1960), 285-305.
28. F. W. Dorr, "The direct solution of the discrete Poisson equation on a rectangle," SIAM Rev., 12 (1970), 278-287.
29. L. Bauer, H. B. Keller and E. L. Reiss, "Multiple eigenvalues lead to secondary bifurcation," SIAM Rev., 17 (1975), 101-122.
30. E. L. Reiss, "Cascading bifurcations," in preparation.

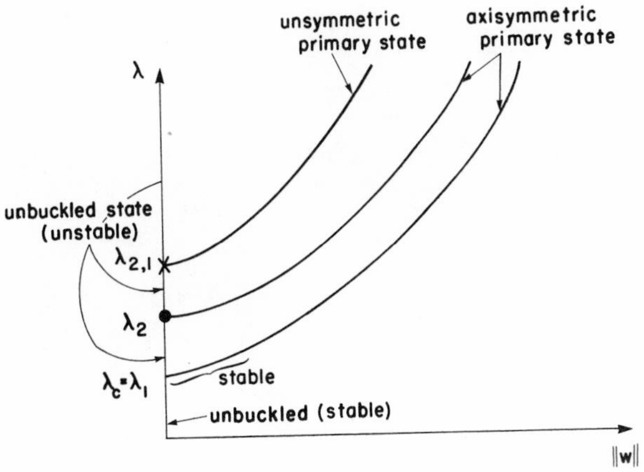

Figure 1

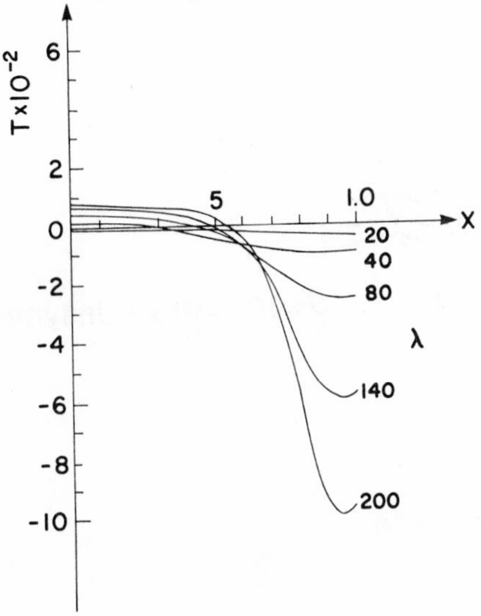

Figure 2

The dimensionless, axisymmetric, circumferential membrane stress for an increasing sequence of values of λ .

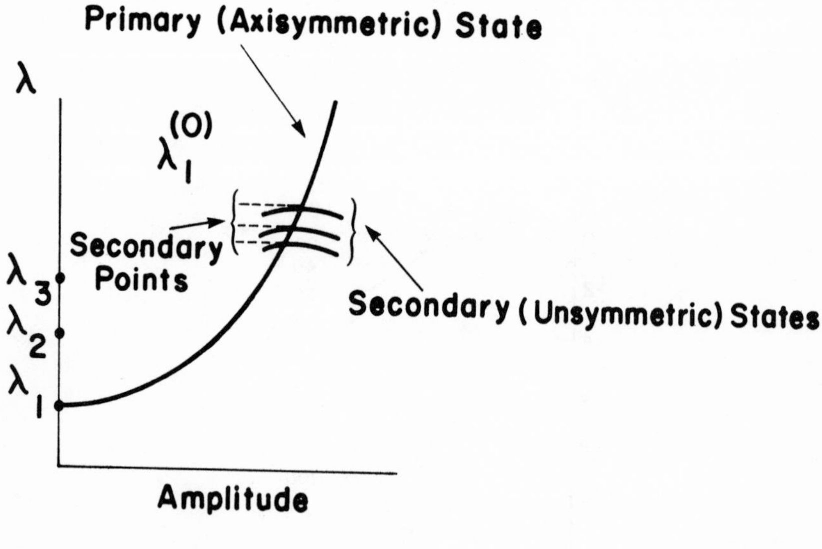
Primary (Axisymmetric) State
λ
$\lambda_1^{(0)}$
Secondary Points
λ_3
λ_2
λ_1
Secondary (Unsymmetric) States
Amplitude

Figure 3

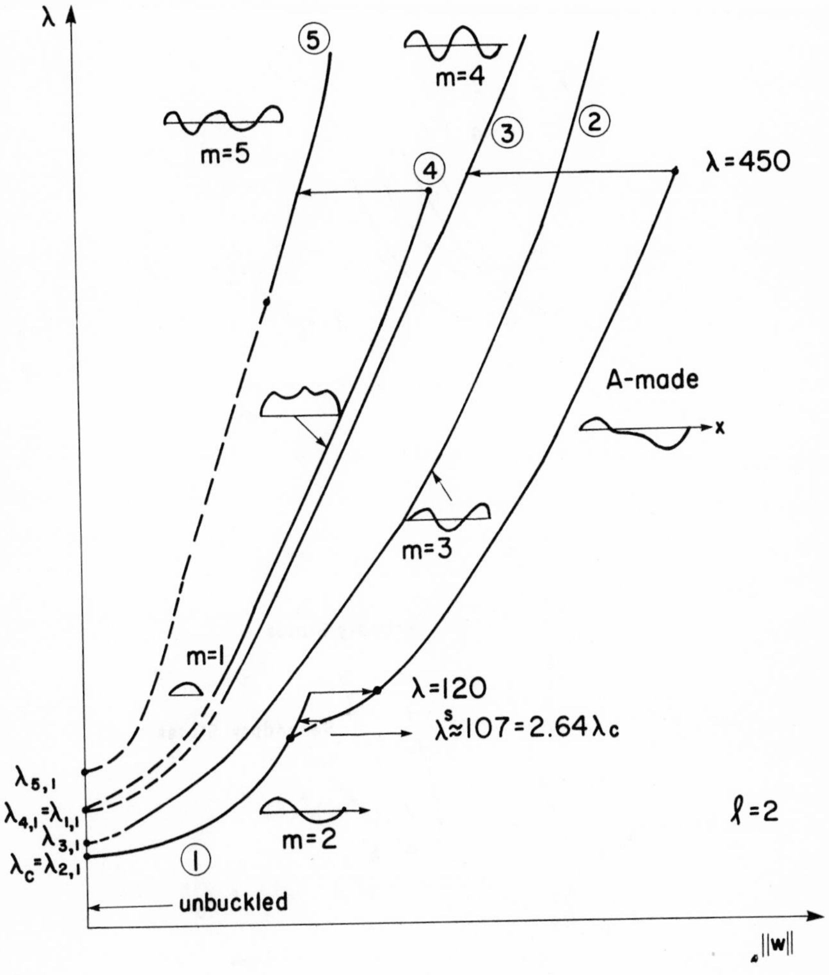

Figure 4

Sketch of the bifurcation diagram for the buckled rectangular plate with $\ell = 2$. The sketches show the deflected shapes of the corresponding states. The minimum eigenvalue is $\lambda_c = \lambda_{2,1}$. The dashed curves are conjectured. The iteration procedure did not converge for the ranges of λ indicated by the dashed lines.

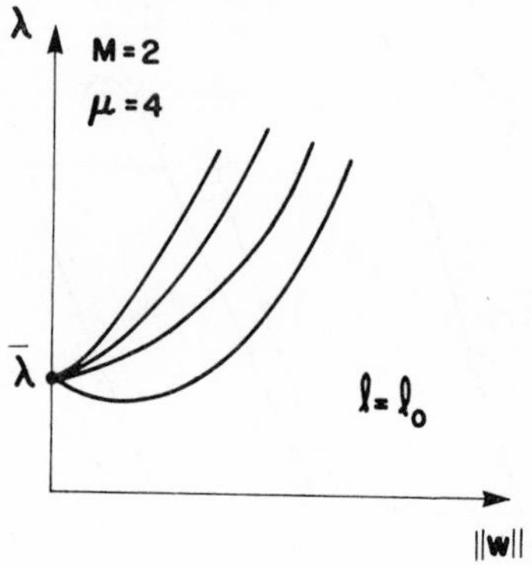

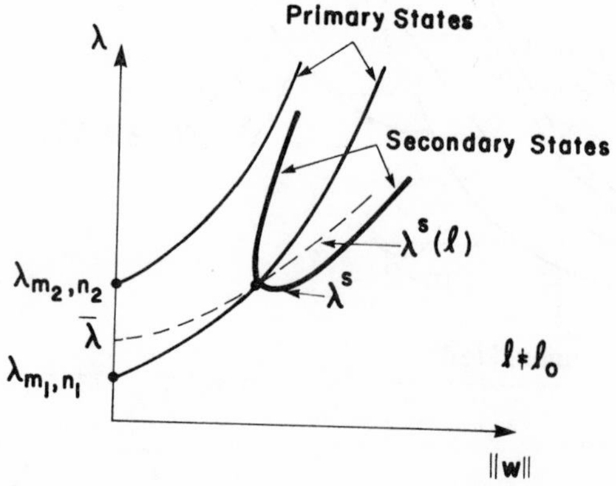

Figure 5

Splitting of a multiple eigenvalue with $M = 2$ and $\mu = 4$ at $\ell = \ell_o$ into component eigenvalues λ_{m_1,n_1} and λ_{m_2,n_2} and secondary point λ^s . The dotted line gives the locus of the secondary points as ℓ varies.

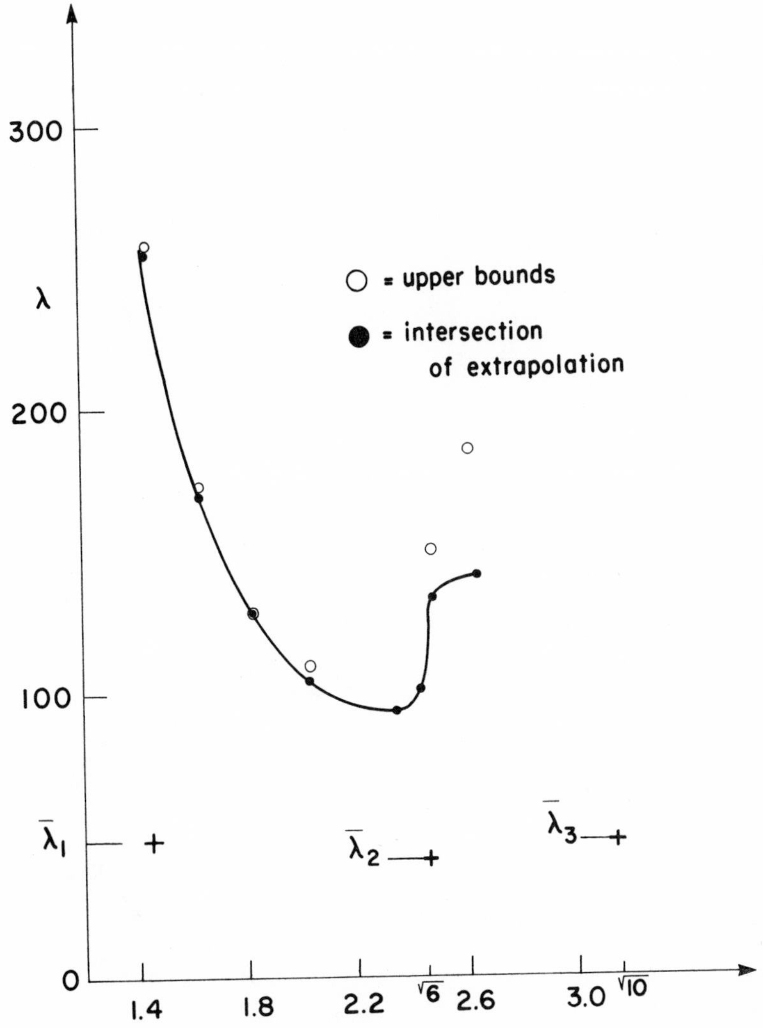

Figure 6

The numerically determined values of $\lambda^s(\ell)$. The open circles are numerical upper bounds. These are the values of λ from which the iterations from the A-mode jump to primary State 1. The solid dots are the intersections of the least-square extrapolations of the A-modes with State 1. The deviations between these values and the upper bounds increase rapidly for $\ell > \sqrt{6}$.

CONVERGENCE IN ENERGY FOR ELLIPTIC OPERATORS

Sergio Spagnolo*

1. Introduction

We discuss here the principal results on G - convergence, i.e. the convergence of the Green's operators for Dirichlet's and other boundary problems.

The study of G - convergence had its starting point with an example of DE GIORGI (see [24], page 661) concerning a sequence of ordinary linear differential operators of the second order whose coefficients rapidly vary.

Several phenomena observed in this example can be extended to a suitable class of elliptic or parabolic partial differential equations. A systematic study in this direction has been effected in the series of papers [24], [25], [11], and [7].

In these papers we often referred to parabolic equations, even to prove some exclusively elliptic results.

The present exposition is, on the contrary, entirely confined to the domain of elliptic equations. For this purpose we give a direct proof of the compactness and locality properties of G - convergence (Theor. 5, 6, 7), which avoids recourse to parabolic equations and uses elliptic regularization.

We note that the results of §4 which deal with the non-Dirichlet boundary value problems are not contained in the former papers.

* University of Pisa

G - convergence has several connections with other theories so that it is difficult to supply an exhaustive bibliography about this subject.

We may however note: SBORDONE ([19]) for the G - convergence for elliptic operators with lower order terms; JOLY ([8]), MARCELLINI ([9], [10]) and MOSCO ([13]) for the abstract convergence of convex functionals; MURAT ([14]), TARTAR ([27]) and ZOLEZZI ([28]) for applications to optimal control theory; BOCCARDO - MARCELLINI ([3]) and BENSOUSSAN - LIONS - PAPANICOLAU ([2]) for the convergence of solutions of variational inequalities; SENATOROV ([22]) for the convergence of eigenvalues and eigenfunctions; RESHETNIAK ([15]) for a convergence of variational problems related to the theory of quasi-conformal mappings (see also [26]); SANCHEZ PALENCIA ([16], [17], and [18]) for the study of periodic problems and many other problems in Mechanics related to G - convergence; BABUŠKA ([1]) for homogeneization problems; DE GIORGI ([5]) for the convergence of integrals of area - type; DE GIORGI - FRANZONI ([6]) for an abstract definition including both the integral of energy - type and of area-type; CRISTIANO ([4]) and SBORDONE ([20]) for some applications of the De Giorgi - Franzoni scheme.

2. The abstract scheme

Let V be a reflexive and separable real Banach space and let V' be the dual of V. We shall denote by the same symbol $\| \|$ both the norm on V and its adjoint norm on V'.

Let us denote by $E(V)$ the class of linear operators

$$A : V \to V'$$

which are symmetric and positive isomorphisms, i.e. such that, for any u and v in V,

(1) $$\langle Au,v\rangle = \langle Av,u\rangle$$

(2) $$\lambda_A\|u\|^2 \le \langle Au,u\rangle \le \Lambda_A\|u\|^2 \quad (0 < \lambda_A \le \Lambda_A) \quad .$$

Let us finally denote by $E_{\lambda_0,\Lambda_0}(V)$, with $0 < \lambda_0 < \Lambda_0$, the class of operators A in $E(V)$ which verify (2) for $\lambda_A \ge \lambda_0$ and $\Lambda_A \le \Lambda_0$.

Definition

Let A_k and A be in $E(V)$, $k \in N$. We say that $\{A_k\}$ is G - convergent to A (in symbols $A_k \xrightarrow{G} A$) for $k \longrightarrow \infty$ if, for any f and g in V' ,

$$\lim_{k\to\infty} \langle g,A_k^{-1}f\rangle = \langle g,A^{-1}f\rangle \quad .$$

Remark 1.

If $A_k \xrightarrow{G} A$, then $\langle A_k\, u,u\rangle \ge \lambda_0\|u\|^2$, $\forall u \in V$, $\forall k$, for some constant $\lambda_0 > 0$.

Remark 2.

G - convergence is the sequence - convergence for the topology defined on $E(V)$ by the family of semi-norms

$$\{A \mapsto \langle f,A^{-1}g\rangle \mid f,g \in V'\} \quad .$$

Remark 3.

Let φ and ψ be two functions defined on V with values in $[0,+\infty]$ such that

$$\{u_k\} \to u \text{ weakly in } V \Rightarrow \varphi(u) \le \min\lim_{k\to\infty} \varphi(u_k) \quad .$$

Then the class of operators A in $E(V)$ such that

$$\varphi(u) \leq \langle A\, u,u\rangle \leq \psi(u) \ , \ \forall u \in V \ ,$$

is closed in $E(V)$ with respect to G - convergence.

Remark 4.

$E_{\lambda_0,\Lambda_0}(V)$ is a compact metric space with respect to G - convergence.

Besides G - convergence, we can consider on $E(V)$ the topologies of uniform, of strong, and of weak convergence, i.e. respectively

$$\{A_k\, u\} \longrightarrow A\, u \quad \text{in} \quad V' \ , \text{ uniformly for } \ \|u\| \leq 1 \ ,$$

$$\{A_k\, u\} \longrightarrow A\, u \quad \text{in} \quad V' \ , \ \forall u \in V \ ,$$

$$\{\langle A_k\, u,v\rangle\} \longrightarrow \langle A\, u,v\rangle \ , \ \forall u,v \in V \ .$$

$E_{\lambda_0,\Lambda_0}(V)$ is a complete metric space for anyone of these topologies and it is compact with respect to weak convergence.

Some relations between these convergences and G - convergence are described in the following Remarks.

Remark 5.

If A_k, A and B, in $E(V)$, $k \in N$, are such that $\{A_k\}$ is G - convergent to A and weakly convergent to B, then $A \leq B$.

Remark 6.

Let A_k and A be in $E_{\lambda_0,\Lambda_0}(V)$, $k \in N$. Then $\{A_k\}$ is strongly convergent to A if and only if $\{A_k\} \longrightarrow A$ simultaneously for G - convergence and weak convergence.

To obtain a closer description of the case of differential operators we must refine our abstract scheme by introducing a linear operator

$$I : V \to V'$$

which is assumed to be compact, symmetric and strictly positive (i.e. $\langle Iu, u\rangle > 0$ for $u \neq 0$).

We can therefore construct an Hilbert space X and two compact embeddings, $j : V \to X$, $\tilde{j} : X \to V'$, such that $\tilde{j} \circ j = I$ and $(ju,jv)_X = \langle Iu,v\rangle$, $\forall u,v$ in V .

Consequently we have the scheme

$$V \xrightarrow{j} X \xrightarrow{\tilde{j}} V'$$

and the G - convergence of $\{A_k\}$ to A is equivalent to

$$j\, A_k^{-1}\, f \to j\, A^{-1}\, f \quad \text{in } X\ ,\ \forall f \in V' .$$

Moreover we have

Theor. 1.

Let A_k and A be in $E(V)$, $k \in N$, and $\lambda \in R$ such that

$$\langle(\lambda I + A_k)u,u\rangle \geq \lambda_0\|u\|^2 \quad , \quad (\lambda_0 > 0)\ \forall u \in V ,$$

$$A_k \xrightarrow{G} A \qquad \text{for} \qquad k \longrightarrow \infty \ .$$

Then we have

$$(\lambda\, I + A_k) \xrightarrow{G} \lambda\, I + A \ .$$

Corollary

Let A_k and A be in $E_{\lambda_0,\Lambda_0}(V)$ and $\lambda > 0$. Then $A_k \xrightarrow{G} A$ if and only if $(\lambda I + A_k) \xrightarrow{G} \lambda I + A$.

The existence of the compact embedding I enables us to approximate any A in $E(V)$ by its resolvent - regularisation

$\{A^{(\varepsilon)}, \varepsilon > 0\}$.

Such an approximation is very useful in the study of G - convergence since it transforms this convergence to strong convergence (see Th. 3 below).

For any A in $E(V)$ let us define the operator

$$A^{(\varepsilon)} : V \to V' \quad (\varepsilon > 0)$$

by means of the equalities

$$(3) \quad A^{(\varepsilon)} \equiv \varepsilon^{-1}[I - I(I+\varepsilon A)^{-1}I] = I(I+\varepsilon A)^{-1}A = A(I+\varepsilon A)^{-1}I \quad .$$

<u>Theor. 2</u>.

<u>Let</u> A <u>be in</u> $E(V)$ <u>and</u> $A^{(\varepsilon)}$ <u>be defined as in</u> (3).

<u>Then</u> $A^{(\varepsilon)}$ <u>is a compact, symmetric and positive linear operator such that</u> $A^{(\varepsilon)} \leq A$.

<u>Moreover</u>

$$A^{(\varepsilon)}u \to A\,u \ \underline{\text{in}}\ V' \ , \ \underline{\text{for}}\ \varepsilon \to 0 \ , \ \forall u \in V \ .$$

<u>Theor. 3</u>.

<u>Let</u> A_k <u>and</u> A <u>be in</u> $E(V)$, $k \in N$, <u>and let</u> $A_k^{(\varepsilon)}$ <u>and</u> $A^{(\varepsilon)}$ <u>be the corresponding operators defined as in</u> (3).

<u>Then if</u> $A_k \xrightarrow{G} A$ <u>we have, for</u> $k \to \infty$,

$$A_k^{(\varepsilon)}u \to A^{(\varepsilon)}\,u \ \underline{\text{in}}\ V' \ , \ \forall u \in V \ , \ \forall \varepsilon > 0 \ .$$

3. <u>Dirichlet's problem</u>

The abstract scheme of §2 can be applied to study the convergence in $L^2(\Omega)$ of the solutions of Dirichlet's boundary problems for elliptic equations on a bounded open set Ω in R^n .

Denoting by $H_0^1(\Omega)$ the closure of $\mathcal{D}(\Omega)$ in $H^1(\Omega)$ (space of real functions $u \in L^2(\Omega)$ whose derivatives in the sense of distributions $D_1 u, \ldots, D_n u$ also belong to $L^2(\Omega)$), equipped with the norm

$$\|u\|_{H_0^1(\Omega)} = [\int_\Omega |Du|^2 dx]^{1/2} ,$$

we can choose V and I as follows:

$$V = H_0^1(\Omega)$$

$$<Iu,v> = \int_\Omega u \; v \; dx \quad .$$

We shall identify V' with $H^{-1}(\Omega)$, the space of distributions g on Ω such that

$$g = f_0 + \sum_1^n {}_i \, D_i \; f_i \quad (f_0, f_1, \ldots, f_n \in L^2(\Omega) \;) \; ,$$

so that any u in $H_0^1(\Omega)$ will be identified with $I\,u \in H^{-1}(\Omega)$.

As a matter of fact we are interested in the study of G - convergence on a proper subclass of $E_{\lambda_0,\Lambda_0}(H_0^1(\Omega))$; namely the class, which we denote by $E(\lambda_0,\Lambda_0;\Omega)$, of operators $A : H_0^1(\Omega) \to H^{-1}(\Omega)$ such that

$$A = - \sum_1^n {}_{i,j} \, D_i(a_{ij}(x) \; D_j)$$

for some real measurable functions a_{ij} on Ω such that

$$\begin{cases} a_{ij} = a_{ji} \\ \lambda_0 \le \sum_1^n {}_{i,j} a_{ij}(x)\xi_i\xi_j|\xi|^{-2} \le \Lambda_0 \quad , \quad \forall \xi \in R^n \quad . \end{cases}$$

The functions a_{ij} are called the coefficients of A .

Let A_k and A be in $E(\lambda_0,\Lambda_0;\Omega)$, $0 < \lambda_0 \le \Lambda_0$, $k \in N$, with coefficients $a_{ij,k}$ and a_{ij} respectively, and let us denote by $u_k(f)$ and $u(f)$, for $f \in H^{-1}(\Omega)$, the solutions u_k and u of Dirichlet's problems

$$\begin{cases} A_k\, u_k = f & \text{on } \Omega \\ u_{k|\partial\Omega} = 0 & \end{cases} \qquad \begin{cases} A\, u = f & \text{on } \Omega \\ u_{|\partial\Omega} = 0 & \end{cases} .$$

The G - convergence of $\{A_k\}$ to A means that

$$u_k(f) \to u(f) \quad , \quad \forall f \in H^{-1}(\Omega) \quad , \quad k \to \infty \quad ,$$

in $L^2(\Omega)$, or, equivalently, in $H_0^1(\Omega)$ - weakly (when f is in $L^\infty(\Omega)$ we have also the uniform convergence on any compact subset of Ω , using the De Giorgi - Nash theorem).

In the following remarks we make a comparison between G - convergence, weak convergence, strong convergence and uniform convergence on $E(\lambda_0,\Lambda_0;\Omega)$ and, on the other hand, the convergence of corresponding coefficients (for $k \to \infty$) .

Remark 7.

$$A_k \longrightarrow A \quad \text{uniformly}$$
$$\text{if and only if}$$
$$a_{ij,k} \longrightarrow a_{ij} \quad \text{in } L^\infty(\Omega) \ , \ \forall i,j \quad .$$

Remark 8.

$$A_k \to A \quad \text{strongly}$$
$$\text{if and only if}$$

$$\begin{cases} a_{ij,k} \to a_{ij} \quad \text{weakly in } L^2(\Omega) \quad , \quad \forall i,j \quad , \\ \qquad\qquad \text{and} \\ \sum_1^n {}_i\, D_i\, a_{ij,k} \to \sum_1^n {}_i\, D_i\, a_{ij} \quad \text{in } H^{-1}_{loc}(\Omega) \ , \ \forall_j \quad . \end{cases}$$

Remark 9.

$$A_k \to A \quad \text{weakly}$$
$$\text{if and only if}$$
$$a_{ij,k} \to a_{ij} \quad \text{weakly in } L^2(\Omega) \ , \ \forall i,j \ .$$

Remark 10.

The strong convergence $A_k \to A$ is equivalent to the convergence (with the same notations as above)

$$u_k(f) \to u(f) \quad \text{in} \quad H_0^1(\Omega) \quad , \quad \forall f \in H^{-1}(\Omega) \quad .$$

In particular strong convergence implies G - convergence.

Remark 11.

The convergence of the coefficients in $L^1_{loc}(\Omega)$,

$$a_{ij,k} \to a_{ij} \quad , \quad \forall i,j \quad ,$$

implies the strong convergence $A_k \to A$.

Contrary to uniform, strong, and weak convergence, the G - convergence $\{A_k\} \to A$ is not equivalent to any convergence of the corresponding coefficients $\{a_{ij,k}\} \to a_{ij}$, $\forall i,j$.

This is made evident by the following result:

Theor. 4.

There exists a constant $c(n) > 0$ such that for any A in $\mathcal{E}(\lambda_0,\Lambda_0;\Omega)$ it is possible to find a sequence of isotropic operators $(k \in N)$

$$A_k \equiv - \sum_1^n {}_i \ D_i(a_k(x) \ D_i) \ , \ c(n)\lambda_0 \le a_k(x) \le \frac{\Lambda_0}{c(n)} \quad ,$$

such that

$$A_k \xrightarrow{G} A \quad .$$

The one dimensional case $n = 1$ (Ω = open interval of R) is trivial:

Remark 12:

$$\left\{- \frac{d}{dx}\left(a_k(x) \frac{d}{dx}\right)\right\} \xrightarrow{G} - \frac{d}{dx} \left(a(x) \frac{d}{dx}\right) \quad (\lambda_0 \le a_k(x) \le \Lambda_0)$$

if and only if

$$a_k(x)^{-1} \longrightarrow a(x)^{-1} \quad \text{weakly in} \quad L^2(\Omega) \quad .$$

This Remark enables us to obtain immediately for $n = 1$ two important properties of G - convergence: the local nature of it and the compactness of $E(\lambda_0,\Lambda_0;\Omega)$.

In the case $n > 1$ these two properties are also true (Th. 5, Th. 6, and Th. 7 below) but their proofs are more complicate and use the following Lemmas, where spt(g) denotes the support of the distribution g on Ω (i.e. the complement with respect to Ω of the largest open set on which g is zero) and $\{A^{(\varepsilon)}\}$ denotes the resolvent - regularisation of A defined by (3).

Lemma 1.

An operator A , in $E_{\lambda_0,\Lambda_0}(H_0^1(\Omega))$, belongs to $E(\lambda_0,\Lambda_0;\Omega)$ if and only if for any u,v in $H_0^1(\Omega)$,

$$\text{spt}(u) \cap \text{spt}(Dv) = \emptyset \Rightarrow \langle Au, v \rangle = 0 \quad .$$

Lemma 2.

Let Ω_0 be an open subset of Ω and $\vartheta(x) = \text{dist}(x, C\Omega_0)$. Then for any u in $H_0^1(\Omega)$ such that

$$A\,u = 0 \quad \text{on} \quad \Omega_0$$

the following estimate holds:

$$\int_\Omega |A^{(\varepsilon)}u|^2 e^{\lambda\vartheta(x)}dx \le \varepsilon^{-1}\Lambda_0 \int_\Omega |Du|^2 dx$$

for any $\varepsilon , \lambda > 0$ with $\varepsilon \lambda^2 \le \Lambda_0^{-1}$.

Lemma 3.

Let Ω_0 be an open subset of Ω and let u , v in $H_0^1(\Omega)$ be such that

$$A u = 0 \text{ on } \Omega_0 , \quad spt(v) \subset \Omega_0 .$$

Then the following estimate holds with $\delta = dist(spt(v),C\Omega_0) , \varepsilon > 0$, :

$$|\langle A^{(\varepsilon)}u,v\rangle| \le \sqrt{\Lambda_0\varepsilon^{-1}} \quad e^{-\delta/(2\sqrt{\Lambda_0\varepsilon})}\|Du\|_{L^2(\Omega)} \|v\|_{L^2(\Omega)} .$$

Theor. 5.

$E(\lambda_0,\Lambda_0;\Omega)$ is compact with respect to G - convergence.

Theor. 6.

Let Ω_0 be an open subset of Ω and let A_k and A , in $E(\lambda_0,\Lambda_0;\Omega)$, u_k and u in $H^1_{loc}(\Omega_0)$, f_k and f , in $H^{-1}_{loc}(\Omega_0)$, $k \in N$, be such that:

$$\begin{aligned} &A_k u_k = f_k \quad \text{on } \Omega_0 , && \forall k , \\ &A_k \xrightarrow{G} A , && \text{for } k \to \infty , \\ &u_k \to u \text{ in } L^2_{loc}(\Omega_0) , && \text{" " "} , \\ &f_k \to f \text{ in } H^{-1}_{loc}(\Omega_0) , && \text{" " "} , \end{aligned}$$

Then we have

$$A u = f \text{ on } \Omega_0 .$$

In the rest of this paper we use the notation

$$\mathcal{E}(\lambda_0,\Lambda_0) \equiv \mathcal{E}(\lambda_0,\Lambda_0;R^n)$$

and we denote by

$$A_{|\Omega} : H_0^1(\Omega) \to H^{-1}(\Omega)$$

the restriction to Ω of some $A \in \mathcal{E}(\lambda_0,\Lambda_0)$.

We shall say that

$$A_k \xrightarrow{G} A \quad \text{on} \quad \Omega$$

if

$$A_{k|\Omega} \xrightarrow{G} A_{|\Omega} \quad (\text{in} \quad \mathcal{E}(\lambda_0,\Lambda_0;\Omega)) \ .$$

Theor. 7.

Let Ω and Ω_h , $h \in N$, be bounded open sets in R^n such that

$$\Omega_h \subseteq \Omega \ , \ \forall h \quad ; \quad \Omega_h \cap \Omega_p = \emptyset \quad \text{for} \quad h \neq p \quad ;$$

$$\text{meas}(\Omega \setminus \bigcup_{h=1}^{\infty} \Omega_h) = 0 \quad ,$$

and let A_k and A be in $\mathcal{E}(\lambda_0,\Lambda_0)$, $k \in N$.

Then

$$A_k \xrightarrow{G} A \quad \text{on} \quad \Omega$$

if and only if

$$A_k \xrightarrow{G} A \quad \text{on any} \quad \Omega_h \ , \ \forall h \in N \ .$$

4. Other boundary problems

In this section we shall consider operators A_k , with coefficients $a_{ij,k}$, and A , with coefficients a_{ij} , belonging to $E(\lambda_0,\Lambda_0)$, $k \in N$.

Moreover we shall use the matrix notation

$$A_k \equiv - \operatorname{div}(a_k(x)D) \quad , \quad A \equiv - \operatorname{div}(a(x)D) \quad .$$

where $a_k(x)$ in the matrix of the coefficients of A_k :

$$a_k(x) = [a_{ij,k}]_{i,j=1,\dots,n} \quad .$$

The function

$$x \longmapsto \langle a(x)\ Du(x)\ ,\ Du(x)\rangle$$

defined for u in $H^1_{loc}(R^n)$, is called the A-energy of u .

Let u_k and u be the solutions of Dirichlet's problem

$$\begin{cases} A_k\ u_k = f & \text{on } \Omega \\ u_{k|\partial\Omega} = 0 \end{cases} \qquad \begin{cases} Au = f & \text{on } \Omega \\ u_{|\partial\Omega} = 0 \end{cases}$$

where Ω is bounded open set in R^n and $f \in H^{-1}(\Omega)$.

It follows directly that, if $A_k \xrightarrow{G} A$ on Ω , the integral on Ω of the A_k - energy of u_k converges to the integral on Ω of the A-energy of u .

On the contrary, it is not trivial to prove the convergence of the A_k-energy of u_k in the interior of Ω , or to extend the former result to the case of non-zero boundary values.

Theor. 8.

Let Ω be a bounded open set in R^n and let us assume that A_k, A, in $E(\lambda_0,\Lambda_0)$, u_k, u, in $H^1_{loc}(\Omega)$, and f in $H^{-1}_{loc}(\Omega)$, $k \in N$, be such that, for $k \to \infty$,

$$A_k \xrightarrow{G} A \quad \text{on } \Omega ,$$

$$u_k \longrightarrow u \quad \text{in } L^2_{loc}(\Omega) ,$$

$$A_k u_k \longrightarrow f \quad \text{in } H^{-1}_{loc}(\Omega) .$$

Then $A\,u = f$ on Ω and, for any $S \subset\subset \Omega$,

$$\int_S \langle a_k Du_k , Du_k\rangle\, dx \longrightarrow \int_S \langle a\, Du , Du\rangle\, dx .$$

Theor. 9.

Let Ω be a bounded open set in R^n with Lipschitz boundary and let A_k, A, in $E(\lambda_0,\Lambda_0)$, u_k, u, in $H^1(\Omega)$, f, in $H^{-1}(\Omega)$, $k \in N$, and w, in $H^1(\Omega)$, be such that

$$\begin{cases} A_k\, u_k = f & \text{on } \Omega \\ u_k - w \in H^1_0(\Omega) \end{cases} \qquad \begin{cases} A\, u = f & \text{on } \Omega \\ u - w \in H^1_0(\Omega) \end{cases} .$$

Then, if $A_k \xrightarrow{G} A$ on Ω, we have

$$u_k \longrightarrow u \quad \text{in } L^2(\Omega) ,$$

$$\int_S \langle a_k Du_k , Du_k\rangle\, dx \longrightarrow \int_S \langle a\, Du , Du\rangle\, dx , \ \forall S \subseteq \Omega .$$

We will now study some more general boundary problems, which will be posed in the variational formulation.

More exactly, let Ω be a bounded open set in R^n with Lipschitz boundary and let V be a closed linear subspace of $H^1(\Omega)$ such that

$$H_0^1(\Omega) \subseteq V \subseteq H^1(\Omega) \quad .$$

Then, given A in $E(\lambda_0,\Lambda_0)$, f in $L^2(\Omega)$, w in $H^1(\Omega)$ and $\lambda > 0$, we pose the following problem:

Problem 1.

To find the functions u^* , in the class

$$w + V \equiv \{u \in H^1(\Omega) : u - w \in V\} \quad ,$$

such that the functional

$$F : u \longmapsto \int_\Omega <a\, Du , Du> dx + \lambda \int_\Omega u^2 dx - 2 \int_\Omega f\, u\, dx \quad ,$$

restricted to $w + V$, takes its minimum value at u^* .

It is known that this problem has one and only one solution u^* . This solution verifies the equation $A\, u^* + \lambda\, u^* = f$ on Ω , with some boundary condition whose nature depends on the choice of V (e.g. Dirichlet's condition for $V = H_0^1(\Omega)$, Neumann's condition for $V = H^1(\Omega)$) .

Now let us consider the functional

$$F_k(u) = \int_\Omega <a_k Du , Du> dx + \lambda \int_\Omega u^2 dx - 2 \int_\Omega f\, u\, dx$$

where $A_k \in E(\lambda_0,\Lambda_0)$, $k \in N$, $f \in L^2(\Omega)$ and $\lambda > 0$, and let us denote by u_k^* the solution of Problem 1 relative to F_k .

Theor. 10.

If A_k is G - convergent to A on Ω , we have (for $k \longrightarrow \infty$)

$$u_k^* \to u^* \quad \text{in} \quad L^2(\Omega) \; , \; \text{and weakly in} \quad H^1(\Omega) \; ,$$

$$\int_S \langle a_k Du_k \, , \, Du_k \rangle \, dx \to \int_S \langle a \, Du \, , \, Du \rangle \, dx$$

for any $S \subset\subset \Omega$ and also for $S = \Omega$.

If we take $f = 0$, we can also consider Problem 1 for $\lambda = 0$. In this case Problem 1 has a unique solution if V does not contain any constant function $\neq 0$, otherwise there exists a unique solution u^* of Problem 1 such that $\int_\Omega u^* \, dx = 0$ (the other solutions being of the form $u^* + \text{const.}$).

With this limitation Theorem 10 can be easily extended to the case $f = 0$, $\lambda = 0$.

Let us now specialize Problem 1 taking $f = \lambda = 0$, Ω equal to some open cube of R^n and w equal to linear function, $w(x) = \langle \xi, x \rangle$, for $\xi \in R^n$.

We put $|\Omega| = \text{meas}(\Omega)$ and

$$\gamma_A(V,\Omega,\xi) = |\Omega|^{-1} \, \text{Min}\left\{\int_\Omega \langle a(x) Du \, , \, Du \rangle \, dx : u - \langle \xi, x \rangle \in V\right\} .$$

Remark 13.

For any A , V and Ω , $\gamma_A(V,\Omega,\xi)$ is a quadratic form in ξ , $\xi \in R^n$, such that

$$\gamma_A(V,\Omega,\xi) \leq |\Omega|^{-1} \int_\Omega \langle a(x)\xi, \xi \rangle \, dx \leq \Lambda_0 |\xi|^2 \quad .$$

The forms γ_A are not, in general, coercive, e.g., for $V = H^1(\Omega)$, $\gamma_A(V,\Omega,\xi) = 0$. However we have the coerciveness if V is small in the following sense:

$$H_0^1(\Omega) \subseteq V \subseteq \tilde{H}^1(\Omega) \tag{4}$$

where

$$\tilde{H}^1(\Omega) = \{u \in H^1(\Omega) : \int_\Omega D \, u \, dx = 0\} \quad .$$

Remark 14.

If V satisfies (4) we have the estimate

$$\gamma_A(V,\Omega,\xi) \geq |\xi|^4 [|\Omega|^{-1} \int_\Omega <a(x)^{-1}\xi,\xi> dx]^{-1} \geq \lambda_0 |\xi|^2 \quad .$$

Remark 15.

In the one-dimensional case, $n = 1$, we have $H_0^1(\Omega) = \tilde{H}^1(\Omega)$ and

$$\gamma_A(H_0^1(\Omega),\Omega,\xi) = |\Omega| [\int_\Omega a(x)^{-1} dx]^{-1} \xi^2 \quad .$$

Remark 16.

If the coefficients a_{ij} of A are constants and V satisfies (4), we have the equality

$$\gamma_A(V,\Omega,\xi) = \sum_{i,j}^{n} a_{ij}\xi_i\xi_j \quad .$$

Remark 17.

If $A_k \xrightarrow{G} A$ on Ω, then $\gamma_{A_k}(V,\Omega,\xi) \longrightarrow \gamma_A(V,\Omega,\xi)$.

Remark 18.

If x_0 is a Lebesque point for any coefficient a_{ij} of A and V_k satisfies (4) with $\Omega = \Omega_k$, then

$$\gamma_A(V_k,\Omega_k,\xi) \longrightarrow \sum_{i,j}^{n} a_{ij}(x_0)\xi_i\xi_j$$

for any sequence $\{\Omega_k\}$ of open cubes of R^n such that $x_0 \in \bar{\Omega}_k$ $\forall k$ and $\text{diam}(\Omega_k) \longrightarrow 0$, for $k \longrightarrow \infty$.

An example of a linear space V which satisfies (4) is the following

$$P(\Omega) = \{u_{|\Omega} : u \in H^1_{loc}(R^n) \text{ , } u \text{ is } \Omega\text{-periodic}\} \quad .$$

Finally, let us remark that

$$\gamma_A^1(\Omega,\xi) \le \gamma_A^{per}(\Omega,\xi) \le \gamma_A^0(\Omega,\xi) \quad , \tag{5}$$

where

$$\gamma_A^0(\Omega,\xi) = \gamma_A(H_0^1(\Omega),\Omega,\xi)$$

$$\gamma_A^1(\Omega,\xi) = \gamma_A(\tilde{H}^1(\Omega),\Omega,\xi)$$

$$\gamma_A^{per}(\Omega,\xi) = \gamma_A(P(\Omega),\Omega,\xi) \ .$$

5. <u>The homogeneization problem</u>

A very illuminating case of G - convergence is the following.

Let A be in $E(\lambda_0,\Lambda_0)$ and let us assume that the coefficients a_{ij} of A are periodic functions with <u>period</u> P, where P is a fixed open cube of R^n.

We can then consider the sequence $\{A_k\} \subseteq E(\lambda_0,\Lambda_0)$ defined by

$$A_k = - \sum_{1\ i,j}^{n} D_i(a_{ij}(kx)D_j) \quad , \quad k \in N \ .$$

We have, with the same notations of §4,

<u>Remark 19</u>.

$$\gamma_{A_k}{}^{per}(P,\xi) = \gamma_A{}^{per}(P,\xi) \quad , \quad \forall k \in N \ .$$

<u>Remark 20</u>.

If $\{A_{k_j}\} \xrightarrow{G} B$ for some $\{k_j\} \longrightarrow \infty$, then B is an operator with constants coefficients.

Theor. 11.

If A is an operator, in $E(\lambda_0,\Lambda_0)$, with periodic coefficients a_{ij} , there exists an operator $\tilde{A}$ with constants coefficients $\tilde{a}_{ij}$ such that

$$\{A_k\} \equiv \left\{-\sum_{1\ i,j}^{n} D_i(a_{ij}(kx)D_j)\right\} \xrightarrow{G} \tilde{A} \equiv -\sum_{1\ i,j}^{n} \tilde{a}_{ij}D_iD_j \ .$$

For the calculation of the coefficients of the homogeneized problem, $\tilde{a}_{ij}$, we use the following estimate (which is a consequence of (5) and Remarks 19, 17, and 16).

$$\gamma^1_{A_k}(P,\xi) \le \gamma_A^{per}(P,\xi) = \sum_{1\ i,j}^{n} \tilde{a}_{ij}\xi_i\xi_j \le \gamma^0_{A_k}(P,\xi) \ , \ k \in N \ .$$

Let us finally observe that the extremal terms of this inequality converge to the interior term for $k \to \infty$.

6. Appendix

In this section we give the proofs of the previous statements.

Remark 1: is a consequence of Banach-Steinhaus theorem.

Remark 2: is trivial.

Remark 3: the inequality $\varphi(u) \le \langle Au,u\rangle$, $\forall u \in V$, can be also written $\varphi(A^{-1}f) \le \langle f,A^{-1}f\rangle$, $\forall f \in V'$, and then it is preserved under the G - convergence of $\{A_k\}$ to A .

On the other hand we use the known inequality

$$\text{(6)}\quad \langle f,u\rangle^2 \le \langle f,A^{-1}f\rangle \langle Au,u\rangle \ , \ u \in V \ , \ f \in V' \ , \ A \in E(V) \ .$$

Therefore the inequality $\langle Au,u\rangle \le \psi(u)$, $\forall u \in V$, is equivalent to $\langle f,u\rangle^2 \le \langle f,A^{-1}f\rangle\psi(u)$, $\forall u \in V$ and $f \in V'$, and it is valid for G - convergence.

Remark 4: is a consequence of reflexivity and separability of V.

Remark 5: is a consequence of inequality (6).

Remark 6: using the identity $A^{-1} - B^{-1} = A^{-1}(B-A)B^{-1}$, for A and B in $E(V)$, we can prove that the strong convergence of $\{A_k\}$ to A in $E_{\lambda_0,\Lambda_0}(V)$ implies the strong convergence of $\{A_k^{-1}\}$ to A^{-1} and hence the G-convergence of $\{A_k\}$ to A.

Conversely, we get from (6)

$$2 \langle Au,u\rangle \leq \langle Au,A_k^{-1}Au\rangle + \langle A_k u,u\rangle \quad ,$$

so that the symmetric operator $B_k \equiv AA_k^{-1}A + A_k - 2A$ is positive.

But if $A_k \to A$ in G - convergence and in weak convergence, we obtain $B_k \to 0$ weakly, and hence $B_k \to 0$ strongly.

As a consequence, $(AA_k^{-1}-I)^2 \equiv B_k A_k^{-1} \to 0$ strongly, so that $(AA_k^{-1}-I) \to 0$ strongly and finally $A_k \to A$ strongly.

Theor. 1: we use the notations $R(\lambda) = (\lambda I + A)^{-1}$, $R_k(\lambda) = (\lambda I + A_k)^{-1}$, and the following identity:

$$R_k(\lambda)(\lambda I + A)A^{-1} = A_k^{-1} - \lambda R_k(\lambda)I(A_k^{-1} - A^{-1}) \quad .$$

Since $A_k^{-1} \to A^{-1}$ weakly, I is a compact operator and $\{R_k(\lambda)\}$ is equibounded, we obtain that $\{R_k(\lambda)(\lambda I+A)A^{-1}\}$ weakly converges to A^{-1}; hence $R_k(\lambda) \to R(\lambda)$ weakly.

Theor. 2: the compactness and the symmetry of $A^{(\varepsilon)}$ are immediate consequence of (3).

To prove the other assertions of Th. 2 let us put

$$u_\varepsilon = (I + \varepsilon A)^{-1}Iu \; , \; v_\varepsilon = (I + \varepsilon A)^{-1}Au \; (u \in V \; , \; \varepsilon > 0) \; .$$

We have then

$$Au_\varepsilon = Iv_\varepsilon = A^{(\varepsilon)}u \tag{7}$$

$$v_\varepsilon = \varepsilon^{-1}(u - u_\varepsilon) \tag{8}$$

$$Iu_\varepsilon + \varepsilon\, Au_\varepsilon = Iu \tag{9}$$

$$Iv_\varepsilon + \varepsilon\, Av_\varepsilon = Au\ . \tag{10}$$

Multiplying (10) by u_ε we obtain

$$<Iv_\varepsilon,\ u_\varepsilon> + \varepsilon <Av_\varepsilon,\ u_\varepsilon> = <Au,\ u_\varepsilon>$$

or equivalently, for (7),

$$<A\ u_\varepsilon, u_\varepsilon> + \varepsilon <Iv_\varepsilon, v_\varepsilon> = <A^{(\varepsilon)}u, u> \quad ,$$

which implies the positivity of $A^{(\varepsilon)}$.

Multiplying (10) by v_ε we obtain

$$<Iv_\varepsilon,\ v_\varepsilon> + \varepsilon <A\ v_\varepsilon,\ v_\varepsilon> = <Au,\ v_\varepsilon>\ , \tag{11}$$

which implies that $<Au, v_\varepsilon> \geq 0$ and therefore, for (8),

$$<Au, u> \geq <Au, u_\varepsilon> = <A^{(\varepsilon)}u, u>\ ,\ \text{i.e.}\ \ A \geq A^{(\varepsilon)}\ .$$

It remains to prove that $A^{(\varepsilon)} \to A$ strongly for $\varepsilon \to 0$ or equivalently, since $A^{(\varepsilon)} \leq A$, $\forall\varepsilon$, that $A^{(\varepsilon)} \to A$ weakly for $\varepsilon \to 0$. Now from (11) it follows that $\varepsilon <A\ v_\varepsilon,\ v_\varepsilon> \leq <Au, v_\varepsilon>$ and therefore, using the Schwartz inequality,

$$<Av_\varepsilon, v_\varepsilon> \leq \varepsilon^{-2} <Au, u>\ . \tag{12}$$

From (12) and (11) it follows that $<Iv_\varepsilon, v_\varepsilon> \leq$ $\leq <Au,u>^{1/2}[\varepsilon^{-2}<Au,u>]^{1/2} = \varepsilon^{-1}<Au,u>$, i.e.

(13) $$<I(u-u_\varepsilon) , u - u_\varepsilon> \leq \varepsilon <Au,u> .$$

On the other hand (12), using (8), implies that $\{u-u_\varepsilon\}$ is bounded in V and hence weakly relatively compact, so that $I(u-u_\varepsilon)$ is in a compact of V' .

This fact, together with (13), assures us that $(u-u_\varepsilon) \to 0$ weakly in V ; hence $A^{(\varepsilon)}u \to Au$ weakly in V' , for $\varepsilon \to 0$.

Theor. 3: Th. 1 gives $(I + \varepsilon A_k)^{-1}Iu \to (I + \varepsilon A)^{-1}Iu$ weakly in V and therefore, using (3) and the compactness of I , we get $A_k^{(\varepsilon)}u \to A^{(\varepsilon)}u$ in V' , for $k \to \infty$, $\forall\varepsilon > 0$.

Remarks 7,8, and 9: are easy to verify (see e.g. [25], page 587).

Remark 10: see the proof of Remark 6.

Remark 11: is a consequence of Remark 8.

Theor. 4: see [11].

Remark 12: in the case $n = 1$ it is possible to find the explicit solution u of the problem $(a(x)u')' = f$ on $[\alpha,\beta]$ with the boundary condition $u(\alpha) = u(\beta) = 0$. Remark 12 then follows easily from examination of this solution (see e.g. [25] page 596.).

Lemma 1: see [23].

Lemma 2: let u be in $H_0^1(\Omega)$ such that $Au \equiv -\text{div}(a(x)Du) = 0$ on Ω_0 and let v_ε be as in the proof of Th. 2

$(A^{(\varepsilon)}u = I\ v_\varepsilon \equiv v_\varepsilon)$. Since $v_\varepsilon(e^{\lambda\vartheta} - 1)$ belongs to $H_0^1(\Omega_0)$, we have

$$\int_\Omega \langle a(x)Du,\ D(v_\varepsilon e^{\lambda\vartheta})\rangle\ dx = \int_\Omega \langle a(x)Du,\ Dv_\varepsilon\rangle\ dx\ .$$

Multiplying equation (10) by $v_\varepsilon e^{\lambda\vartheta}$ and integrating on Ω we obtain then the equality

$$\int_\Omega v_\varepsilon^2 e^{\lambda\vartheta}dx + \varepsilon \int_\Omega \langle a\ Dv_\varepsilon, Dv_\varepsilon\rangle\ e^{\lambda\vartheta}dx + \varepsilon\lambda \int_\Omega \langle a\ Dv_\varepsilon, D\vartheta\rangle\ v_\varepsilon e^{\lambda\vartheta}dx =$$

$$= \int_\Omega \langle a\ Du, Dv_\varepsilon\rangle\ dx\ .$$

Using the inequality $|D\vartheta| \le 1$ and the Schwartz inequality we get

$$\int_\Omega v_\varepsilon^2 e^{\lambda\vartheta}dx + \varepsilon \int_\Omega \langle a\ Dv_\varepsilon, Dv_\varepsilon\rangle\ e^{\lambda\vartheta}dx \le$$

$$\le \Big[\int_\Omega \langle a\ Dv_\varepsilon, Dv_\varepsilon\rangle\ e^{\lambda\vartheta}dx\Big]^{1/2} \cdot \Big[\varepsilon\ \lambda\Big(\Lambda_0 \int_\Omega v_\varepsilon^2 e^{\lambda\vartheta}dx\Big)^{1/2} +$$

$$+ \Big(\Lambda_0 \int_\Omega |Du|^2 dx\Big)^{1/2}\Big]\ .$$

From here (using the fact that $x^2 + \varepsilon y^2 \le y\ z$ implies $x^2 \le z^2/4\varepsilon$) we obtain

$$\int_\Omega v_\varepsilon^2 e^{\lambda\vartheta}dx \le \frac{\Lambda_0}{4\varepsilon}\Big[\varepsilon\lambda\Big(\int_\Omega v_\varepsilon^2 e^{\lambda\vartheta}dx\Big)^{1/2} + \Big(\int_\Omega |Du|^2 dx\Big)^{1/2}\Big]^2$$

$$\le \frac{\Lambda_0}{2\varepsilon}\Big[\varepsilon^2\lambda^2 \int_\Omega v_\varepsilon^2 e^{\lambda\vartheta}\ dx + \int_\Omega |Du|^2 dx\Big]$$

and hence, if $\Lambda_0 \varepsilon \lambda^2 \le 1$, the result of Lemma 2.

Lemma 3: since $Au = 0$ on Ω_0 and $spt(v) \subset \Omega_0$, (and therefore $e^{-\lambda\vartheta(x)} \le e^{-\lambda\delta}$ on $spt(v)$) we have, $\forall\lambda > 0$,

$$|\int_\Omega A^{(\varepsilon)}u\cdot v\,dx| \le |\int_\Omega A^{(\varepsilon)}u\, e^{\lambda\vartheta/2}\, v\, e^{-\lambda\vartheta/2}dx| \le$$

$$\le [\int_\Omega |A^{(\varepsilon)}u|^2 e^{\lambda\vartheta}dx]^{1/2}[\int_\Omega v^2dx]^{1/2}e^{-\lambda\delta/2}.$$

Hence, using Lemma 2 with $\lambda = (\varepsilon\Lambda_0)^{-1/2}$, we obtain the result of Lemma 3.

Theor. 5: using Remark 4 we have only to prove that the G-limit A of some sequence $\{A_k\} \subset E(\lambda_0,\Lambda_0;\Omega)$ also belongs to $E(\lambda_0,\Lambda_0;\Omega)$, i.e. (Lemma 1) that $\langle Au,v\rangle = 0$ if $spt(Du) \cap spt(v) = 0$.

Now, for u and v as above, we can find an open neighborhood Ω_0 of $spt(v)$ such that $Du \equiv 0$ on Ω_0, and hence $A_k u = 0$ on Ω_0, $\forall k$.

In virtue of Lemma 3 we get then the inequality

$$|\langle A_k^{(\varepsilon)}u,v\rangle| \le \sqrt{\Lambda_0\varepsilon^{-1}}\, e^{-\delta/(2\sqrt{\Lambda_0\varepsilon})}\|Du\|_{L^2(\Omega)}\|v\|_{L^2(\Omega)}$$

for any k and ε; hence passing to the limit for $k \to \infty$ and then for $\varepsilon \to 0$ (and using Th. 3 and Th. 2) we obtain $\langle Au,v\rangle = 0$.

Theor. 6:

Caccioppoli's estimate on solutions of elliptic equations assure us that $\{u_k\}$ is a bounded sequence in $H^1_{loc}(\Omega_0)$. Therefore we can consider, for any $\tilde{\Omega}_0 \subset\subset \Omega_0$, a function ψ

in $\mathcal{D}(\Omega_0)$ such that $\psi \equiv 1$ on $\tilde{\Omega}_0$, and we can replace u_k by ψu_k and Ω_0 by $\tilde{\Omega}_0$; so that we can assume that $u_k \to u$ weakly in $H_0^1(\Omega)$.

Let us now introduce the solutions w_k and w, in $H_0^1(\Omega)$, of equations $A_k w_k = f_k$ and $Aw = f$ on Ω and set $\tilde{u}_k = u_k - w_k$, $\tilde{u} = u - w$. Since $A_k \xrightarrow{G} A$ and $f_k \to f$ in $H^{-1}(\Omega)$ we have $w_k \to w$, and hence $\tilde{u}_k \to \tilde{u}$, weakly in $H_0^1(\Omega)$. Moreover $A_k \tilde{u}_k = 0$ on Ω_0.

We wish to prove that $A\tilde{u} = 0$ on Ω_0.

Now Lemma 3 gives us the inequality, $\forall k$, $\forall \varepsilon$, :

$$|<A_k^{(\varepsilon)}\tilde{u}_k, v>| \leq \sqrt{\Lambda_0 \varepsilon^{-1}}\, e^{-\delta/(2\sqrt{\Lambda_0 \varepsilon})} \|D\tilde{u}_k\|_{L^2(\Omega)} \|v\|_{L^2(\Omega)}$$

for $v \in \mathcal{D}(\Omega_0)$, with $\delta = \text{dist}(\text{spt}(v), (\Omega_0)$.

Therefore passing to the limit, in this inequality, for $k \to \infty$ and then for $\varepsilon \to 0$ (and using Th. 3 and Th. 2) we obtain $<A\tilde{u}, v> = 0$, $\forall v \in \mathcal{D}(\Omega_0)$.

<u>Theor. 7</u>: in virtue of compactness of G - convergence we need only prove that, for $\Omega_0 \subseteq \Omega$, :

$$\left\{A_k \xrightarrow{G} A \text{ on } \Omega,\ A_k \xrightarrow{G} B \text{ on } \Omega_0\right\} \Rightarrow A_{|\Omega_0} = B_{|\Omega_0}.$$

Now these hypotheses imply, $\forall f \in H^{-1}(\Omega)$,

$$\begin{cases} (A_{k|\Omega})^{-1} f \to (A_{|\Omega})^{-1} f \text{ in } L^2(\Omega) \\ \qquad\qquad\qquad\qquad\qquad\qquad (k \to \infty) \\ (A_{k|\Omega_0})^{-1} f \to (B_{|\Omega_0})^{-1} f \text{ in } L^2(\Omega_0), \end{cases}$$

so that the sequence $\{u_k\} \equiv \{(A_{k|\Omega})^{-1} f - (A_{k|\Omega_0})^{-1} f\}$ is convergent in $L^2(\Omega_0)$ to $(A_{|\Omega})^{-1}f - (B_{|\Omega_0})^{-1}f$.

But $A_k\, u_k = 0$ on Ω_0 , $\forall k$, and hence (Th. 6)

$$A[(A_{|\Omega})^{-1}f - (B_{|\Omega_0})^{-1}f] = 0 \text{ on } \Omega_0 \text{ , i.e.}$$

$$f = A[(B_{|\Omega_0})^{-1}f] \text{ on } \Omega_0 \text{ .}$$

This equality for $f = Bu$, $u \in H^1_0(\Omega_0)$, becomes $Bu = Au$ on Ω_0 , i.e. $B_{|\Omega_0} = A_{|\Omega_0}$.

Theor. 8: see [7], page 400.

Theor. 9: see [7], page 401.

Theor. 10: since for the solution of Problem 1 we can get easily an "a priori" estimate, we can assume that, for $k \to \infty$,

(14) $$F_k(u^*_k) \to \eta$$

(15) $$u^*_k \to u_0 \text{ , in } L^2(\Omega) \text{ and weakly in } H^1(\Omega) \text{ ,}$$

for some function u_0 such that $u_0 - w \in V$.

Now, for a fixed u such that $u - w \in V$, we will prove that $F(u_0) \le F(u)$. To this end, let us introduce the solution $\tilde{u}_k$ of Dirichlet's problem $A_k\, \tilde{u}_k = Au$ on Ω , $(\tilde{u}_k - u)_{|\partial\Omega} = 0$, and let us observe that $\tilde{u}_k - w$ belongs to V .

We have then, for any $S \subseteq \Omega$, the inequality:

(16) $$\int_S \langle a_k\, Du^*_k, Du^*_k\rangle\, dx + \lambda \int_\Omega u^{*2}_k dx - 2\int_\Omega f\, u^*_k\, dx \le F_k(u^*_k) \le F_k(\tilde{u}_k).$$

Now, for $k \to \infty$, $u_k^* \to u_0$ in $L^2(\Omega)$ and $A_k u_k^* \to (f - \lambda u_0)$ in $L^2(\Omega)$ (indeed $A_k u_k^* + \lambda u_k^* = f$ on Ω) ; hence (Th. 8)

$$(17) \qquad \int_S <a_k Du_k^*, Du_k^*> dx \to \int_S <a Du_0, Du_0> dx \ , \ \forall S \subset\subset \Omega \ .$$

On the other hand $\tilde{u}_k \to u$ in $L^2(\Omega)$ (since $A_k \xrightarrow{G} A$) and $\int_\Omega <a_k D\tilde{u}_k, D\tilde{u}_k> dx \to \int_\Omega <a\, Du, Du> dx$ (Th.9) so that $F_k(\tilde{u}_k) \to F(u)$.

Therefore passing to the limit for $k \to \infty$ in (16) we get

$$\int_S <a\, Du_0, Du_0> dx + \lambda \int_\Omega u_0^2 dx - 2 \int_\Omega f\, u_0 dx \le \eta \le F(u)$$

$\forall S \subset\subset \Omega$, hence $F(u_0) \le \eta \le F(u)$.

But since u is arbitrary u $(u - w \in V)$ this means that $\eta = F(u_0)$ and $u_0 = u_*$; hence (15), (17) and (14) coincide with the result of Th. 10.

Remark 13: is trivial; indeed $x \mapsto <\xi, x>$ is an admissible function.

Remark 14: for any $u \in H^1(\Omega)$, we have (see (6))

$$<\xi, Du> \le <a(x)Du, Du>^{1/2} <a(x)^{-1}\xi, \xi>^{1/2} \ .$$

Hence, integrating on Ω and using Schwartz's inequality, if $\int_\Omega Du\, dx = \xi$ we get the result.

Remark 15: it follows from the explicit solution of Dirichlet's problem.

Remark 16: for any $u = \langle\xi,x\rangle + v$, with $\int_\Omega Dv\,dx = 0$, we have

$$\sum_{1\,ij}^{n} a_{ij}\int_\Omega D_i u\, D_j u\, dx = |\Omega| \sum_{1\,ij}^{n} a_{ij}\xi_i\xi_j + \sum_{1\,ij}^{n} a_{ij}\int_\Omega D_i v\, D_j v\, dx +$$

$$+ 2\sum_{1\,ij}^{n} a_{ij}\xi_i \int_\Omega D_j v\, dx \geq \sum_{1\,ij}^{n} a_{ij}\xi_i\xi_j|\Omega| .$$

The converse inequality follows from Remark 13.

Remark 17: the proof is quite similar to that of Th. 10.

Remark 18: it suffices to prove that $\lim\limits_{k\to\infty} \gamma_A(V_k,\Omega_k,\xi) =$ $= \sum_{1\,ij}^{n} a_{ij}(x_0)\xi_i\xi_j$ for $V_k = H_0^1(\Omega_k), \forall k$, and also for $V_k = \tilde{H}^1(\Omega_k)$, $\forall k$. In both cases, by an appropriate change of variables, we can assume that $x_0 = 0$ and $\Omega_k = \delta_k\Omega$ ($\Omega =$ = unitary cube centered in the origin, $\lim\limits_{k\to\infty}\delta_k = 0$). Now by a dilatation we are reduced to proving that $\gamma_{A_k}(V,\Omega,\xi) \to \gamma_A(V,\Omega,\xi)$, with $V = H_0^1(\Omega)$ or $V = \tilde{H}^1(\Omega)$, and $A_k \equiv -\operatorname{div}(a(\frac{x}{k})\, D)$.

But since x_0 is a Lebesgue point for $a_{ij}(x)$, $\forall_{i,j}$, the coefficients of A_k converge in $L^1(\Omega)$ to the coefficients of A and this (Remark 17 and Remark 11) completes the proof.

Remark 19: is a simple computation.

Remark 20: if we denote by $\tau_b(A)$, $b \in R^n$, the operator obtained from A by a b-translation of coefficients, we can prove that, if $\{b_j\} \to b$ in R^n and $A_j \xrightarrow{G} A$, then $\tau_{b_j}(A_j) \xrightarrow{G} \tau_b(A)$.

Now for any b in R^n there exists $\{b_j\} \to b$ such that $\tau_{b_j}(A_{k_j}) = A_{k_j}$, $\forall j$. Hence $\tau_b(B) = B$, $\forall b$; i.e.

B is a constant operator.

Theor. 11: using the compactness of G - convergence we must only prove that, for any G - limit B_1 and B_2 of two G - convergent subsequences of $\{A_k\}$, the equality $B_1 = B_2$ holds. Now Remark 20 assures us that B_1 and B_2 are constant operators and Remark 19 gives $\gamma_{B_1}^{per}(P,\xi) = \gamma_{B_2}^{per}(P,\xi)$. Therefore using Remark 16 we have $B_1 = B_2$.

REFERENCES

1. Babuška, I., "Solutions of Interface Problem by Homogeneization, I and II," Institute for Fluid Dynamics, University of Maryland, (1974).
2. Bensoussan, A., Lions, J. L., and Papanicolaou, G., "Some asymptotic results for solutions of variational inequalities with highly oscillatory periodic coefficients," to appear.
3. Boccardo, L. and Marcellini, P., "Sulla convergenza delle soluzioni di disequazioni variazionali," to appear on Ann. di Mat. Pura e Appl.
4. Cristiano, L.C., "Γ -convergenza di funzioni a gradiente limitato," to appear.
5. De Giorgi, E., "Sulla convergenza di alcuni funzionali del tipo dell-area," Rend. Matematica, 8 (1975), 277-294.
6. De Giorgi, E. and Franzoni, T., "Su un tipo di convergenza variazionale," to appear on Atti Accad. Naz. Lincei.
7. De Giorgi, E. and Spagnolo, S., "Sulla convergenza degli integrali dell'energia per operatori ellittici del secondo ordine," Boll. U.M.I., 8 (1973), 391-411.
8. Joly, J.L., "Une famille de topologies sur l'ensemble des functions convexes pour lesquelles la polarite' est bicontinue," J. Math. Pure Appl., 52 (1973), 421-441.
9. Marcellini, P., "Su una convergenza di funzioni convesse," Boll. U.M.I., 8 (1973), 137-158.
10. Marcellini, P., "Un teorema di passaggio al limite per la somma di funzioni convesse," Boll. U.M.I., 11 (1975).
11. Marino, A. and Spagnolo, S., "Un tipo di approssimazione dell'operatore $\Sigma D_i(a_{ij}D_j)$ con operatori $\Sigma D_j(b\, D_j)$," Ann. Scu. Norm. Pisa., 23 (1969), 657-673.
12. Meyers, N.G., "An L^p-estimate for the gradient of solutions of second order elliptic divergence equations," Ann. Scu. Norm. Pisa, 17 (1963), 189-200.
13. Mosco, U., "Convergence of convex sets and of solutions of variational inequalities," Advances in Math., 3 (1969), 510-585.

14. Murat, F., "Théorèmes de non-existence pour des problèmes de contrôle dans les coefficients," C.R. Acad. Sc. Paris, 274 (1972), 395-398.
15. Reshetnyak, Y., "General theorems on semicontinuity and on convergence with a functional," Sib. Mat. Ẑ, 8 (1967), 1051-1069.
16. Sanchez-Palencia, E., "Solutions périodiques par rapport aux variables d'espace et applications," C.R. Acad. Sc. Paris, 271 (1970), 1129-1132.
17. Sanchez-Palencia, E., "Equations aux dérivées partielles dans un type de milieux hétérogènes," C.R. Acad. Sc. Paris, 272 (1971), 1410-1411.
18. Sanchez-Palencia, E., "Comportement local et macroscopique d'un type de milieux physiques héterogenes," Int. J. Enging. Sci., 12 (1974), 331-351.
19. Sbordone, C., "Sulla G-convergenza di equazioni ellittiche e paraboliche," Ricerche di Mat., 24 (1975), 76-136.
20. Sbordone, C., "Su alcune applicazioni di un tipo di convergenza variazionale," to appear on Ann. Scn. Norm. Pisa.
21. Senatorov, P.K., "The stability of a solution of Dirichlet's problem for an elliptic equation with respect to perturbations in measure of its coefficients," Diff. Urav., 6 (1970), 1725-1726.
22. Senatorov, P.K., "The stability of the eigenvalues and eigenfunctions of a Sturm-Liouville problem," Diff. Urav., 7 (1971), 1667-1671.
23. Spagnolo, S., "Una caratterizzazione degli operatori differenziali autoaggiunti del 2° ordine a coefficienti misurabili e limitati," Rend. Semi Mat. Padova, 39 (1967), 56-64.
24. Spagnolo, S., "Sul limite delle soluzioni di problemi di Cauchy relativi all'equazione del calore," Ann. Scu. Norm. Pisa, 21 (1967), 657-699.
25. Spagnolo, S., "Sulla convergenza di soluzioni di equazioni paraboliche ed ellittiche," Ann. Scu. Norm. Pisa, 22 (1968), 577-597.
26. Spagnolo, S., "Some convergence problems," to appear on Symp. Math. (1975) (Conv. Alta Matematica Roma, Marzo 1974).
27. Tartar, L., "Problèmes de contrôle des coefficients dans les equations aux derivees partielles," to appear.
28. Zolezzi, T., "On convergence of minima," Boll. U.M.I., 8 (1973), 246-257.

Short Communications

Berger, A. E., Ciment, M., and Rogers, J. C. W., "An Alternating Phase Truncation Method for the Stefan Problem"

Carey, G. F., Knight, C. J., and Oates, G. C., "Finite Element Analysis of the Compressible Throughflow Problem in Turbomachines"

Chung, T. J., "Computational Stability of Nonlinear Time Dependent Finite Element Equations"

Dey, S. K., "Numerical Studies of the Stability and Convergence Properties of Navier-Stokes' Equations"

Ghil, M., "A Nonlinear Parabolic Equation with Applications to Climate Theory"

Harten, A., "On Finite-Difference Approximations to Generalized Shocks"

Harten, A., "The Artificial Compression Method"

Hirt, C. W., and Romero, N. C., "Numerical Simulation of Transient Two-Phase Flow"

Hsu, Y. K., "Non-Steady Flow Through a Heavily Loaded Actuator Disk"

Kentzer, C. P., "Shocked Potential Flows"

Patel, V. A., "Time-Dependent Solutions of the Viscous Imcompressible Flow Past a Circular Cylinder by the Method of Series Truncation"

Polk, J. F., "Singular Perturbation Analysis in Diffusion and Heat Conduction Problems"

Sermer, P., "Least Squares Methods for Equations of Mixed Type"

Trangenstein, J., and More, J., "Finite Element Methods for the Tricomi Equation"

Wahlbin, L. B., and Schatz, A. H., "Error Estimates in the Maximum Norm for Finite Element Methods for Poisson's Equation in Plane Polygonal Domains"

Wang, K. C., "Numerical Solution of "Multi-Time" Parabolic Equations"